The Brilliance of Black Children in Mathematics:
Beyond the Numbers and Toward New Discourse

The Brilliance of Black Children in Mathematics:
Beyond the Numbers and Toward New Discourse

Edited by

Jacqueline Leonard
Director, Science and Mathematics Teaching Center
University of Wyoming, Laramie, WY

Danny B. Martin
University of Illinois at Chicago

INFORMATION AGE PUBLISHING, INC.
Charlotte, NC • www.infoagepub.com

Library of Congress Cataloging-in-Publication Data

The brilliance of black children in mathematics : beyond the numbers and toward new discourse / edited by Jacqueline Leonard, director, Science and Mathematics Teaching Center Laramie, WY, Danny B. Martin University of Illinois at Chicago, James Cook University, Townsville, Australia.
   pages cm
 Includes bibliographical references and index.
 ISBN 978-1-62396-080-3 (hardcover) – ISBN 978-1-62396-079-7 (pbk.) – ISBN (invalid) 978-1-62396-081-0 (ebook) 1. Mathematics–Study and teaching–United States. 2. African American children–Education. 3. Mathematical ability in children. 4. Educational equalization–United States. I. Leonard, Jacqueline, editor of compilation.. II. Martin, Danny Bernard, editor of compilation.
 QA13.B755 2013
 371.829'68073–dc23

2012041276

CONTENTS

SECTION I

CULTURAL-HISTORICAL PERSPECTIVES

SECTION V

PREPARING TEACHERS TO EMBRACE THE BRILLIANCE OF BLACK CHILDREN

FOREWORD

Richard Kitchen

In New Mexico, the state where I live, a brief examination of the history of education demonstrates that the Native American and Hispano peoples who live here have been denied access to a quality education for centuries. Similar to how Jim Crow laws in the South guaranteed that Blacks received inferior schooling, schools in New Mexico, particularly on Indian reservations, have a long history of being staffed by poorly trained teachers with little understanding of and respect for the unique cultural heritages of their students (Mondragón & Stapleton, 2005). For instance, non-native peoples may have little understanding of indigenous epistemology, which emphasizes the realization of individual agency through service to and caring for your community (Cajete, 2000). Interestingly, New Mexico was not granted statehood until January, 1912 under the Taft administration, despite having been a territory of the United States since 1848 (Roberts & Roberts, 2004). One of the primary reasons why statehood was denied for so many decades was because of the religious and racial prejudice that existed against both Native American and citizens of Spanish descent (Mondragón & Stapleton, 2005).

I bring up New Mexican history to underscore the historic tragedy of racism in this country, and how the legacy of racism against Blacks, Native

The Brilliance of Black Children in Mathematics:
Beyond the Numbers and Toward New Discourse, pages ix–xiv.
Copyright © 2013 by Information Age Publishing

Americans, and Hispanos, in particular, provide the backdrop for the impoverished educational system that we find today in urban and highly rural districts that primarily serve students of color and the poor. Put bluntly, the sad state of education currently found in many parts of the U.S. are a result of historical racism and White privilege that is manifest in the lack of resources provided urban and highly rural schools and the lack of political will needed to significantly improve education in these schools.

Until only recently in the mathematics education research community, there has been minimal attention paid to issues of equity and diversity needed to improve the mathematics education of students of color and the poor. Instead, there has been a fixation with cognition and how students learn. Many of these studies have been undertaken in suburban schools that predominantly serve middle class Whites. Unknowingly, these studies may promote colorblind racial policies, given the preponderance of the notion that the race and ethnicity of the research subjects are unimportant since the brain and student culture are taken as mutually exclusive. At this point in history, when the achievement gap is becoming a taken-for-granted aspect of the educational landscape in the U.S., we desperately need scholarly contributions that challenge such perspectives, while providing guidance to those working for more just and humane approaches in mathematics education.

Given the historical reality that African American children have consistently been denied access to a quality education (see, for example, Darling-Hammond, 1996; Dreeben, 1987), there must be a sense of urgency among progressives in the mathematics education community to develop new policies and practices to reform mathematics education for the betterment of African American students. Mainstream mathematics education research has, in fact, done little to help policymakers and practitioners address the achievement gap (Martin, Gholson, & Leonard, 2010; Tate, 1997). To improve the mathematics education of African Americans, the mathematics education community needs more scholarly work that challenges taken-for-granted notions about who can and who cannot do mathematics. Such scholarship will also support the transformation of education, in general, for all groups who have historically been denied access to a quality education.

The Brilliance of Black Children in Mathematics: Beyond the Numbers and Toward a New Discourse is a book that will make the sort of significant contribution needed to help promote and guide the transformation of the mathematical experiences of African American children. This is a book about possibilities and hope, in which the successes and resiliency of Black children are highlighted. There is an intense focus throughout the book on the brilliance of children who, because of the nation's obsession with the achievement gap, are frequently portrayed as inherently less capable.

It also draws attention to the beauty of African American culture and the marginalization of that culture within mainstream U.S. culture. The authors of the chapters in this unique book offer narratives rarely included in the mainstream mathematics education literature. Moreover, the content of this book supports the ongoing formation of a new paradigm in the research literature in which the potentialities of all students are valued and nurtured in schools.

The Brilliance of Black Children addresses the educational debt that has been rapidly escalating for African American children since separate and unequal schools were decreed for Black students (Ladson-Billings, 2006). Student culture, identity and issues of power are all addressed in this book. The authors examine issues related to the mathematics curriculum, culturally relevant instruction, and assessment that are fundamental to advance the mathematics education of African American children. Most importantly, this book is about advancing agency, for teachers, students, scholars, and anyone interested in addressing the historical injustices associated with the (under)education of Black children.

This book is a critically important contribution to the work underway to transform schooling for students who have historically been denied access to a quality education, specifically African American children. The first section of the book provides some historical perspective critical to understanding the current state of education in the U.S., specifically for the education of African American children. The following sections include chapters on policy, learning, ethnomathematics, student identity, and teacher preparation as it relates to the mathematical education of Black children. Through offering "counterstories" about mathematically successful Black youth, advocating for a curriculum that is grounded in African American culture and ways of thinking, providing shining examples of the brilliance of Black students, and promoting high expectations for all rather than situating students as the problem, the authors of this book provide powerful insights related to the teaching and learning of mathematics for African American students. As is made evident in this book, effective teaching involves much more than just engaging students in inquiry-based pedagogy (Kitchen, 2003). The chapters offered in this book demonstrate how mathematics instruction for African American students needs to take into account historical marginalization and present-day policies that do harm to Black students (Kunjufu, 2005). Empowering mathematics instruction for African American students needs to take into consideration and promote students' cultural, spiritual, and historical identities. Furthermore, mathematics instruction for African American students should create opportunities for students to express themselves and the needs of their communities as a means to promote social justice both within their classrooms and communities.

The ideals promoted in this book are beyond those articulated in the equity vision put forth in the *Principles and Standards for School Mathematics* (*PSSM*) document (National Council of Teachers of Mathematics, 2000), which largely supports learning dominant, albeit reform-based, mathematics (Gutiérrez, 2002; Rodriguez & Kitchen, 2005) with little attention given to issues of culture and social criticism. In general, more scholarship needs to be done to explore sociopolitical issues relevant in African American communities and to the teaching and learning of mathematics among Black children. This is another contribution of this book. There is also an urgent need for more research in mathematics education that gives attention to the role of African American parents' and students' experiences and the cultural knowledge they bring to the classroom, as well as their views vis-à-vis the role and importance of mathematics education in their lives. Much more scholarly work needs to be done that includes the voices of marginalized students, families and communities. This void in the research literature is addressed in several of the book's chapters.

It is important that we do not lose sight of how many Black children have historically been denied access to a quality education. Much emphasis needs to be placed on pursuing policies that ensure that poor students of color have access to rigorous mathematics curricula that is heavy on algebra, both fostering algebraic ways of thinking (Driscoll, 1999), as well as making traditional algebraic skills (e.g., solving algebraic equations) a primary focus that will help students succeed in higher education. Chapters in this book address the need for high-quality mathematics instruction for Black students by, for example, examining the Algebra Project (Moses & Cobb, 2001) as a practical educational intervention that supports algebraic literacy for African American students. Programs such as the Algebra Project need to be undertaken specifically for African American students in critically important areas such as sense-making related to understanding multiplication and division and proportional reasoning. Such interventions will make a difference in terms of providing students with opportunities in the future while promoting positive student identity in mathematics precisely because students are developing mathematical competence and know it.

Finally, much more attention needs to be given within the mathematics education research community to studies that explore how best to prepare teachers to work in schools that primarily serve students of color, specifically African American students. It is clear that inner-city schools as well as schools in highly rural areas of the U.S. (e.g., reservation schools) tend to attract poorly trained teachers (Darling-Hammond, 1996). Kitchen, DePree, Celedón-Pattichis, and Brinkerhoff (2007) found that "turn around" schools that served poor and diverse communities focused heavily on making high-quality professional development opportunities available for their teachers. At one of the schools studied, the community served by the school developed tremendous agency focused specifically on improving the opportunities afforded their students and neighborhood. Prospective teachers need opportunities to observe, study and teach in schools such as these to develop both beliefs and practices about what is possible in schools that serve students of color and the poor. It has been my experience that so often prospective teachers are not placed in such schools for a variety of reasons unrelated to doing what is best for prospective teachers. It is my belief that we need to focus more on what is needed to transform the beliefs and practices of prospective teachers regarding the brilliance of all students, while empowering prospective teachers about how to deal with legislative mandates (e.g., NCLB) and the many obstacles that they may face daily to teach all students well (e.g., colleagues with low expectations). This requires a commitment to approaches in teacher education that emphasize what is possible as opposed to just what is.

I'd like to conclude by thanking Drs. Leonard and Martin for giving me the honor of writing the foreword to their book. They have done the mathematics education community a great service by bringing together so many fine scholars who have contributed much needed ideas related to improving the mathematics education of Black children. I have little doubt that this book will have a significant impact to the benefit of African American students throughout the U.S. Delight in this book and be sure to share it widely with colleagues and students!

NOTES

1. "Hispano" or "Hispanic" rather than "Latino/a," "Mexican-American" or "Chicano/a" is the preferred term generally used in New Mexico as a means to distinguish a common Spanish ancestry.

REFERENCES

Cajete, G. (2000). *Native science: Natural laws of interdependence.* Santa Fe, NM: Clear Light Publishers.

Darling-Hammond, L. (1996). The right to learn and the advancement of teaching: Research, policy, and practice for democratic education. *Educational Researcher, 25*(6), 5–17.

Dreeben, R. (1987). Comments on "tomorrow's teachers." *Teachers College Record, 88*(3), 359–365.

Driscoll, M. (1999). *Fostering algebraic thinking: A guide for teachers, grades 6–10.* Portsmouth, NH: Heinemann.

Gutiérrez, R. (2002). Beyond essentialism: The complexity of language in teaching mathematics to Latina/o students. *American Educational Research Journal, 39,* 1047–1088.

Kitchen, R. S. (2003). Getting real about mathematics education reform in high poverty communities. *For the Learning of Mathematics, 23*(3), 16–22.

Kitchen, R. S., DePree, J., Celedón-Pattichis, S., & Brinkerhoff, J. (2007). *Mathematics education at highly effective schools that serve the poor: Strategies for change.* Mahwah, NJ: Lawrence Erlbaum Associates.

Kunjufu, J. (2005). *Keeping black boys out of special education.* Chicago, IL: African American Images.

Ladson-Billings, G. (2006). From the achievement gap to the education debt: Understanding achievement in U.S. schools (2006 Presidential Address). *Educational Researcher, 35*(7), 3–12.

Martin, D. B., Gholson, M. L., & Leonard, J. (2010). Mathematics as gatekeeper: Power and privilege in the production of knowledge. *Journal of Urban Mathematics Education, 3*(2), 12–24.

Mondragón, J., B. & Stapleton, E. S. (2005). *Public education in New Mexico.* Albuquerque, NM: University of New Mexico Press.

Moses, R. P. & Cobb, C. E. (2001). *Radical equations: Math literacy and civil rights.* Boston, MA: Beacon Press.

National Council of Teachers of Mathematics. (2000). *Principles and standards for school mathematics.* Reston, VA: Author.

Roberts, C. A. & Roberts, S. A. (2004). *A history of New Mexico* (3rd ed.). Albuquerque, NM: University of New Mexico Press.

Rodriguez, A. J. & Kitchen, R. S. (Eds.). (2005). *Preparing mathematics and science teachers for diverse classrooms: Promising strategies for transformative pedagogy.* Mahwah, NJ: Lawrence Erlbaum Associates.

Tate, W. F. (1997). Race, ethnicity, SES, gender, and language proficiency trends in mathematics achievement: An update. *Journal for Research in Mathematics Education, 28*, 652–680.

PREFACE

This book focuses on the brilliance of Black learners in mathematics. The project is an extension of ongoing research, policy, school, and community-based efforts by the Benjamin Banneker Association (BBA)—the nation's largest organization focusing specifically on the mathematics education of Black children—to meet the needs of Black learners. The theme for the book grew out of a National Science Foundation (NSF) funded conference series that focused on mathematics teaching and learning among Black children.

Authors were charged with the task of generating and supporting an alternative discourse about Black children and mathematics, one that focuses on brilliance. However, they were not asked to prove or justify the assertion that Black children are brilliant. Rather, this assertion is the starting point for the discussions that take place in this volume.

The research and policy literatures are replete with discussions of Black student failure. We believe these discussions have been narrowly focused on this failure and have left little room for discussing alternative conceptualizations of Black learners' competencies. We believe the time is right for changing and broadening the conversation. We believe that others in the field will respond positively and be able to utilize what is known about Black

The Brilliance of Black Children in Mathematics:
Beyond the Numbers and Toward New Discourse, pages xv–xix.
Copyright © 2013 by Information Age Publishing
All rights of reproduction in any form reserved.

learners' brilliance to affect changes in teaching, curriculum, assessment, and policy.

Collectively, the editors and authors reject theories and discourses that suggest Black learners are inferior to other children and that Black culture is deficient. The editors and many of the authors in this text have documented these deficit discourses (e.g., Berry, 2008; Leonard, 2008; Martin, 2000, 2007, 2009a, 2009b, 2012; Martin & McGee, 2009; McGee & Martin, 2011a, 2011b; Stinson, 2008). We also reject research and policy perspectives that ignore Black student success and that frame these successes as counterexamples within mainstream, deficit-oriented ideologies about Black students. Instead, we embrace the cultural background and knowledge that Black learners bring to their school and classroom contexts.

Although there is a growing knowledge base on Black learners and mathematics, this volume is unique in its focus. The authors explicate the experiences of Black learners across contexts, using diverse theoretical and conceptual perspectives, and critically analyze extant research with respect to those experiences. Rather than reify failure, we give attention to Black students' success and resiliency. The conception of brilliance adopted for this volume is not restricted to high test scores or exceptional grades. Many students do not achieve either but are, nevertheless, highly intelligent and capable. They are able to demonstrate their brilliance in non-school contexts and in their ordinary everyday mathematical lives. They are also able to demonstrate their brilliance in schools, but it may be often overlooked. In this volume, we bring it to light.

Within the current political context for mathematics education reform—characterized by shifting standards, stricter and more harsh accountability, and heightened demand for educational equity—the authors in this volume were also asked to address critical policy-oriented themes related to Black learners and how those themes support or undermine Black students' attempts to display their competence. Further, the authors make recommendations for modifications to existing and proposed policies as well as propose new policy directions. It is our belief that the policy context is a key point of leverage for improving mathematics education for Black students and responding to their needs in meaningful ways.

This volume is organized into five sections. Section one takes a cultural-historical perspective on mathematics education as it relates to Black students. In particular, one chapter in this section focuses on the brilliance of Benjamin Banneker, a free Black mathematician and scientist who lived during the colonial period, as a pretext for the brilliance of Black children. Banneker argued in his famous letter to Thomas Jefferson, then Secretary of State, that any dismissing of African talents, skills and abilities by European Americans was a direct consequence of enslavement and not Blacks' inability to perform. Likewise, this volume assumes such a premise. Section two includes two chapters that provide critical commentaries on research methods and policy, respectively. The third section focuses on learning and learning environments and explores mathematics learning among Black learners in particular content areas and learning environments and the impact of culturally relevant pedagogy on mathematics learning. Arguments are made that learning and achievement are not synonymous terms. Furthermore, policies that focus on achieving minimum standards do little to illuminate Black children's brilliance or promote excellence. Studying Black children's achievement without giving due consideration to how they learn, regardless of the school context, is misguided. The fourth section addresses student identity and student success. Too often research on Black children and mathematics has focused on deficits rather than assets. There are many stories of Black students' success that need to be shared, not as isolated cases, to help teachers recognize children's competence and brilliance (Berry, 2008; Ladson-Billings, 2009; Leonard, 2008; Leonard & Dantley, 2002; Martin, 2000, 2009a, 2009b, 2012; McGee & Martin, 2011a, 2011b; Stinson, 2008). The fifth section of the book focuses on preparing teachers to embrace the brilliance of Black children. In order to reform urban mathematics classrooms, teachers must believe that *Black children are brilliant.* The reform movement in education has shifted to producing high-quality teachers. This section also focuses on how teacher educators address pre-service teachers' cultural knowledge bases as well as beliefs and dispositions about Black children, which are important to realizing the brilliance of Black children.

In the last thirty five years, there have only been *six* other books dedicated to the mathematical education of African Americans. The National Council of Teachers of Mathematics (NCTM) published two of these books. The first, *Challenges in the Mathematics Education of African American Children, Proceedings of the Benjamin Banneker Association Leadership Conference,* appeared in 1997 and was edited by Carol Malloy and Laura Brader-Araje. The second, *Changing the Faces of Mathematics: African American Perspectives,* appeared in 2001 as part of the NCTM *Changing the Faces of Mathematics* series and was edited by Marilyn Strutchens, Martin Johnson, and William Tate. The third and fourth, *Mathematics Success and Failure Among African-American Youth* and *Mathematics Teaching, Learning, and Liberation in the Lives of Black Children,* appeared in the years 2000 and 2009, respectively, as part of the *Studies in Mathematical Thinking and Learning* series and were authored and edited by Danny Martin. Another important book, *Culturally Specific Pedagogy in the Mathematics Classroom: Strategies for Teachers and Students,* authored by Jacqueline Leonard (2008), takes up issues of culturally responsive pedagogy and race. All five of these books have been well received, the last three because of their simultaneous appeal to practitioners and their theoretical and methodological contributions. More recently, a text authored by Erica

Walker and published by Teachers College Press in May 2012 addresses mathematics learning among high school students: *Building Mathematics Learning Communities: Improving Outcomes in Urban High Schools.* However, none of these texts foreground Black brilliance as the central theme.

This volume will help to shape research and practice as well as enrich discussions about Black students and mathematics. Audiences who will find this book helpful include policymakers, mathematics education researchers and graduate students, teacher educators, school administrators, teachers, and parents of Black children.

—Jacqueline Leonard and Danny Bernard Martin

REFERENCES

Berry III, R. Q. (2008). Access to upper-level mathematics: The stories of African American middle school boys who are successful with school mathematics. Journal for Research in Mathematics Education, 39(5), 464–488.

Ladson-Billings, G. (2009). *The Dreamkeepers* (2nd ed.). San Francisco: Jossey-Bass.

Leonard, J. (2008). *Culturally specific pedagogy in the mathematics classroom: Strategies for teachers and students.* New York, NY: Routledge.

Leonard J. & Dantley, S. J. (2002, Winter). Why Malik can do math: Race and status in integrated classrooms. *Trotter Review, 14*(1), 61–78.

Malloy, C. & Brader-Araje, L. (Eds.). (1997). *Challenges in the mathematics education of African American children: Proceedings of the Benjamin Banneker Association leadership conference.* Reston, VA: National Council of Teachers of Mathematics.

Martin, D. (2000). Mathematics success and failure among African American youth: The roles of sociohistorical context, community forces, school influence, and individual agency. Mahwah, NJ: Lawrence Erlbaum Associates.

Martin, D. (2007). Beyond missionaries or cannibals: Who should teach mathematics to African American children? *The High School Journal, 91*(1), 6–28.

Martin, D. (2009a). Researching race in mathematics education. *Teachers College Record, 111*(2), 295–338.

Martin, D. (2009b). *Mathematics teaching, learning, and liberation in the lives of Black children.* London: Routledge.

Martin. D. (2012). Learning mathematics while Black. *The Journal of Educational Foundations, 26*(1–2), 47–66. Retrieved from http://education.uic.edu/images/documents/userfiles/file/DBM-learningmathwhileBlack.pdf

Martin, D. & McGee, E. (2009). Mathematics literacy for liberation: Reframing mathematics education for African American children. In B. Greer, S. Mukhophadhay, S. Nelson-Barber, & A. Powell (Eds.), *Culturally responsive mathematics education* (pp. 207–238). New York, NY: Routledge. Retrieved from http://education.uic.edu/images/documents/userfiles/file/Culturally Responsive Education0001.pdf

McGee, E. & Martin, D. (2011a). From the hood to being hooded: A case study of a Black male PhD. *Journal of African American Males in Education, 2*(1), 46–65.

McGee, E. & Martin, D. (2011b). You would not believe what I have to go through to prove my intellectual value! Stereotype management among successful Black college mathematics and engineering students. *American Educational Research Journal, 48*(6), 1347–1389. Retrieved from http://education.uic.edu/images/documents/userfiles/file/martinmcgee2011.pdf

Stinson, D. (2008). Negotiating sociocultural discourses: The counter-storytelling of academically (and mathematically) successful African American male students. *American Educational Research Journal, 45*(4), 975–1010.

Strutchens, M., Johnson, M., & Tate, W. (Eds.). (2001). *Changing the faces of mathematics: African American perspectives.* Reston, VA: National Council of Teachers of Mathematics.

Walker, E. N. (2012). *Building mathematics learning communities: Improving outcomes in urban high schools.* New York, NY: Teachers College Press.

ACKNOWLEDGMENTS

We wish to thank our graduate advisors, who continue to be supportive and encouraging in all our endeavors: Martin L. Johnson and Alan H. Schoenfeld, respectively. We also appreciate Richard Kitchen for writing the foreword, which poignantly expresses the challenges associated with receiving equitable educational opportunities in the United States for Native Americans, Hispanos, and African Americans, the latter which are the focus of this volume. We also extend a special thanks to our colleagues, who served as external reviewers for the chapters and whose comments made the manuscripts much stronger: Celia Keiko Anderson, Michaele Chappell, Marta Civil, Sandra Crespo, Geneva Gay, Kara J. Jackson, Heather Johnson, Sarah Lubienski, Na'ilah Suad Nasir, Arthur Powell, Joi Spencer, Marilyn Strutchens, Edd Taylor, La Mont Terry, and Erica Walker. Finally, we acknowledge Cara M. Moore for editing a couple of the book chapters, Floyd Vaughn for front cover photography, and Cydnei M. Quinn (pictured on front cover) for inspiration.

This material is based upon work supported by the National Science Foundation under Grant No. 0907896. Any opinions, findings, and conclusions or recommendations expressed in this material are those of the author(s) and do not necessarily reflect the views of the National Science Foundation.

—Jacqueline and Danny

SECTION I

CULTURAL-HISTORICAL PERSPECTIVES

CHAPTER 1

THE HISTORY, BRILLIANCE, AND LEGACY OF BENJAMIN BANNEKER REVISITED

Jacqueline Leonard and Cheryl Lewis Beverly

INTRODUCTION

The Benjamin Banneker Association, Inc. (BBA), founded in 1986, is the nation's largest organization focusing on the mathematical education of Black children to specifically meet the needs of Black children. Named for the famous mathematician and scientist, Benjamin Banneker, one of the activities of the BBA is to recognize outstanding students of color who have achieved in mathematics in the city that hosts the National Council of Teachers of Mathematics (NCTM) Annual Meeting. In May 2004, Leonard (first author) traveled to Masterman Middle and High School in Philadelphia, Pennsylvania, to bestow the Benjamin Banneker Award on a high school senior during the awards assembly. Prior to presenting the award, Leonard overheard the recipient telling another student, "I'm getting this Banneker Award, and I have no idea who Banneker is or what I did to de-

The Brilliance of Black Children in Mathematics:
Beyond the Numbers and Toward New Discourse, pages 3–21.
Copyright © 2013 by Information Age Publishing
All rights of reproduction in any form reserved.

serve it." Needless to say, too many Americans have limited knowledge of Benjamin Banneker, and, because of racial bias, many of his achievements have been underrated or glossed over entirely.

The purpose of this chapter is threefold: 1) to consolidate historical data and biographical narratives about Banneker to honor his genius; 2) to show his humanity as an advocate for social justice; and 3) to show the legacy of Black achievement in mathematics and science to enhance Banneker's argument that Black brilliance in mathematics and science is just as much a part of our cultural heritage as individuals from other racial and ethnic backgrounds. While more than 100 narratives have been written about Benjamin Banneker (Perot, 2008), none have accomplished all three of the aforementioned purposes.

While Banneker's life is a singular story and Black brilliance is often unrecognized or challenged, there is a need to understand, more broadly, the historical and social forces that perpetuate underachievement and marginalize success among Black students in mathematics. Once these forces are understood, they can be challenged to help Black children realize their mathematics identity (Nasir, 2005) and mathematical brilliance. Simply put, too few examples of African Americans excelling in mathematics and engaging in careers that utilize mathematics, critical thinking, and reasoning have been shared. The film *The Pursuit of Happyness* (Muccino, 2006) starring Will Smith is one of the few that demonstrates the mathematical and reasoning ability of an African American man. Little is known about Black mathematicians, historically. For example, Thomas Fuller (1710–1790), known as the Virginia Calculator, "was a brilliant calculating prodigy with amazing arithmetic skills and extraordinary computational power" (Chahine, this volume, p. 195). Yet, very few know Fuller's story. However, the story of Benjamin Banneker (1731–1806) can be used as a backdrop to make the case for Black brilliance in mathematics.

BANNEKER'S ENGLISH AND AFRICAN ANCESTRY

Benjamin Banneker's ancestry has been controversial. In one account, McHenry provided an ambiguous introduction of Banneker in his letter, claiming:

> Benjamin Banneker, a free negro, has calculated an almanac, for the ensuing year 1792....This man is about fifty-nine years of age; he was born in Baltimore County; his father was an African, and his mother the offspring of African parents. (McHenry as cited in Perot, 2008, p. 45)

What is more certain is what Banneker said of himself in a letter to Thomas Jefferson, where he described himself as a man of "African descent" and of "the deepest dye" (Perot, 2008), as well as what was recorded in the family Bible: "Benjamin Banneker was born November the 9th, the year of the Lord God, 1731, and Robert Banneker departed this life July the 10th, 1759" (Bedini, as cited in Perot, 2008). While Banneker made reference to his African father and his African grandfather in his writings, information about his mother and his grandmother is indistinct. Yet, a story of his ancestry may be created from interviews and legal documents (marriage records, land purchases), 17th and 18th century narratives, Benjamin Banneker's writings (1791), as well as 20th and 21st century accounts of his life.

The most consistent story of Banneker's maternal ancestry purports that he is the grandson of Molly Welsh (some records have Molly Walsh), who was born about 1666. Tyson (as cited in Perot, 2008) conducted an interview with Banneker's first cousin, John Hendon. This interview revealed Banneker's grandmother, Molly Welsh, was "not only a *white woman*, but had a remarkably fair complexion" (Tyson as cited in Perot, 2008, p. 33). Molly immigrated to Maryland from England as an indentured servant as punishment for stealing a pail of milk (Baker, 1918). Such crimes were punishable by death, but Molly may have been spared because she was able to read and there was a shortage of farm labor and suitable women for companionship in the colonies (Tyson as cited in Perot, 2008). While no transportation records exist for such an indentured servant, records do exist for several women named Mary (another name for Molly), who set sail for Maryland during the middle of the 17th century (Perot, 2008). Thus, as Perot (2008) illustrates vividly in her thesis, the story of Molly Welsh is quite credible.

After working as an indentured servant, Welsh was paid a sum of money or given a small plot of land (25–50 acres) for her labor, which was customary during this time period (Perot, 2008). Baker's (1918) account claims that Welsh purchased two slaves, one of whom was named BannaKa (some spellings include Bannka, Bannaky), who arrived in 1692 on a ship that docked in the Chesapeake Bay near Annapolis, Maryland, to help her work on a small farm that she purchased near Baltimore. BannaKa was described as "a man of much intelligence and fine temper, with a very agreeable presence, dignified manner and contemplative habits" (Tyson as cited in Baker, 1918, p. 102). Having gained her favor, BannaKa later married Molly.

However, this account is perplexing because Molly Welsh had been an indentured servant—a second-class citizen. Although paid for her service, she was forced to work as an indentured servant for four to seven years, depending on her age when she arrived in Maryland (Perot, 2008). Having suffered under a system that dehumanized her, it is puzzling why she would engage in the slave trade—a horrible system of chattel slavery where Blacks worked in perpetual servitude. A conversation conducted via email with Mr. Robert Lett, a descendant of Benjamin Banneker, revealed the following explanation:

The account of Molly Walsh acquiring two slaves and then going into Maryland wilderness is difficult to perceive based upon our perceptions of slavery. Further, the concept of a White but "freed" indentured servant—woman taking these steps and...adjourning in the wilderness is nearly beyond comprehension. However...I temper these thoughts with the fact that the "ruling [class]" or "those in leadership" within the colonies would have looked at a White indentured woman brought to this country particularly for an alleged offense of thievery, as a second-class woman and not desirable. My sense of things from numerous sources is that particularly in Maryland a sub-culture existed which blurred the lines between slave and indentured people.... No doubt the White gentry viewed Molly's quest as bold or fool hearty, but then again, if you were the slave sold to Molly, and she was able to convey the fact that she only wanted farm hands and that she treated them well in gratitude for their work, a new dynamic arises. (Email communication from Mr. Robert Lett on November 13, 2011)

Mr. Lett's assumptions about women and indentured servants during colonial times are supported by Perot's (2008) thesis on Molly Welsh:

One of the most complicated issues concerning Molly Welsh's story is the interesting problem of race. Molly Welsh, oral tradition captured in the nineteenth century tells us, was a white Englishwoman who worked as an indentured servant. The same tradition has it that she owned slaves, although she is said to have married (or formed a union with) one of them. (p. 3)

Further, Perot (2008) suggested that such an alignment between White females who were indentured servants and Black male slaves was common, even though Maryland law prohibited interracial marriage. Yet, there would be no need for a law if such behavior were not occurring. Given the "minority" status of White female indentured servants, it is not difficult to imagine that close working conditions with Black male slaves would result in friendship and marriage (Perot, 2008).

If...the narrative...is true that she [Welsh] acquired slaves, it is likely that Bannka "kept alive important strands of African consciousness" and shared with her his African traditions in song, dance, farming and perhaps most importantly his respect for her as a person, Molly might have been attracted to his approach to life as much as his physical being. Thus Molly's marriage to one of her own slaves seems a plausible choice that did not necessarily cause her moral strife. (Tyson as cited in Perot, 2008, pp. 12–13)

To Molly's credit, she emancipated both slaves after a number of years because of their joint success in cultivating the farm (Baker, 1918). When Molly married BannaKa (there are no records of the marriage), she took his name.

It is reported that BannaKa was caught and enslaved in Africa and transported via the Middle Passage to America. Several sources also report that

BannaKa was the son of an African king (Baker, 1918; Cromwell, 1914; Perot, 2008). Cerami (2002) suggests BannaKa was a Dogon prince who possessed knowledge about astronomy and the cosmos. Together, BannaKa and Molly lived a low-profile life along the Patapsco River where land was plentiful and affordable and raised four daughters: Mary, Katherine, Esther, and Jemima (Perot, 2008). BannaKa died prematurely, leaving Molly to raise their four daughters alone (Perot, 2008).

The oldest daughter, Mary Banneker (note change of spelling, also spelled Banneky in some sources) was born around 1710. While Mary was a free woman, records show she was sentenced to serve Thomas Harwood on November 13, 1728, when she admitted to the Prince George's County, Maryland, court that she had a mulatto child. Because Maryland law[1] barred White women who had been indentured servants (law could have been binding on the children since Mary's mother was indentured) from marrying Blacks, one can only assume that Mary was mulatto as well and passed for White. Mary was sentenced to seven years of servitude, and her son, Henry (2 months old), was sentenced to indentured servitude until he was 31 years of age. Recall that Benjamin Banneker was born in 1731. However, Mary would have only completed three years of servitude at that point in time. In 1736, records show Mary was married to Robert, and they were using the name surname of Banneker. Robert, a native African, had been a slave on a neighboring farm. Robert was emancipated by his owner after he was baptized in the Anglican Church (Baker, 1918). Having no surname, he took Mary's surname when they married (Perot, 2008). Like her parents, there are no records of Mary's marriage. In 1737, they purchased 100 acres of land with 7,000 pounds of tobacco (Baker, 1918; Perot, 2008). Records show that Benjamin was listed as the heir of the farm at the age of six.

In summary, Benjamin Banneker was a third generation African—son of an African father from Guinea and grandson of an African grandfather, who was possibly a Dogon prince. His mother was the mulatto offspring of English indentured servant, Molly Walsh, and a former African slave, BannaKa. Interestingly, despite laws prohibiting interracial marriages in the state of Maryland, three generations of this family lived secluded lives on their respective farms as free Blacks.

BANNEKER'S EARLY YEARS

Benjamin was a precocious young lad. As a child, he was considered a prodigy. At age five or six, Cerami (2002) states that Banneker was reported by his neighbors to be a phenomenon because of his ability to read the Bible, write, and do math. Several sources credit Banneker's grandmother, Molly, with teaching him to read using the Bible as a text (Baker, 1918). Records found in the Maryland Historical Society also noted that Banneker attended "'an obscure and distant country school'…attended by both white and colored children" (Latrobe as cited in Baker, 1918). McHenry's letter concurred, stating "…His father and mother having obtained their freedom were enabled to send him to school, where he learned, when a boy, reading, writing, and arithmetic, as far as the double position" (McHenry as cited in Perot, 2008, p. 45). However, the school was only open in the winter (Leonard, 2002), so whatever education Banneker routinely received could be attributed to home schooling.

According to Baker (1918), Banneker delighted himself with reading the books he could acquire and was a close observer of his natural surroundings. In addition to studying, he also spent a great deal of time helping his father cultivate the farm. As he matured, he displayed a remarkable ability to predict the weather, ensuring that the farm flourished from his knowledge of planting seasons. The farm was said to be "one of the best kept farms in his neighborhood" (Baker, 1918, p. 104).

Banneker's scientific genius has been *reluctantly* attributed to his African ancestry (Cerami, 2002). Oral tradition revealed BannaKa, Benjamin Banneker's grandfather, was an African prince (Baker, 1918; Perot, 2008), whose knowledge of the cosmos linked him to the Dogons (Cerami, 2002). The Dogons' secretly held knowledge of the universe was transferred through the centuries to successive generations through their patriarchal line. While Cerami (2002) acknowledged the mysterious knowledge of the solar system displayed by the Dogon people, he attributed Banneker's acquisition of knowledge of it to his mulatto mother, Mary, along with his grandmother, Molly Welsh. Recall that Bedini (1999) claimed that BannaKa died at an early age, leaving his wife, Molly, to raise their four daughters alone. In this scenario, the grandfather could not have taught young Benjamin if he had died before he was born. Other sources report that Banneker's father, Robert, was from Guinea (Baker, 1918; Perot, 2008). While it is doubtful that Robert was also Dogon, we are unsure what contributions he may have made to Banneker's knowledge of astronomy. It is possible, since Robert was a slave on a neighboring farm, that he could have interacted with BannaKa and received some scientific knowledge from him that was later passed on to his son, Benjamin.

The aforementioned oral traditions leave us to ponder who may have influenced Banneker's astronomical knowledge. Since Dogon tradition asserted that knowledge can only be transmitted through the patriarchal line (van Beek, 2001), which specifically excluded women, there is some doubt that BannaKa would have shared his knowledge with Molly and Mary. Another possibility is that Banneker was self-taught. Evidence suggested he used the books and instruments that he borrowed from George Ellicott—his friend and neighbor—and made several discoveries on his own. Another theory is that the second slave Molly freed may have stayed on the farm since it was so successful. It is doubtful that Molly could have managed a 50-acre farm and raised four girls entirely on her own. Since only White women who were single or widowed could own land, it is very possible that she never remarried. Thus, it is plausible that Molly's other African slave could have learned from BannaKa and passed the tradition down.

When it came to farming, we are more certain that Benjamin learned what he knew from his father, who was wise and frugal enough to purchase his own farm at a time when few Blacks were free, let alone land owners (Baker, 1918; Cromwell, 1914; Perot, 2008). The Banneker farm was greatly admired and acclaimed by neighbors as one of the best in the area. It became renowned for its vegetables, fruits, poultry goods, and honey. Banneker's system of irrigation techniques helped the farm to flourish. After his father's death in 1759, Banneker and his mother inherited 72 acres (his sisters received the other 28 acres) of the farm, and they were able to live comfortably from its produce (Baker, 1918).

In summary, as a free Black, Banneker was privileged to early childhood education. He was able to attend school with White children where he was able to read and do mathematics. This educational experience may have been one of the first accounts of Black children attending an integrated school. His love for reading led him to acquire a great deal of general knowledge as well as knowledge about astronomy. He was intrigued with mathematical puzzles, and people came from far and wide to visit and engage in conversations with him (Baker, 1918; Perot, 2008).

THE GENIUS OF BANNEKER

Benjamin Banneker was a gifted prodigy. He became an inventor, mathematician, surveyor, engineer, and an astronomer. His most notable contributions to American society are his clock, his mathematical puzzles, his survey of Washington, DC, and his almanacs, which derived from his work in astronomy.

The Wooden Clock

In 1753, at age 21, Benjamin Banneker constructed a wooden clock. After examining the gears of a pocket watch, Banneker made the clock entirely out of wood, including the wheels, springs, and balances (Cromwell, 1914). This clock was the first of its kind to be produced in America. It was a reliable timepiece for more than 20 years (Leonard, 2002) and continued to work for 40 years (Cromwell, 1914). The clock was lost in a fire that consumed Banneker's home after his death in 1806.

Mathematical Puzzles

In one of Banneker's famous puzzles simply known as *Trigonometry*, he revealed his knowledge of logarithms in the solution (Fasanelli, Jagger,

&Lumpkin, 1992). Although he admitted that the problem could have been solved without logarithms, Banneker used the Law of Sines to find the hypotenuse and side of a right triangle with a base of 26. To find the hypotenuse, Banneker set up a proportion: sin C/c = sin B/b. Since the sine complement of angle A was sin 60, Banneker derived the following problem: $x = 26 \sin 90/\sin 60$. Using logarithms, he then solved the following problem: $\log x = \log 26 + \log \sin 90 - \log \sin 60$, which is equal to $1.41497 + 10 - 9.93753 = 1.47744$. To get the answer, Banneker needed antilogarithm tables, which did not exist at the time. Instead he used his logarithm table in reverse to come up with an answer of 30 for the hypotenuse. To find the remaining side, Banneker used the Law of Sines again to obtain $\log x = 1.7641$. The antilog of this number is approximately 15. With this puzzle, Banneker clearly demonstrated his mathematical prowess.

Survey of Washington, DC

One of Banneker's most distinguished honors came from an invitation to serve on the commission appointed by President George Washington at the recommendation of George Ellicott to survey the land for the nation's capital in what is now known as Washington, DC. Some credit French engineer, Major Pierre Charles L'Enfant with creating the plans. However, Major Andrew Ellicott reported the following in his letters written a year apart to his cousin, George Ellicott.

February 1791
Dear George Ellicott,

Today Mr. Banneker was asked by Thomas Jefferson to survey the land where the District of Columbia and nation's capital would be placed. Another person, Pierre L'Enfant, was fired and was the one who had the plans. Mr. L'Enfant also took the plans with him! Surprisingly, Mr. Banneker recreated the plans from memory. He had seen the plans when we first started. I'm amazed by Mr. Banneker's memory.

Sincerely,
Maj. Andrew Ellicott

***

February 3, 1792
Dear George Ellicott,

I'm sorry that it's been so long since I replied to your letter. We have been very busy with constructing the new state capital. I'm not sure if Mr. Banneker's plans were the same as Mr. L'Enfant's. I think Mr. L'Enfant's plans were something like the Versailles of Paris. Is there anything else that you think I might find interesting?

Sincerely,
Maj. Andrew Ellicott

The above communication clearly showed that it was Banneker's plans that were used to construct the nation's capital. The letters suggested that Banneker's photographic memory and creativity led to a unique design that was similar to and yet different from L'Enfant's plans. However, it is L'Enfant whose name is a location in Washington, DC (L'Enfant Plaza) and not Banneker's. It is no wonder Banneker became a recluse after completing his work with the commission in Washington, DC. Banneker returned to his home in Maryland and focused his time and effort on publishing his almanac, the first of which appeared in 1792.

The Almanacs (1792–1797)

At the age of 58, Banneker formally studied astronomy using books loaned to him by George Ellicott (Baker, 1918). This study, along with his other astronomical knowledge, resulted in the production of an almanac in 1791, his most prized accomplishment. However, Banneker had problems getting his almanac published. He solicited the help of Dr. James McHenry, a patriot and statesman (Perot, 2008). McHenry's letter illustrated some of the prejudices Banneker faced in his effort to get the almanac published:

> …fully satisfied myself, with respect to his title and this kind of authorship, if you can agree with him for the price of the work, I may venture to assure you, it will do you credit as editors, while it will afford you opportunity to encourage talents that have thus far surmounted the most discouraging circumstances and prejudices. (McHenry as cited in Perot, 2008, p. 44)

With McHenry's letter as an introduction, Goddard and Angell published Banneker's first almanac after he recalculated the tables for 1792. Yearly almanacs were published until 1797 and widely distributed in Pennsylvania, Delaware, Virginia, and Kentucky. These almanacs predicted solar and lunar eclipses and even corrected the miscalculations of more well-known astronomers: James Ferguson's solar eclipse schedule and Charles Leadbetter's astronomical tables. In Banneker's almanacs, he predicted snowfall, rainfall, and tidal occurrences for the Chesapeake Bay area. He provided information to farmers about seasonal planting and the care of livestock (McKissack & McKissack, 1994). Furthermore, he included political abolitionist and humanitarian writings, including his 1791 letter to Thomas Jefferson on the intelligence of the African, the poetry of Phyllis Wheatley and others, recipes, and the medicinal recipes that his mother Mary was renowned for. His almanacs proved to be comprehensive and entertaining, as well as an accurate body of work that was better than any other available at the time. Banneker's almanacs were the first known scientific books produced by an American of African descent.

BANNEKER AS A SOCIAL JUSTICE ADVOCATE

The foregoing historical accounts not only reveal the mathematical brilliance of Banneker, but of Africans, which affirms and celebrates the mathematics identity of Blacks. Assumptions that Black children are not the *right* kind of mathematics students are simply false (Kitchen, Depree, Celedón-Pattichis, & Brinkerhoff, 2007; Martin, 2007; Stiff & Harvey, 1988). Banneker argued that Blacks were indeed intelligent and capable when he wrote his famous letter to Thomas Jefferson, then Secretary of State, on August 19, 1791, regarding the intellectual capacity of Blacks. In his letter, Banneker urgently requested that slavery be abolished, arguing passionately and authoritatively on behalf of Blacks because Jefferson had been consistently critical of the *intelligence* of Negroes (Perot, 2008). Banneker suggested that any dismissing of African talents, skills and abilities by European Americans was a direct consequence of enslavement and not Black's inability to perform.

Supporting Banneker's claim of equal intelligence was an article in the *Georgetown Weekly Ledger* on March 12, 1791: "Benjamin Banneker, an Ethiopian, whose abilities as a surveyor and an astronomer already prove that Mr. [Thomas] Jefferson's concluding that that race of men were void of mental endowment was without foundation"(Weekly Ledger, as cited in Baker, 1918, p. 112). The following year, Maryland legislator James McHenry also wrote "I consider this Negro as fresh proof that the powers of mind are disconnected with the color of skin...in every civilized country we shall find thousands of whites liberally educated, and who have enjoyed greater opportunities of instance than this Negro" (McHenry as cited in Perot, 2008, p. 45). However, many who were less talented than Banneker intellectually did not have the human characteristics needed to respect the contributions of a Black man (Brodie, 1993).

Banneker's quest to inform the world and America of African talents and skills was routinely dismissed and generally ignored by the elite of his day. Unfortunately, all too often, negative perceptions of Blacks in mathematics are still held today. While Benjamin Banneker was a product of centuries of intellectual and spiritual African gifts, he was considered an anomaly. The accomplishments in science and mathematics of many others of African descent, while providing examples of brilliance in these fields, are still viewed as outliers. We must also be cognizant of the fact that Blacks must often "negotiate the most difficult and oppressive life circumstances" (Martin, 2009, p. 22) while trying to obtain mathematics education. Like

Benjamin Banneker, these African Americans are given little recognition and acknowledgement. The preface to Banneker's 1796 Almanac read, "To whom do you think that you are indebted to for this entertainment? Why, to a Black man—Strange! Is a Black capable of compiling an almanac? Indeed, it is no less strange than true; and a clear wise long-headed Black he is. Why so strange?"

Banneker wanted the world to know that Americans of African descent, as a group, shared his brilliance, which could not be fully realized by involuntary servitude, slavery, and the slave trade. As quoted in the *Federal Gazette* (1806) three weeks after his death, "Mr. Banneker is a prominent instance to prove that a descendant of Africa is susceptible of as great mental improvement and deep knowledge into the mysteries of nature as that of any other nation." Yet, there continues to be a need to not only revisit Banneker's work and his genius, but to promote his legacy, which includes the recognition of Black brilliance.

FROM MALI TO AMERICA: BANNEKER'S LEGACY

Cerami (2002) believed that Banneker's legacy extended from grandfather BannaKa's Dogon ancestry. The Dogons are African people who reside in the west central area of Africa in southern Mali in the Bandiagara Escarpment, an isolated area south of the Niger River and Tombucto (Timbuktu). The Dogon systems of irrigation, architecture, farming and knowledge of the cosmos were known and understood by BannaKa and later possessed and successfully used by Benjamin. His absorption with mathematical structure, the cosmos and numerology (Cerami, 2002), which emboldened him to correct the inaccuracies of others, was African; it was Dogon.

The Dogon Legacy

New evidence in the form of an artifact known as the Covenant stone supports Cerami's claim that Benjamin Banneker was of Dogon ancestry. The Covenant stone was asserted to have been passed down from generation to generation through the patriarchal line. Most likely it was brought to this country by BannaKa and was given to Banneker by his grandmother. The stone has inscribed on its surface a snake, which is quite possibly symbolic of the Nommos serpent (Dogon mythology). Moreover, the stone bears a mathematical symbol similar to infinity or the number eight, which is the base number of the Dogon numerical system. The number eight may also represent the eight civilizing skills of Dogon culture and the eight ancestral Dogon tribes.

Serendipitously, the Dogon stone (see Figure 1.1) was found by Mr. Brian Deese, an insurance investigator, who has done extensive research on Banneker since finding the stone. The Covenant stone not only shows the

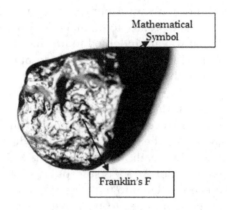

FIGURE 1.1.   Covenant Stone

link between Banneker and the Dogon culture but also a relationship be-tween Benjamin Banneker and Benjamin Franklin. The assumption is that Banneker gave Franklin the stone, onto which Banneker forged Franklin's "F," perhaps as a pact to fight against slavery. It is believed that Franklin lost the stone, which Deese found on May 3, 2011, in Martha's Vineyard near the home of a relative that Franklin frequently visited during his lifetime. A relationship between Franklin and Banneker is credible because Banneker, while espousing no religion, often worshipped with the Quakers at the Society of Friends (Baker, 1918); his famous clock, built in 1753, drew visitors from miles around (Baker, 1918; Cromwell, 1914); and his almanac, which was sold in Pennsylvania, would have been noticed by Franklin.

The intricate carvings of a snake and a mathematical symbol, as well as the balance and triangular shape of the stone, reveal the creativity and mathematical genius of the Dogons. Recall that the Dogons were known for their complex mythology and cosmology, which included an understanding of the solar system and the rare understanding of the double star Sirius well before the same discoveries were made by Europeans (Cerami, 2002). Banneker's knowledge of the cosmos was evident when he proclaimed Sirius as his favorite star and his lucky star. Identifying the unknown and unseen double star of Sirius, invisible to the naked eye and only scientifically reported in the 1930s by Europeans Marcel Griaule and Germaine Dieterlen, Banneker was not credited with discovering it in the 1770s. Nevertheless, Benjamin Banneker's gift of solar knowledge was verified by Frenchman Marcel Griaule's research of the Dogon people between 1931 and 1956, with Sirius B confirmed by telescope in 1970 and Sirius C confirmed in 1995, a fact that the Dogon people have known for centuries (van Beek, 2001).

The belief that Banneker's interest in Sirius and his skills in architecture and irrigation were passed down to him by his Dogon ancestors is plausible and challenges Bedini's characterization of Banneker as an amateur astronomer (Bedini, 1999). Communication between George Ellicott and his cousin Major Andrew Ellicott after Banneker started surveying Washington, DC, for the new capital support this theory:

March 20, 1791
Dear George Ellicott,

Yes, I have heard about the clock that Mr. Banneker made. I think it is a mechanical wonder. How on Earth did he make it? Has he borrowed anything from you? Has he published any books? Please write back as soon as you get this letter.

Sincerely,
Maj. Andrew Ellicott

***

April 3, 1791
Dear Maj. Andrew Ellicott,

Mr. Banneker borrowed a pocket watch from a friend, and then he took it apart. He took a look at the pieces and wrote all the essential facts down in his journal, and he drew a picture of every piece. He set a marker on the day that he made the clock. He was the first person to make a clock in America. Yes, he has borrowed some items of mine. He borrowed most of my astronomy equipment. He tried madly to get his first almanac published, but it was already the New Year. Then he successfully got his first almanac in 1792. If you have any other questions, please include them in your next letter.

Your cousin,
George Ellicott

Memoirs of Mali

In 1999, Beverly (second author) visited Mali and the Bandiagara cliffs of the Dogon people. Their cultural environment appeared to be highly "untainted" by Western influence and/or knowledge. Beverly can attest that their "preferred" location was one of complete isolation from other African and Western influences. The houses consisted of hundreds of mud brick and grass huts. The houses exhibited the design of a geodesic dome (see Figure 1.2). To create the homes, knowledge of mathematics and geometric principles were needed. These homes and other cultural specifics were uniquely identifiable and consistent from village area to village area.

Beverly also noted there was little evidence of western clothing on hundreds of people, which was so commonly seen in other tribes. Beverly recalled, as she walked through the village, that it felt like a large step back

FIGURE 1.2. Mali Homes

into time, suggesting the Dogon method and philosophy of living had remained the same over the centuries. Even in the late 20th century, the Dogon people were still, from her observation, authentic. Beverly observed that farming was conducted outside of the village; farms were enclosed behind a system of wooden fences unlike those observed in other African villages where farms were incorporated within the village walls. Large plots of land were planted and tended to by the men and male children of the village. Rows of produce were separated by gullies for irrigation. The tradition of male farming and the system of irrigation are remarkably similar to BannaKa's, Robert's, and Benjamin Banneker's methods, adding credence to the theory that BannaKa may have been a Dogon prince (Bedini, 1999).

Banneker's Progeny

While Banneker never married or had offspring, his three younger sisters married and had children. There are records of two of these sisters' marriages (Perot, 2008). No records connecting the Banneker family to his brother Henry have been found (Perot, 2008). From the three younger sisters' offspring, a host of nieces and nephews became Banneker's progeny. Additional information about Banneker's legacy was obtained from an interview with one of Banneker's descendants, Robert Lett. Benjamin Banneker was Robert's seventh uncle on his father's side. Robert's father is a direct descendant of Banneker's sister, Jemima.

> I had general knowledge that my family background was special with a cornucopia of family activists. Charles Henry Lett of Zanesville, Ohio, is the family oral historian. Professor George Simpson of Wilberforce University has papers dating to the 1880s of interviews with family members as recorded history.

> We are very spiritually committed to the Christian principles of having a responsibility to help and support others in the community. The Lett family was held with great respect by the White community. We were always made aware of our mixed race background and the African and American links.

> I learned of the Dogon culture in college at California State University in Los Angeles as part of my degree in Pan African Studies. It is not taught or relayed in my family. We were all raised, however, with the belief that we were "blessed" as if guided. We were made to feel "royal" without boundaries on what we could achieve.

The above excerpts speak to the power of not only knowing who you are but also the power of instilling positive beliefs about one's ability. While the family did not acknowledge the Dogon legacy, the notion of royalty was passed down throughout the generations. Knowing who you are and where you come from is critical to developing a strong sense of identity. High expectations and role models helped to perpetuate family pride and success. What would happen if parents, teachers, and community leaders believed that *all* children, especially poor Black children, had no limits or boundaries to what they could achieve? Yet, regardless of family history, Black students must understand that their family legacy can begin with them. There are no limits on brilliance. This was the message that Banneker tried to convey when he claimed that God had given Blacks equal mental endowments.

When asked about Banneker's farm and the fire, Robert Lett responded:

> Benjamin sold the land to the Ellicotts and when they wanted to sell it they were petitioned along with the county by the Black community to have it acquired by the county and turn into a park.

> Nothing survived the fire. We have only oral history that was recorded and written down in the late 1880s and is the property of Charles Henry Lett, acclaimed family historian, or griot. There is no consensus within the family as to how the fire started. There is the general suspicion that it was arson. White people in the area had little respect for Benjamin. He was frequently shot at. The family migrated after his death.

Robert Lett's account about the land and the fire are consistent with the historical record. Baker (1918) contended that Banneker lived alone on the farm in his later years and sold it to one of the Ellicotts under the condition that Banneker could continue to live on the property and receive an annuity of 12 pounds for the rest of his life. Banneker, however, miscalculated his lifespan and lived eight years longer than he projected. It is reported that this was the only miscalculation Banneker made, ironically in his favor. After all, who can accurately predict his or her death? Nevertheless, Ellicott fulfilled the terms of the contract. The contractual arrangement allowed Banneker the opportunity to focus on his intellectual interests rather than the farm.

After Banneker's death, his sisters, Minta and Mollie, disposed of his personal effects per Banneker's request: returning Ellicott's books, instruments, and table and giving Ellicott his volume of almanacs, Jefferson's letter and reply (Baker, 1918). A feather bed was also reported to have been removed and preserved by one of Banneker's sisters. Everything else burned in a fire while Banneker's funeral was in progress. As previously stated, the fire consumed Banneker's famous clock and the rest of his belongings. Arson was a reasonable conclusion.

When asked if there were relatives with an affinity for mathematics or science, Lett responded:

> My son is very gifted in mathematics and science and is in the gifted and talented education (GATE) program. I have cousins who are doctors. One is a PhD in neuro-molecular research in medicine at the University of Wisconsin.

When asked what he thought the Banneker legacy was, Lett replied:

> I think the legacy is one of our expectation and responsibility to take care of ourselves and our community. Further, it is expected that we stand up for what is right even at a personal cost or loss. It is moral to do the right thing.

This final excerpt is a tribute to Banneker and his family's ideals concerning social justice. While he worked as a surveyor in Washington, DC, Banneker refused to eat at the same table as the other men who were commissioned for the work, although he was invited on numerous occasions. While some may think Banneker did not esteem himself worthy to eat with Major Ellicott, Major L'Enfant, and others, we believe he may have chosen not to eat with them in silent protest. Until all of God's children can sit at the table, Banneker chose not to sit at the table either.

SUMMARY

Beginning with Thomas Fuller, the human calculator, and Benjamin Banneker there have been many outstanding achievements in mathematics by African Americans (Williams, 2008). Charles Reason (1814–1893) was the first Black to serve as college faculty in mathematics. Alexander Bouchet was the first African American to earn a PhD in Physics from Yale University (1878), and Kelly Miller was the first to study mathematics at the graduate level at Johns Hopkins University (1886). From 1923 to 1947, there were twelve Blacks who earned PhDs in mathematics. The first African American to earn the PhD was Elbert Frank Cox from Cornell University in 1925. Dudley Woodard was awarded the second PhD in mathematics from the University of Pennsylvania in 1928. From 1943 to 1969, thirteen African American women earned a PhD in mathematics. The first was Euphemia Lofton Haynes, who earned her degree from Catholic University in 1943. The second woman to earn her PhD in mathematics was Evelyn Boyd Granville from Yale University in 1949. These prolific men and women are the giants upon whose shoulders contemporary Black mathematicians and mathematics educators stand.

Contemporary achievements in mathematics include Professor C. Dwight Lahr, who was the first African American to become a tenured (1981) and full professor (1984) of mathematics at an Ivy League institution. Other notable accomplished African American mathematicians include President Freeman A. Hrabowksi, III (University of Maryland, Baltimore County), also named one of America's Best Leaders by *U.S. News & World Report* in 2008, and Senior Vice President and Provost Howard C. Johnson (Medgar Evers College, CUNY), who has also served as past-president of the Benjamin Banneker Association (BBA) (1995–1997). Other outstanding Black mathematicians include William A. Massey, Edwin S. Wilsey Professor of Operations Research and Financial Engineering (Princeton University), and Professor Raymond Johnson (Rice University), the first African American to earn the PhD in Mathematics at Rice. Professors Massey and Johnson also organized the first Conference for African American Researchers in the Mathematical Sciences (CAARMS) in 1995. Promoted to full professor in 2011, Danny B. Martin (University of Illinois Chicago) holds joint appointments in the Department of Mathematics, Statistics, & Computer Science and the Department of Curriculum & Instruction, where he served as chair (2006–2011).

In addition to mathematicians, there are many African American mathematics educators in higher education in the U.S. Several have also reached the rank of full professor at prestigious private, state, and land grant institutions: Irvin Vance (emeritus, Michigan State University), also past-president of BBA (1993–1995); Martin L. Johnson (emeritus, University of Maryland), also founder of the Maryland Institute for Minority Achievement & Urban Education; Lee Stiff (North Carolina State University), also past-president of the National Council of Teachers of Mathematics (2000–2002); William F. Tate, IV (Washington University in St. Louis), also chair and Edward Mallinckrodt Distinguished Professor and past-president of the American Educational Research Association (2007–2008); Marilyn Strutchens (Auburn University), also president of the Association of Mathematics Teacher Educators (2011–2013); Michaele Chappell (Middle Tennessee State University); Rodney E. McNair (Delaware State University); M. Bernadette Carter (emerita, Cheyney University); and Jacqueline Leonard, director of the Science and Mathematics Teaching Center (University of Wyoming), also past-president of BBA (2009–2011). We recognize that this is not an exhaustive list. Many other African Americans have helped to shape the field of mathematics in the past quarter century.

African Americans have always been *achievers* in mathematics and science. Throughout American history, we have demonstrated our talent, whether enslaved (Fuller) or free (Banneker). Often self-taught, relying largely on our own native intelligence and creativity, we have persisted to attain the highest degrees. Given almost 300 years of history, we salute and give tribute to Benjamin Banneker, who was destined to be brilliant, and to all who have contributed to the education and legacy of Black children in mathematics. We believe that mathematics is in our blood because it was in the blood of our ancestors, who did the mathematics in their heads long before it was written down on paper. Thus, we acknowledge, embrace, and celebrate the brilliance and endowment of blacks in mathematics.

NOTES

1. A 1661 Maryland preamble stated: "…Whatsoever free-born woman shall intermarry with any slave, form and after the last day of the present assembly, shall serve the master of such slave during the life of her husband; and that all the issues of such free-born women, so married, shall be slaves as their fathers were….And be it further enacted: That all the issues of English, or other free-born women, that have already married negroes, shall serve the master of their parents, till they be thirty years of age and no longer." Archives of Maryland, Proceedings of the General Assembly, 1637-1664, pp. 533–534.
2. Some sources have 17,000 pounds (see Baker, 1918).
3. "Letter exchange between George Ellicott and Major Andrew Ellicott." Source: *Dear Mr. Banneker* by Blake, retrieved from http://www.kyrene.org/schools/brisas/sunda/bhistory/banneker/banneker.pdf
4. *Federal Gazette and Baltimore Daily Advertiser,* October 28, 1806.
5. Interview with Brian Deese by Jacqueline Leonard, August 23, 2011, Denver, Colorado.
6. "Letter exchange between George Ellicott and Major Andrew Ellicott": Source: *Dear Mr. Banneker* by Blake, retrieved from http://www.kyrene.org/schools/brisas/sunda/bhistory/banneker/banneker.pdf

7. Robert Lett interview with Cheryl Beverly, January 17, 2012, Pasadena, California.

REFERENCES

Baker, H. E. (1918). Benjamin Banneker, the Negro mathematician and astronomer. *The Journal of Negro History, 3*(2), 99–118.

Bedini, S. A. (1999). *The life of Benjamin Banneker: The first African-American man of science* (2nd ed.). Baltimore, MD: Maryland Historical Society.

Brodie, J. (1993). *Created equal: The lives and ideas of Black American innovators.* New York, NY: William Morrow.

Cerami, C. A. (2002). *Benjamin Banneker: Surveyor, astronomer, publisher, patriot.* New York, NY: John Wiley & Sons.

Cromwell, J. W. (1914). *Benjamin Banneker. The Negro in American history.* Washington, DC: The American Negro Academy.

Fasanelli, F., Jagger, G., & Lumpkin. B. (2012). *Benjamin Banneker's trigonometry puzzle.* Retrieved on September 24, 2012 from http://mathdl.maa.org/mathDL/23/?pa=content&sa=viewDocument&nodeId=212

Kitchen, R., Depree, J., Celedón-Pattichis, S., & Brinkerhoff, J. (2007). *Mathematics education at highly effective schools that serve the poor: Strategies for change.* Mahwah, NJ: Lawrence Erlbaum.

Leonard, J. (2008). *Culturally specific pedagogy in the mathematics classroom: Strategies for teachers and students.* New York, NY: Routledge.

Martin, D. B. (2007). Beyond missionaries or cannibals: Who should teach mathematics to African American children? *The High School Journal, 91*(1), 6–28.

Martin, D. B. (2009). Little Black boys and little Black girls: How mathematics education and research treat them? In S. L. Swars, D. W. Stinson, & S. Lemons-Smith (Eds.), *Proceedings of the 31st Annual Meeting of the North American Chapter of the International Group for the Psychology of Mathematics Education* (pp. 22–41). Atlanta, GA: Georgia State University.

McKissack, P. & McKissack, F. (1994). *African-American inventors.* Brookfield, CT: Millbrook Press.

Muccino, G. (Director). (2006). *The Pursuit of Happyness.* Los Angeles, CA: Columbia Pictures.

Nasir, N. S. (2005). Individual cognitive structuring and the sociocultural context: Strategy shifts in the game of dominoes. *The Journal of the Learning Sciences, 14*(1), 5–34.

Perot, S. (2008). *Reconstructing Molly Welsh: Race, memory and the story of Benjamin Banneker's grandmother.* Unpublished thesis, University of Massachusetts Amherst.

Stiff, L. & Harvey, W. B. (1988). On the education of black children in mathematics. *Journal of Black Studies, 19*(2), 190–203.

Williams, S. W. (2008). Mathematicians of the African Diaspora. *A modern history of Blacks in mathematics.* Retrieved on January 3, 2012 from http://www.math.buffalo.edu/mad/madhist.html

van Beek, W. E. A. (2001). *Dogon—Africa's peoples of the cliffs.* New York, NY: Harry N. Abrams.

CHAPTER 2

A CRITICAL REVIEW OF AMERICAN K–12 MATHEMATICS EDUCATION, 1900–PRESENT

Implications for the Experiences and Achievement of Black Children

Robert Q. Berry, III, Holly Henderson Pinter, and Oren L. McClain

INTRODUCTION

Examination of the history of mathematics education in the United States suggests that there have always been, and remain, tensions in conceptualizing the aims and goals of mathematics teaching and learning. Within the larger social and political landscapes of American society, these tensions have focused on what mathematics children should learn (content), how students should be taught (pedagogy), and who is qualified to teach them (quality). Throughout this history, researchers and policymakers have provided evidence which, on one hand, suggests that mathematics is best learned through drill and practice and, on the other hand, suggests that

The Brilliance of Black Children in Mathematics:
Beyond the Numbers and Toward New Discourse, pages 23–53.
Copyright © 2013 by Information Age Publishing
All rights of reproduction in any form reserved.

mathematics teaching and learning should have practical value and should build from the experiences of children. These debates have culminated in a number of "Math Wars." Also, we find that shifts in mathematics content and debates about mathematics pedagogy and quality have typically been aligned with political and social agendas, especially national security. Moreover, throughout this history, research and policy have frequently been used to make claims about what is best for *all* students.

In this chapter, we will critically examine the historical timeline of mathematics education research, policy, and reform to highlight the impact on *Black* children. Specifically, we will examine the mathematics education timeline through a lens that simultaneously considers mathematics education research, policy, and the social conditions of Black people at historically significant points within the timeline. The primary aim of this chapter is to provide a lens on Black children's mathematical experiences from the nadir to the present. We examined research articles, book chapters, and policy documents along with venues for social commentary (e.g. newspaper and magazines articles) to provide an analysis of the contexts in which Black children's experiences with mathematics teaching and learning have taken place and the consequences of those experiences. Given the historical shifts described above, we raise several questions when examining the amount and type of research on Black children:

1. Given the larger social and political contexts, how do issues of power and authority impact the mathematics experiences of Black children?
2. What are the structural and institutional variables that affect the mathematics experiences of Black children?
3. In what ways do policy and reforms affect the mathematics experiences of Black children?
4. What variables are researchers investigating that impact Black children's mathematics achievement?
5. What other variables beyond mathematics achievement should be considered when describing outcomes for Black children?

The scope of these five questions is discussed along the historical timelines presented throughout this chapter. Additionally, these questions will be revisited at the end of this chapter.

NEW INDUSTRIALISM AND THE BLACK NADIR (1900S–1940)

The beginning of the twentieth century saw much change in the character of the United States. Science, industry, and transportation were making strides in this period, placing new demands on schools. The new industrialism, involving factory systems and mass production, required workers to have at minimum, elementary grade competencies (Martin, 1964). Additionally, compulsory attendance laws brought more children into schools. Compulsory attendance laws were also passed to deal with child labor issues; however, these laws were irregularly enforced. While the primary intent of the compulsory attendance laws was to protect White children, they also led to increases in primary school attendance for Black children (Du Bois, 1915; Martin, 1964). The increase in Black children attending primary schools also led to increases in the number of Black schools. However, for many Black children in the South, economic and social forces limited school attendance (Anderson, 1988; Martin, 1964). It is important to note that during this period Black children were either denied education or were provided an education in a separate, segregated environment. The advent of new industrialism placed demands on schools to determine what mathematics should be taught and how it should be taught. The mathematics taught in school during this period focused on mathematics used in the everyday work requirements of adults, based on the set of needs determined by new industrialism (DeVault & Weaver, 1970). That is, mathematics content and teaching focused on increasing productivity in various industries such as agriculture, mechanics, carpentry, factory systems, and others.

For Black Americans, the period from the end of Reconstruction through the early 20th century, is the "nadir" (Logan, 1954). The nadir is when racism is deemed to have been worse than in any other period after the American Civil War. During this period, Blacks lost many civil rights protections afforded by the fourteenth and fifteenth amendments. Anti-Black violence, lynching, segregation, legal racial discrimination, and expressions of White supremacy increased. One such protection provided prior to the nadir was the Freedmen's Bureau bill. The Freedmen's Bureau provided aid to those formerly enslaved through legal supports, food, housing, civil rights protections, education, health care, and employment contracts with private landowners. The Freedmen's Bureau also provided supports for Black children to attend schools (Anderson, 1988). However, the Reconstruction Act of 1867 required the removal of citizen agents' support by the Freedmen's Bureau, thus marking the approximate beginning of the nadir. After the elimination of the Freedmen's Bureau, philanthropic organizations contributed substantially to the creation of or continuance of Black schools (Standish, 2006). The Peabody Education Foundation assisted the South in establishing schools after the financial ruin resulting from the Civil War (Smith, 1950). The Southern Education Foundation assisted schools and communities in the South by building infrastructure and providing teacher training. The Anna T. Jeanes Foundation focused its efforts on rural schools and vocational training (Pincham, 2005). In 1913 Julius Rosenwald developed a foundation to build over 5,000 Black schools in rural southern areas of the South between 1913 and 1932 (Anderson, 1988; Hughes, Gordon, & Hillman, 1980).

Booker T. Washington and W. E. B. Du Bois emerged as two prominent educators and activists during the nadir. Both Washington and Du Bois were advocates for Blacks gaining economic stability and self-sufficiency through training and education, respectively. Washington stressed vocational and industrial training as a way for Blacks to gain equal social rights. In his "Atlanta Compromise" speech in 1895, Washington's vision of the role of education for Black Southerners called on them to "cast down their buckets...in agriculture, in mechanics, in commerce, in domestic service and in other professions" (Hill et al., 1998, p. 682). Because Washington stressed vocational and industrial training for Black southerners, his views for mathematics content and teaching focused on preparing Black learners to use mathematics that was required for their daily work lives and to meet the needs of industries. In his autobiography, *Up from Slavery,* Washington (1901) referred several times to mathematics for everyday purposes by describing the classrooms he observed:

> The students who came first seemed to be fond of memorizing long and complicated "rules" in grammar and mathematics, but had little thought or knowledge of applying these rules to their everyday affairs of their life. One subject which they liked to talk about, and tell me that they had mastered, in arithmetic, was "banking and discount," but I soon found out that neither they nor almost anyone in the neighbourhood in which they had lived had ever had a bank account. (p. 122)

While Washington focused on learning that connected to the lives of Black children, he recognized the brilliance of Black children. It appeared that he struggled with finding a balance between what he described as "high-sounding things" and vocational studies. Through his observations, we see that he acknowledges the brilliance of Black learners' capacities to know and understand "high-sounding things," while at the same time stressing the importance of vocational and everyday ideals:

> I have never seen a more earnest and willing company of young men and women than these students were. They were all willing to learn the right thing as soon as it was shown them what was right. I was determined to start them off on a solid and thorough foundation, so far as their books were concerned. I soon learned that most of them had the merest smattering of the high-sounding things that they had studied. While they could locate the Desert of Sahara or the capital of China on an artificial globe, I found out that the girls could not locate the proper places for the knives and forks on an actual dinner table, or the places on which the bread and meat should be set....I had to summon a good deal of courage to take a student who had been studying cube root and "banking and discount," and explain to him that the wis-

est thing for him to do first was thoroughly master the multiplication table. (Washington, 1901, p. 122)

We see Washington's tension as he described a student who is studying cube root and "banking and discount." Many in mathematics would expect that a student who is studying cube root and banking would be skillful with the multiplication table, proficient with arithmetic, and competent in other areas of mathematics. The tension lies in the connection between "high-sounding" mathematics and the use of mathematics for vocational uplift. However, Washington indirectly acknowledges the brilliance of Black learners as being capable of knowing rigorous mathematics topics and other "high-sounding things."

W. E. B. Du Bois challenged the position of primarily training Black people for vocations and industries. Du Bois (1915) argued that vocational and industrial education as the primary educational program for Black people does little to prepare for the rapidly changing pace of industry. In essence, training one for industry leads to displacement as technologies of industries evolve to take the place of laborers. Du Bois (1915) asserted, "While we teach men to earn a living, that teaching is incidental and subordinate to the larger training of intelligence in human beings and to the largest development of self-realization in men" (p. 133). Expounding on Du Bois' views, Alridge (1999) described six educational principles in Du Bois' work:

1. African-American-centered education—this focuses on understanding the social realities of what Du Bois described as a "double consciousness": knowing what it means to be African and American;
2. Communal education—this focuses on advocacy for social and economic cooperation;
3. Broad-based education—this focuses on providing a comprehensive education to meet the demand of technological changes;
4. Group leadership education—this focuses on preparing all Black children for success and leadership, which diverges from Du Bois' early assertions of the "talented tenth;"
5. Pan-Africanist education—this focuses on understanding the philosophies, histories and worldviews of the African diaspora; and
6. Global education—this focuses on a perspective that encourages understanding the American democracy in a global context.

Within the context of these six educational principles, it is plausible to consider that Du Bois would view mathematics teaching as preparing Black children to be thinkers and use mathematics to critique the world in which they lived. From this point of view, mathematics is not simply a mental exercise, but mathematics is a tool to analyze social and economic issues, critique power dynamics, and to build advocacy within the Black community. While Du Bois' six education principles suggest that he supported democratic education and creativity in the mathematics curriculum, his work also suggests that he supported rote pedagogies for mathematics. For example, Du Bois stated:

> The learning of the multiplication table cannot be done by inspiration or exhortation. It is a matter of *blunt, hard, exercise of memory, done so repeatedly and for so many years, that it becomes second nature, so it cannot be forgotten* [emphasis Du Bois']. (as cited in Aldridge, 2008, p. 104)

Du Bois argued that children needed a broad and balanced curriculum that would teach the particulars of reading, writing, and arithmetic. Aldridge (2008) described Du Bois' educational views as a dualism for advocacy of democratic education, creative, and interactive learning along with an advocacy for rote memorization. Part of Du Bois' dualism is reflected in the mainstream mathematics education during this time period.

The mainstream history of mathematics education tells us that the work of Edward L. Thorndike had a tremendous influence on mathematics education at the beginning of the twentieth century. Thorndike's work is reflective of the new industrialism of the early twentieth century because it focused on efficiency and sorting students into mathematics courses suited to their perceived future needs (Gould, 1996; Henriques, Hollway, Urwin, Venn, & Walkerdine, 1998; Thayer, 1928). Thorndike's Stimulus-Response Bond theory had a profound influence on the teaching and learning of mathematics (English & Halford, 1995; Willoughby, 2000). Thorndike contended that mathematics is learned through drill and practice and viewed mathematics as a "hierarchy of mental habits or connections" (Thorndike, 1923, p. 52) that must be carefully sequenced, explicitly taught, and then practiced with much repetition in order for learning to occur.

Thorndike's theory about mathematics education converges with Booker T. Washington's and W. E. B. Du Bois' views. Washington appears to support mastery of facts, which is consistent with Thorndike's theory. While it is clear that Du Bois supported drill and practice for mathematics teaching and learning, it is not clear whether Washington supported drill and practice. However, it appears that Washington would support efficient mathematics teaching. Many schools educated Black children by teaching them to memorize facts (Martin, 1964). Du Bois situated mathematics teaching as a "mental habit," while also contending that Black folks be liberated through communal, broad-based, global and Pan-Africanist education. Du Bois' dualism supported the acquisition of basic mathematics skills and affirmed teaching mathematics to understand issues in the context of children's lives but also learning the classical studies of mathematics. Du Bois' contemporary, Carter G. Woodson, author of *The Mis-education of the Negro*

(1933/1999), called for an approach for mathematics teaching reflecting the experiences and traditions of Black people. Woodson argued, "And even in the certitude of science or mathematics it has been unfortunate that the approach to the Negro has been borrowed from a 'foreign' method" (p. 4). In mathematics education, a "foreign" pedagogy ignores the issues that are relevant to the experiences of Black children. Both Du Bois and Woodson would contend that education built strictly on the thinking and experiences of White children was inappropriate for Black children (Tate, 1994). While mathematics education was heavily influenced by the work of Thorndike, both Washington and Du Bois supported curricula that prepared Black children to meet the needs of their communities.

BROWN, SPUTNIK, AND NEW MATH (1940S–1960S)

Lessons learned from World War II had a great influence on the elementary school mathematics curriculum (DeVault & Weaver, 1970). Many World War II recruits failed to exhibit minimal competencies in mathematics, which was viewed as a threat to national security. As a result, during the late 1940s and early 1950s, increased attention was given to the mathematics curriculum (DeVault & Weaver, 1970). Congress established the National Science Foundation in 1950, which contributed greatly to the development of new mathematics curricula projects of the 1950s and 1960s. Parallel to these activities, the "new math" reform movement began in the early 1950s with the initiation of many new mathematics curricula projects led by university mathematicians (Walmsley, 2003). The launch of the first artificial satellite, *Sputnik*, on October 4, 1957, by the USSR gave impetus to the drive to improve mathematics education in America. The launching of *Sputnik* brought heightened concern about America's national security as well as concern that children in America were lagging behind the Russians in mathematics and science. Shortly, after *Sputnik* launched, more federal funds for mathematics and mathematics education became available through the U.S. National Defense Education Act of 1958 (Walmsley, 2003). These funds were provided for improvements in mathematics and science education and to identify America's "best and brightest" young scientific minds.

Approximately three years prior to the launching of *Sputnik*, the United States Supreme Court issued the landmark ruling in *Brown v Board of Education of Topeka, Kansas* (*Brown I*) on May 17, 1954. The Supreme Court's ruling in the *Brown I* case revoked the "separate but equal" doctrine, which legally sanctioned segregation in public education and all aspects of daily life among Black Americans and White Americans. There was much opposition to the Brown decision, and it took further rulings from the federal courts (including the *Brown II* decision handed down in 1955) to instruct the states to act with "all deliberate speed" with respect to desegregating public schools. Critics of the Brown decisions thought the decisions fell short because the vague language did not provide specific timetables; thus, segregationists delayed implementation of the *Brown I* decision (Mayo, 2007). The Supreme Court and other federal courts began to force meaningful school desegregation in the South in the mid to late 1960s and early 1970s (Feagin & Barnett, 2004). From the beginning, Black parents and community leaders sought desegregation primarily to secure greater access to educational and related resources. They did not seek desegregation because they felt that Black children needed to sit with White children to be educated. The assumption was that better school resources accompanied schools where White children were taught, and Black parents believed that better resources provided greater opportunities.

The *Brown* decisions occurred in the midst of large-scale efforts to reform what mathematics should be taught and how it should be taught. As stated earlier, these efforts were initiated in response to *Sputnik*; the "new math" reform offered new content as well as new teaching approaches (Walmsley, 2003). The new content consisted of abstract algebra, topology, symbolic logic, set theory, and Boolean algebra taught in conjunction with much of the traditional curriculum. Set theory often became synonymous with "new math," but the public thought set theory was being taught instead of arithmetic. One main idea of "new math" was to abandon the drill and practice approach with approaches where students could develop conceptual understanding of the mathematics. These new approaches included the use of hands-on models (manipulatives), discovery/guided-discovery learning teaching practices, and the spiral curriculum (Walmsley, 2003; Willoughby, 2000).

The reforms of "new math" did very little to address the concerns Black parents had for their children (Tate, 2000). Those responsible for the "new math" reforms were most concerned with identifying the "best and brightest" scientific minds rather than mathematics for all children (Tate, 2000). Only a select few communities and students were to be served by the "new math" reform. The appeal to limit the mathematics reform to the perceived "best and brightest" was built on a political philosophy focused on protecting the national security and interests of America (Tate, 1997). This meant that the mathematics experiences of Black children were largely ignored and pushed aside within the larger discussions of *Sputnik*. In fact, one could argue that Black children had virtually no access to the "new math" curricula and pedagogy that resulted from this reform. The approach to mathematics teaching remained similar to that described during the new industrialism and nadir periods, which Woodson described as a "foreign" pedagogy focusing on drill, practice, and memorization of facts. Similarly, Tate (2000) characterized the mathematics reform in the 1950s and 1960s for Black children as "benign neglect" (p. 201).

Snipes and Waters (2005) provided a lens on the experiences of Black children's mathematics experiences in the 1960s in segregated schools in North Carolina. Snipes and Waters presented a case study of a former state-level mathematics curriculum specialist in North Carolina, Mr. Smith. Mr. Smith's primary responsibility was to determine the mathematics needs of Black, Native American, and White schools. Mr. Smith observed that Black students in segregated schools received used textbooks handed down from White schools, and these texts served as the curriculum. Because Black schools used old textbooks and the textbook was viewed as the curriculum, the reforms of the "new math" were not envisioned for Black children in North Carolina. That is, Black children did not receive access to new content nor experience the new approaches to teaching associated with the "new math" reform movement. Mr. Smith described mathematics instruction at Black schools as follows: "There was too much lecture, show and tell, and kids sitting pretty much idly by in all the schools that I observed. The kids were just not involved" (Snipes & Waters, 2005, p. 116).

Standish (2006) described similar circumstances in her work focusing on Black teachers in Texas during the pre- and post-*Brown* eras. She found that Black schools received secondhand books and supplies. However, she observed that Black teachers in segregated schools "made do" with the lack of books, supplies, and poor facilities. One teacher commented:

> Quality of the teachers, now this is just my view, but I think that under the circumstances under which they labored they were superb. Because they were given poor equipment, they were given limited money for supplies. Basic things like the books were old. I think that to even be able to impart enough knowledge to get students into college, under those circumstances, they did a superb job. (Standish, 2006, p. 174)

Making do reflected in the high expectations Black teachers had for themselves and their students. Another teacher in Standish's (2006) work reported, "the quality of teaching was excellent…High expectations for all students and a love for children" (p. 184). A mathematics teacher expressed, "They [teachers] made you believe in yourself. You may not be able to do all of the math problems, but you could do some of them…" (Standish, 2006, p. 186). These reports suggest that Black teachers in segregated schools believed in the brilliance of Black children and "made do" with substandard materials because they expected the children to work hard, and they wanted to improve students' lives.

Offering advanced studies in mathematics and requiring at minimum Algebra I as a graduation requirement suggested that segregated Black high schools had high expectations in mathematics for Black children (Foster, 1997; Snipes & Waters, 2005; Standish, 2006; Walker, 2000). Foster (1997) described a teacher, Everett Dawson, who taught a course containing content beyond geometry and Algebra II at Horton High School, a segregated Black school in North Carolina. Dawson was the first teacher to teach this advanced course in his county. However, when county officials found out, they blocked Dawson from teaching the course. According to Dawson, "They cut it out until the White school could establish the course and catch up with us. That's how determined the White folks were to be better than we were" (p. 5). Similarly, when Dunbar High School in Washington, DC, added calculus to its curriculum, the school board eliminated it from course offerings (Sowell, 1974). In North Carolina, segregated Black high schools had Algebra I as a requirement for graduation before integration, whereas White schools did not have such a requirement (Snipes & Waters, 2005). Similarly, teachers in Standish's (2006) study reported that Black schools in Texas had Algebra I as a graduation requirement. Because Black students in North Carolina and Texas were steered into Algebra I due to the graduation requirement, it appeared that these students benefited from heterogeneous grouping. This requirement suggests that teachers believed that Black children were capable of studying and learning at minimum Algebra I.

After school integration, Algebra I was no longer a requirement for graduation. North Carolina created mathematics courses below the mathematics rigor of Algebra I, which included consumer mathematics, general mathematics and other less rigorous mathematics courses to accommodate the perceived lack of preparation of Black students (Snipes & Waters, 2005). From Snipes and Water's (2005) case study, Mr. Smith reported, "When we desegregated the schools, the kids that might have stayed in a Black school and gone through Algebra I were sent to general math... (p. 118)." In Texas, Black children were primarily placed in what was perceived as low-level mathematics courses (Standish, 2006). The change in graduation requirement and expectations appeared to have an impact on Black children's mathematical experiences that persisted well after integration. This is not to suggest that the mathematical experiences of Black children were better in segregated schools; rather, the process of integration did not attend to or appreciate the social realities and needs of Black children to ensure that these children would receive caring teachers who were vested in their mathematics success. During integration, Black children did not have the same access to Algebra I as they did prior to integration. In segregated schools, many Black children had mathematics teachers who cared about them as people and expected them to be prepared for rigorous mathematics studies. However, in integrated schools, the level of attention and care provided to Black children was lost; consequently, these children were placed in low-level mathematics courses, which did little to adequately prepare these children for rigorous mathematics studies and college. The consequence of lowered expectations coupled with the systematic design to diminish access to rigorous studies in mathematics appears to have had generational negative impacts on Black children. We see these impacts in classrooms today when we observe the racial compositions of mathematics courses, mathematics achievement, and the care that Black children perceive as receiving in mathematics classrooms (Berry, 2008).

THE GREAT SOCIETY, DESEGREGATION, AND BACK TO BASICS (MID 1960s–1980)

During the mid 1960s there was increased spending by the federal government to broaden opportunities for all Americans. Federal programs and legislations provided reforms to eliminate poverty and support civil rights. President Lyndon Johnson had a vision for a "Great Society," which was an effort to address issues of civil rights, poverty, economic inequities, and access to health care, housing, jobs, and education. The civil rights legislations and the development of programs such as Title I, Medicare, and Medicaid were part of President Johnson's vision (Levitan & Taggart, 1976). Title I, a federal compensatory program, was enacted through the Elementary and Secondary Education Act of 1965 (Wong & Nicotera, 2004). The purpose of the Title I program was to allocate extra funds to schools with high concentrations of poverty in order to improve the educational opportunities of poor students. Several civil rights legislations provided greater opportunities for Americans. The Civil Rights Act of 1964 forbade job discrimination and the segregation of public accommodations. The Voting Rights Act of 1965 suspended use of literacy tests and other voter qualification tests, and stopped poll taxes. The Civil Rights Act of 1968 banned housing discrimination and extended constitutional protections to Native Americans on reservations. Critics argued that President Johnson's vision for a "Great Society" was in response to the increasing pressures from civil rights activists of the 1960s.

The work of James Coleman coincided with the Great Society initiatives, and, in 1966, Coleman and colleagues issued a report titled *Equality of Educational Opportunity* commonly referred to as the Coleman Report (Coleman et al., 1966). The report argued that school resources had little effect on student achievement and that student background and socioeconomic status are much more important in determining educational outcomes (Wong & Nicotera, 2004). One finding that received significant attention from policymakers and civil rights activist was that peer effects had a significant impact on student achievement, meaning the background characteristics of other students influenced student achievement. Many interpreted this finding to mean that Black children who attended integrated schools would have higher test scores if a majority of their classmates were White (Wong & Nicotera, 2004). This one finding coupled with the tensions of desegregation was a catalyst for the implementation of the desegregation busing systems that occurred in many places in the United States.

Lessons learned from hindsight inform us that the findings of the Coleman Report should be considered with caution. There are a number of counter-stories to suggest that human and material resources have a significant effect on Black children's achievement and participation in mathematics during segregation and desegregation (Standish, 2006; Walker, 2000). To that end, Walker (2000) identified four common themes valued among Black segregated schools from 1935 to 1969: a) exemplary teachers, b) curriculum and extracurricular activities, c) parental support, and d) leadership of the school principal. Across these themes, we see a significant grounding in relationship between people caring about and connecting to the social realities of Black children. Interestingly, Coleman issued a report in 1975 concluding that busing failed largely because it had prompted "White flight." That is, as White families fled to suburban schools, the report concluded, the opportunity for achieving racial balance evaporated. This implies that significant thought was not given to understanding Black children within the context of who they are; rather, we argue that the Coleman report did not seek to understand Black children and had no regard for who they are and the resources these children bring to schools. The Coleman report only sought to understand Black children within the school context and relative to White children. Consequently, there may be an over-reliance on the peer effect finding.

Busing was one strategy used to desegregate schools to provide equitable opportunities for Black children. Black children were more likely to be bused than White children to promote desegregation; thus, many Black children were displaced from their neighborhoods (Doughty, 1978). The efforts to desegregate schools focused on racial balance in schools rather than improving the academic experiences and life chances of Black children. While busing sought to desegregate at the school level, we must consider what was happening at the classroom level. In schools where significant numbers of Black children were bused, these children experienced resegregation for their mathematics instruction. In fact 70% of school districts had racially identifiable classrooms as a result of ability grouping resegregation (Doughty, 1978). That is, because of the development of and placement in low-level mathematics courses, Black children were in situations that were contrary to Coleman's peer effect findings. Additionally, Black children were more likely to be placed in special education programs. In fact, Doughty (1978) estimated that 91% of Black children in special education programs during this period were incorrectly assigned on the basis of low expectations and inaccuracies in IQ scores. Resegregation during this period is perhaps the foundational legacy of the tracking Black children experience today in schools and in mathematics. Mickelson

(2001) argued that resegregation in classrooms through tracking undermined any potential benefits of school-level desegregation. Further, the effects of tracking on academic outcomes must be considered in assessing the effects of desegregation. In theory, tracking is a meritocratic process that allocates resources and opportunities commensurate with students' prior academic achievement, abilities, and interests (Mickelson, 2001). This meritocratic process failed to consider the interests and abilities of Black children studying mathematics in resegregated classrooms; rather, these spaces served to limit access and opportunities. Tracking placed Black children in a status hierarchy where differential tracks prepared many of these children for paths designated for workers and laborers. Because tracks tend to be homogeneous, students receive little exposure to individuals who differ from themselves. Further, the literature is compelling in proposing that differences in tracked mathematics courses lead to differences in opportunities, engagement, support, rigor, and cognitive demands (Oakes, 1990). When we consider the effects of busing and resegregation, it is plausible to question the validity of the findings from the Coleman Report. It is possible that for many Black children, integration at the classroom level never happened.

While we are critical of aspects of the Coleman Report and desegregation efforts, we do not want to imply that the federal policies, programs, and legislations enacted during the 1960s and 1970s were ineffective and unnecessary. Black people had many economic, political, and academic gains during this period. During the 1960s and 1970s, median Black family income rose 53 percent, Black employment in professional and technical occupations doubled, and the average Black educational attainment increased by four years. The proportion of Blacks families below the poverty line fell from 55 percent in 1960 to 27 percent in 1968, and the Black unemployment rate fell 34 percent (Mintz, 2007). When examining mathematics trends as measured by the National Assessment of Educational Progress (NAEP), Black children at the ages of nine and 13 years old had increases in achievement from 1972 to 1980 (Burton & Jones, 1982). However, Tate (2005) is critical of these gains because they are closely aligned with a basic skills mathematics curriculum. In this sense, Black students' performance reflected achievement in basic arithmetic facts and skills rather than understanding in problem-solving, reasoning, and complex procedures.

President Richard Nixon's 1968 election served as a conservative response to the liberal 1960s policies (Hughes, 2006). During the Nixon administration, schools moved to approaches that emphasized "core curriculum" in mathematics, increased accountability for teachers and students, and the initiation of large-scale achievement tests. In the late 1960s and early 1970s, the "back to basics" reform movement in mathematics emerged in response to the perceived shortcomings of "new math" (Burrill, 2001).

During this period, the National Science Foundation discontinued funding programs focused on "new math," and there was a call to go back to the "core curriculum," which was understood to be basic skills in mathematics. Tate (2000) suggested that this movement was an outgrowth of efforts to achieve equality of educational opportunity through compensatory education. That is, by focusing on the core content in mathematics, students would learn and achieve basic skills. The "back to basics" movement called for teaching basic mathematics procedures and skills and was closely connected to the minimum competency testing movement used extensively by states in the 1970s and 1980s (Resnick, 1980; Tate, 2000).

Testing had a significant impact on the mathematics content that was taught and the methods used to teach mathematics. Typically, students were taught mathematics content deemed important for passing tests. The focus on basic skills dominated textbook publishing through the early 1980s, leading to another generation of Thorndike-like mathematics textbooks (Ellis & Berry, 2005; English & Halford, 1995). Although the emphasis on skills did result in slightly improved standardized test scores for Black children, it was criticized for not adequately preparing these students for mathematics coursework requiring higher levels of cognition and understanding (Tate, 2000; United States Congress Office of Technology Assessment, 1992). At this time, Black students were underrepresented in the upper achievement distribution and in upper-level mathematics courses (Tate, 2000). This phenomenon could be explained by tracking, a narrowing of the curriculum, and low teacher expectations.

When considering the immediate and larger impact of desegregation coupled with an emphasis on testing, we find that Black students had decreased opportunities to study rigorous mathematics and that mathematics pedagogy did not account for the realities of their lives. For Black children, there was no "back to basics" movement because the pedagogy consistently emphasized during this period focused on drill, practice, and the memorization of facts. These pedagogies were constant in many of these children's mathematical experiences. The emphasis on testing provided an additional layer that undermined the experiences of many Black children in schools. Consequently, many schools used testing to legitimize the perception that Black children are not capable of rigorous studies in mathematics.

The "back to basics" movement provided more focus on using achievement tests to pathologize Black children as being inferior, deficit and deviant. We find the first of many research studies focusing on the Black-White "achievement gap" in mathematics during this period (Perry, 2003). If one considers the context of this period and the persistent limited educational opportunities available to Black children, discussion of a Black-White "achievement gap" serves to reinforce an ideology about Black children's intellectual inferiority, and, as such, undermines how Black children are seen by others and themselves. Considering the factors that affect the mathematics experiences of many Black children, there are examples of brilliance in mathematics among Black children during this period. For example, Tate (1994) described his elementary mathematics experiences at an urban Catholic school. He discussed three mathematics experiences in sixth grade. The first was a traditional sixth-grade mathematics course, the second was a combined algebra and geometry course, and the third experience was the informal opportunities that allowed him to develop out-of-school mathematical knowledge. Tate described how he and his classmate built a telephone system for the class using various electronic components. They explored mathematics within the context of technical and economic issues. He also described how he made weekly museum trips, read scientific journals, and shared his knowledge with classmates, a younger brother, and cousins. Tate credited the "good teaching" he received at his Catholic school as a catalyst for not only affecting his intellectual curiosity but also affecting others.

A NATION AT RISK AND THE STANDARDS MOVEMENT (1980–2000)

In 1983, the National Commission on Excellence in Education, appointed by President Ronald Reagan, issued a report titled, *A Nation at Risk: The Imperative for Educational Reform.* This report contributed to the sense that American schools were failing, thus initiating local, state, and federal educational reform efforts. The report suggested that education reform is necessary because competitors throughout the world are overtaking America's preeminence in commerce, industry, science, and technology. Furthermore, the report stated, "If an unfriendly foreign power had attempted to impose on America the mediocre educational performance that exists today, we might well have viewed it as an act of war" (National Commission on Excellence in Education, 1983, p. 1). The inflammatory rhetoric of *A Nation at Risk* heightened concerns about America's national security that schools and colleges were not doing an adequate job of preparation, and America's children were lagging in mathematics and science when compared internationally. *A Nation at Risk* stated that through educational reform, American children's promise of economic, social, and political security in the future would be earned by meritocratic ideals of effort, competence, and informed judgment.

Many states placed Algebra I as a high school graduation requirement as a reaction to *A Nation at Risk.* Between 1982 and 1992, students enrolled in Algebra I increased from 65 to 89 percent, in Algebra II from 35 to 62 percent, and in calculus from five to 11 percent (Raizen, McLead, & Rowe, 1997). Planty, Provasnik, and Daniel (2007) reported that the percentage of graduates who completed a semester or more of Algebra II rose from 40 percent in 1982 to 67 percent in 2004. This evidence suggests that since the early 1980s, the average number of mathematics courses at or above Algebra I taken by high school students has increased. The increased focus on maximizing students' performance on standardized tests has led schools to rethink course-taking patterns (Kelly, 2009). In 1982, the National Assessment of Education Progress (NAEP) reported that 57% of Black 17 year-olds studied Algebra I and 34% studied geometry (National Center for Education Statistics, 2012). The National Science Foundation (2003) reported that in 1990, 56% of Black students graduated from high school with geometry credit, 41% with Algebra II credit, 6% with trigonometry credit, 6% with pre-calculus/analysis credit, and 3% with calculus credit. These enrollment rates are significantly lower than the enrollment patterns of the national sample, with Black students to be more likely to be enrolled in Algebra I and geometry but less likely enrolled in higher-level courses. In the early 1990s, Bob Moses, one of the stalwarts of the civil rights movement, argued that access to algebra is the "new" civil rights (Jetter, 1993). Moses contended that algebra served as a curricular gatekeeper tracking numerous students out of many in-school and out-of-school opportunities. This had an impact on higher-level mathematics enrollment and educational and economic opportunities.

Another reaction to *A Nation at* Risk was that these courses remained quite traditional in spite of reform initiatives to change pedagogical methodologies. The increased enrollment in the upper-level mathematics courses did not influence instructional methodologies to meet the increase in the diverse learning needs of children (Porter, Kirst, Osthoff, Smithson, & Schneider, 1993). That is, for Black children, instruction focused primarily on the acquisition of skills. Additionally, much of the increase in mathematics course enrollment occurred by simply placing students in Algebra I tracks or eliminating low-level courses such as general mathematics. Little considerations were given towards preparing these students for rigorous mathematics, unlike the care given to Black children's preparation for rigorous mathematics during segregation when many Black schools had Algebra I as a minimum graduation requirement. When the increased enrollment in mathematics courses seemed an insufficient means for increasing student achievement, policymakers and reformers began to investigate notions of systemic reform within the larger education system (Raizen et al., 1997). In the fall of 1989, President George H. W. Bush, the nation's governors, and other leaders held an educational summit in Charlottesville, Virginia. One result of this meeting was a call for national standards. Participants at the 1989 National Education Summit made a commitment to make U.S. students first in the world in mathematics and science by the year 2000. The National Education Summit provided the foundation for the National Education Goals Panel (1990) charged with providing a catalyst for change in schools and communities.

In mathematics education, the first set of standards was neither federally sponsored nor considered elitist (Raizen et al., 1997). Recall, that the "new math" movement was led by mathematicians and scientists with the aim to identify the mathematical elite. Rather, the efforts of the 1980s were led by mathematics educators and teacher educators with the support of the National Council of Teachers of Mathematics (NCTM). In 1980, NCTM put forth its *Agenda for Action*, which diverged in some ways from "back to basics." The *Agenda for Action* put forward eight recommendations, that:

1. Problem solving be the focus of school mathematics in the 1980s;
2. Basic skills in mathematics be defined to encompass more than computational facility;
3. Mathematics programs take full advantage of the power of calculators and computers at all grade levels;
4. Stringent standards of both effectiveness and efficiency be applied to the teaching of mathematics;
5. The success of mathematics programs and student learning be evaluated by a wider range of measures than conventional testing;
6. More mathematics study be required for all students and a flexible curriculum with a greater range of options be designed to accommodate the diverse needs of the student population;
7. Mathematics teachers demand of themselves and their colleagues a high level of professionalism;
8. Public support for mathematics instruction be raised to a level commensurate with the importance of mathematical understanding to individuals and society.

The eight recommendations reflected a move away from elitism towards mathematics teaching and learning that broadens the notion of basic skills as the acquisition of skills toward focusing on problem solving, use of technology, broad notions of achievement, and an effort to meet students' diverse needs. While the *Agenda for Action* was not a standards document, it was the foundation for the first standards documents developed by a content area association, NCTM. NCTM provided the framework for a grassroots effort led by mathematics educators, teachers, and teacher educators to develop the *Curriculum and Evaluation Standards for School Mathematics* (*CSSM*) (NCTM, 1989). The *CSSM* consisted of content and process standards for pre-kindergarten through twelfth grades. The NCTM standards were comprised of grade bands, rather than being grade-level specific, and included goals and evaluation standards for mathematics teaching and learning.

Apple (1992) commented that the NCTM standards were vague enough to allow many supporters but specific enough to offer something to educators. *CSSM* was a visionary document for mathematics teaching and learning, providing broad frameworks for mathematics content and processes across grade bands. Emphasis was placed on an inquiry-based approach to mathematics teaching and learning in which students were exposed to real-world problems that help them develop fluency in number sense, reasoning, and problem-solving skills. The inquiry-based approach supported conceptual understanding as a primary goal, and algorithmic fluency would follow once conceptual understanding was developed. Critics of *CSSM* argued that the primary goal of conceptual understanding through an inquiry-based approach did not help children acquire basic skills efficiently nor learn standard algorithms and formulas. They also claimed that there was too much emphasis on constructivist pedagogy, student-centered learning, discovery learning, and real-world problems (Klein, 2003). However, given the timeliness of the *CSSM's* release and the call for national mathematics standards, *CSSM* received massive support from the U.S. Department of Education and the National Science Foundation (NSF). Through the 1990s, NSF supported the creation and development of commercial mathematics curricula aligned to the standards in *CSSM*. As many as 15 elementary, middle grades, and high school curricula projects were developed with NSF support (Klein, 2003). Critics of the curricula objected to the inquiry-based approaches, claiming that not enough emphasis was placed on acquisition of basic skills and general mathematics principles (Klein, 2003). Tension between proponents and opponents of *CSSM* resulted in the "Math Wars." The "Math Wars" started in the 1990s and persist today through public debates, popular press, online forums, and other avenues. It is reasonable to consider that there were proponents for improving mathematics instruction for Black children on both sides of the "Math Wars." The dualism of Du Bois' perspective, discussed earlier, is on both sides of the "Math Wars." The primary tensions focus on mathematics content, pedagogical approaches, and student achievement, with both sides agreeing that reform is necessary for America's economic, technological, and security interests.

The *CSSM* outlined four social goals for schools: a) mathematically literate workers, b) lifelong learning, c) opportunity for all, and d) an informed electorate. These four goals derived from the fact that at the time society was moving towards an increase in technologies. These goals situated social justice issues in school mathematics within the framework of economic competition and national technological interests. For example, the opportunity-for-all goal stated:

> The social injustices of past schooling practices can no longer be tolerated. Current statistics indicate that those who study advanced mathematics are most often white males. Women and most minorities study less mathematics and are seriously underrepresented in careers using science and technology. Creating a just society in which women and various ethnic groups enjoy equal

opportunities and equitable treatment is no longer an issue. Mathematics has become a critical filter for employment and full participation in our society. We cannot afford to have the majority of our population mathematically illiterate: Equity has become an economic necessity. (NCTM, 1989, p. 4)

Positioning social justice in mathematics education within the framework of economic competition and national technological interests situates mathematics education as being primarily utilitarian. That is, from an economic competition perspective, increasing Black children's participation in mathematics education is based on ensuring that America's economic and social interests are met rather than the interest to serve the families and communities of Black children. Consequently, the interests of Black children are not given careful consideration; it is the interests of the broader American contexts that drive implementation of standards. Within this context, mathematics is always situated as a utilitarian area of study, and the focus of mathematics education is on the needs of national interests rather than the needs of a democratic society.

There is a long history indicating that during times of reform, the interests and needs of Black children are in many ways dismissed. The tensions of the "Math Wars" appear to have an underlying narrative focusing on the nation's technological interests, social efficiency, and perpetuation of privilege. There are intense debates focusing on curriculum, teaching, and assessment but little debate focusing on understanding the realities of children's lives. For Black children, issues of race, racism, identity, and conditions are not under consideration in the "Math Wars." There is evidence from the *CSSM* to suggest that the standards were moving towards a democratic vision by including "for all" language. However, critics of the "for all" language argue that the use does not delve into serious considerations of the social and structural realities faced by Black children; rather the language suggests a myopic focus on modifying curricula, classrooms, and school cultures (Martin, 2003). When we consider the experiences and achievement of Black children, it is clear that that these children have not benefited from previous reform efforts. Consequently, the underpinning of the "for all" message has done little to understand the variables that impact mathematics teaching and learning for Black children.

Even within this context, we find examples of the brilliance of Black children. Hrabowski, Maton, and Greif (1998) described the experiences of academically successful Black men who are Meyerhoff Scholars at the University of Maryland at Baltimore County. Meyerhoff Scholars are high-achieving undergraduate students enrolled in programs of study in the fields of mathematics, science and engineering. Hrabowski et al. (1998) collected data for their work from 1994 to1996. Many of the young men and their parents provided reflections of K–12 experiences that influenced their success in mathematics and science. One mother described when she realized her son was brilliant, "I first realized he had a special talent for math when he was in the fourth grade. They were learning about new theories then, and my son came up with a new one." Another mother gives the following account:

> At two-and-a-half, my son wasn't talking…So we went over to Johns Hopkins for a week, going through all kinds of tests. The doctor was in front of me saying, "…Don't look for him to ever go any further—he's mentally retarded…" When he started talking, he started talking in sentences. And he started talking like a computer…But he always did well on tests…Always in the ninety-ninth percentile…when he was in fourth grade his teacher started telling me, "There's something with him with mathematical concepts. We're doing something, and he'll explain some kind of concept that nobody's even told him about…"

Below is a reflection of one young man, realizing his brilliance through his father:

> My father was a carpenter, and over the summer I would have to go to work with him sometimes. I use to hate it for a while because it would be hot outside and everything. But I would just work with him and every now and then there would be a little problem as far as the work goes, and he would have to solve that problem. And you know, he might work a little and then there would be another problem…And I saw, I actually saw math being used…And sometime he would bring home little problems, like something he encountered during the day, and he would tell me about it. And then he would ask, "Well, how would you do that?"…When we first learned area in math class, I remember he [his father] was talking digging a hole for concrete, and he gave me the dimensions…I always knew I didn't want to be a carpenter because I didn't want to work out in the sun, but I wanted to do something like that. And then I found out that engineering was that field.

The stories of these men's lives and others like them are not captured in the traditional contexts in which mathematics education reform and research are conducted. When we delve into understanding Black children within their own realities, we can see brilliance not captured in school contexts, and we see how out-of-school contexts serve as motivators for children to persist with mathematics. When we consider the misdiagnosis of mental retardation for the one male, one must consider the number of Black children who are similarly misdiagnosed, leading to their brilliance never being realized.

NO CHILD LEFT BEHIND AND THE NATIONAL MATHEMATICS
ADVISORY PANEL (2000–PRESENT)

As stated earlier, the "Math Wars" persisted through the 1990s with opponents of the *CSSM* and NSF funded curricula critiquing these works as not having enough focus on basic skills and emphasizing constructivist pedagogy. NCTM revised it standards document in 2000 through the release of the *Principles and Standards for School Mathematics* (*PSSM*). The *PSSM* revision provided slightly greater emphasis on the importance of algorithms and computational fluency. It was less vague by providing guidelines for the grade bands. *PSSM* was received as more balanced than *CSSM*, which led to some calming of but not an ending to the "Math Wars." *PSSM* highlighted equity by making it the first of six principles for school mathematics. According to *PSSM* (NCTM, 2000), equity requires: a) high expectations and worthwhile opportunities, b) accommodating differences to help everyone learn mathematics, and c) resources and support for all classrooms and all students. These points situate equity in a broad context but fail to recognize issues of social justice or understanding social, economic, and political context in which mathematics is learned. Martin (2003) is critical of *PSSM's* Equity Principle for not providing a sense of equity that considers the contexts of students' lives, identities of students, and conditions under which mathematics is taught and learned. He states:

> …The Equity Principle of the *Standards* contains no explicit or particular references to African American, Latino, Native American, and poor students or the conditions they face in their lives outside of school, including the inequitable arrangements of mathematical opportunities in these out of school contexts. I would argue that blanket statements about *all* students signals an uneasiness or unwillingness to grapple with the complexities and particularities of race, minority/marginalized status, differential treatment, underachievement in deference to the assumption that teaching, curriculum, learning, and assessment are all that matter. (Martin, 2003, p. 10)

Too often, race, racism, social justice, contexts, identities, conditions, and others are relegated as issues not appropriate for mathematics education when in fact these issues are central to the learning and teaching of mathematics for all children, specifically Black children. Starting in the mid 1980s through the present, there have been mathematics education researchers conducting studies in areas that challenge the assumptions that mathematics teaching, learning, curriculum, and assessment are the only factors that matter when understanding the mathematical experiences and achievement of children (Johnson, 1984; Malloy & Jones, 1998; Stiff & Harvey 1988; Tate 1995). Although there is a growing body of research challenging these assumptions, policymakers continue to focus on teaching, curriculum, learning, and assessment as the primary variables to develop policies about mathematics education.

On January 8, 2002, President George W. Bush signed the Elementary and Secondary Education Act, better known as the No Child Left Behind Act (NCLB), into law with the declared intention of helping "all students meet high academic standards" (U.S. Department of Education, 2004). NCLB is a standards-based educational reform document with the premise that setting high standards and establishing measurable goals will improve student achievement and change American school culture (Jones & Hancock, 2005). NCLB required states to implement student testing, collect and disseminate subgroup results, ensure that teachers are highly qualified, and guarantee that all students achieve academic proficiency by 2014. States were required to use sanctions to hold schools and districts accountable for their success in meeting adequate yearly progress (AYP) goals, set by the states, for both overall performance and performance in each subgroup. NCLB was supposed to help improve mathematics education by 1) creating mathematics and science partnerships, 2) increasing the ranks and pay of teachers of mathematics and science, and 3) funding research to determine the best way to teach mathematics and science and measure students' progress in mathematics (U.S. Department of Education, 2004).

Similar to previous reforms, NCLB motives are cast in the national interest rather than aimed at developing a democratic society. NCLB assumes that solutions to issues of student achievement are technical solutions requiring increased attention to teaching, curriculum, learning, and assessment. NCLB narrows the definition of achievement, thus focusing on measurable outcomes and applying a technical solution to a complex issue of student learning. For Black children, it is likely that instruction focused primarily on the acquisition of knowledge by content covered on achievement tests. Rather than focusing on the children's needs, the solutions derived from NCLB focused on inputs (i.e. teaching and curriculum) and outputs (i.e. assessment) rather than what actual learning may look like.

By reporting achievement by subgroups, NCLB assumes that members within a subgroup have static identities that are quantifiable in term of race, class, gender, language, and so on (Gutiérrez, 2008). Critiques of NCLB have focused on the standardization of highly qualified teachers "for all" children (Martin, 2007). Martin (2007) questioned whether the standard of highly qualified teachers of mathematics is applicable to all children by examining the term highly qualified through two lenses: the achievement lens and the experience lens. The achievement lens privileges standardized outcomes such as degree, exam performance, and certification as proxies for teacher quality. The experience lens offers greater clarity to the idea that achievement outcomes among Black children are indicators of the ways they experience mathematics learning and participation. This suggests that teachers must understand how Black children make meaning of their own experiences. NCLB implies that teachers who are knowledgeable in content, teaching, and learning are qualified to teach all children while ignoring the realities of Black children's lives. We contend that highly qualified teachers need also to understand contexts, cultural backgrounds, and identities when teaching Black children.

On April 18, 2006, President George W. Bush created the National Mathematics Advisory Panel (NMAP). The NMAP was part of the President Bush's plan to strengthen mathematics education so that America's students could receive the tools and skills necessary for success in the 21st century. The Panel was charged with providing recommendations on the best use of scientifically-based research to advance the teaching and learning of mathematics. Implicit in this charge is the assumption that there are technical solutions to advancing the teaching and learning of mathematics for all children. Consequently, the charge excluded considerations of advancing mathematics teaching and learning along non-technical issues such as race, racism, contexts, identities, and conditions. The absence of non-technical issues in the NMAP final report suggests an unwillingness to grapple with issues that are significant to the mathematics experiences of Black children, even though there is a growing body of research focusing on these issues.

Efforts such as NCLB and NMAP often drive research agendas that situate Black children as deficient. The research efforts often use comparison groups and take on an achievement gap lens. The implicit message is that Black children are not worth studying in their own right and that a comparison group is necessary. Such framing situates Whiteness as the norm, positioning Black children and Black culture as deviant (Gutiérrez, 2008). However, there is a growing body of research that positions Black children as brilliant (Berry, 2005, 2008; Jett, 2010; Martin, 2000, 2006, 2008; Noble, 2011; Stinson, 2010; Thompson & Lewis, 2005; Walker, 2006). This body of research considers issues of race, racism, contexts, identities, and conditions as variables that impact the mathematical experiences of Black children. This growing body of research is relatively small but an important contribution when considering the broader body of mathematics education research. This body of research challenges the dominant discourse and pushes the field of mathematics education to consider sociological, anthropological, and critical theories. It encourages researchers to consider outcomes other than achievement as the primary measure of success and brilliance.

To understand the broader body of research focusing on Black learners in mathematics education and to examine trends in research methodologies, we conducted searches in four databases focusing on research articles since 1970: a) Google Scholar, b) ERIC, c) Education Retro & Full-text, and d) Sociological Abstracts. The protocols included peer-reviewed journals using (math OR mathematics) AND (Black OR African American) as the search

TABLE 2.1. Mathematics Education Research Focusing on Black Children from 1970–2011

Years	Google Scholar	Eric	Educational Retro & Full Text	Sociological Abstracts
1970–1979	64	2	9	11
1980–1989	284	75	25	25
1990–1999	3,070	164	1,776	55
2000–2009	15,000+	399	6.408	299
2010–2011*	4,700+	113	941	119

*Note: Only two years of data compared to 10 in all other categories

criteria. Table 2.1 shows the result from the searches, which suggests that research on Black people and mathematics has grown significantly since 1970.

This discussion will focus on the findings and trends from the Sociological Abstract database to determine topics and research methodologies used from 1970 to 2011. From 1970 through 1979, seven of the eleven articles focused on comparing the achievement of Black children to the achievement of White children, thus positioning Black children as deficient. These articles primarily focused on the technical dimensions of mathematics education. From 1980 through 1989, trends in research on Blacks show a continued focus on achievement disparities but also include issues related to course-taking patterns, testing bias, workforce development, and gender research comparing Black boys and girls. Also during the 1980s, there was an emerging body of qualitative research focused on Black learners. From 1990 through 1999, there was a focus on achievement gap research using varying methodologies that were different from previous time intervals. The achievement gap research used large sets such as NAEP, NELS:88, and TIMSS. Additionally, there was a focus on exploring gaps between urban schools and other schools, race, and gender. Qualitative research involving Black teachers and students focusing on pedagogies (i.e., culturally relevant pedagogy), access issues, and curriculum arose during the mid to late 1990s. During this period, there are case studies of effective teachers using different curricula in urban classrooms and case studies of Black learners in mathematics classrooms and other settings. Also, in the 1990s there was a growing body of mathematics education research using sociological, anthropological and critical theories. From 2000 through 2009, achievement gap/comparison research was still significant; however, there was a substantial increase in research focusing on issues of race, identities, language, attitudes, success, and resiliency. These topics represent a shift from examining issues in mathematics education as being technical to examining the realities of people's lives. From 2010 to 2011, the same trends of the early

2000s persisted. However, there were studies using large data sets focusing on Black learners without using a comparison group, suggesting a shift towards research aimed at understanding the complexities within Black learners (Joe & Davis, 2009; Lewis, James, Hancock, & Hill-Jackson, 2008).

CONCLUSION: REVISITING THE FIVE QUESTIONS

The five questions raised at the beginning of this chapter provided a framework for examining the historical unfolding of the issues, contexts, policies, reforms, and variables that impacted the experiences of African American children in mathematics education. This conclusion discusses the questions raised and highlights a few key points. The discussion of the questions is not exhaustive, nor is the discussion exclusive to any single question. Rather, the intent is to extend the conversation and to make inferences about points in the timeline that are relevant to Black children's mathematics experiences today.

Resegregation addresses questions one and two. Resegregation, in the form of tracking, has had a tremendous impact on the mathematics education of Black children and still persists today. The process of resegregating (tracking) students at the classroom level after school desegregation is an example of how power and authority was enacted to limit access and opportunities in mathematics education for Black children. While schools had to desegregate by law, it is plausible to suggest that many schools resegregated at the classroom level as a form of resistance to desegregation, thus enacting power and placing Black children in a status hierarchy where their interests and abilities were largely ignored. If one accepts the Coleman Report's peer effect findings, then one must consider that resegregation served to undermine this finding. The process of resegregation continues today at both the school and classroom levels. In a broad context, Black children's limited access to high quality human and material resources negatively impacts their mathematical experiences in schools.

Question three in the beginning of this chapter asked, "In what ways do policy and reforms affect the mathematics experiences of Black children?" Policies and reforms have typically not attended to or appreciated the social realities and needs of Black children in ways that lead to improvements in their life circumstances. This critique suggests that Black children's voices and experiences within a broader context are either missing or are situated within deficit perspectives in mathematics education research, policy, and reform. In fact, patterns over time suggest that Black children have experienced mathematics instruction as a "foreign" pedagogy, focusing on the acquisition of facts over the entire history discussed in this chapter. Thus, policies and reform have had little to no impact on the type of instruction these children receive, when we consider the nadir period of instruction, which focused on preparation for the work force and basic skills. After *Sput-*

nik was launched in 1957, the "race to space" movement began with a significant amount of money allocated to mathematics and science education to identify and develop the "best and brightest" mathematicians and scientists to serve the national interests. When considering the social conditions and experiences of Black children in mathematics around the time of *Sputnik,* one could argue that the intent of identifying and developing the "best and brightest" actually meant identifying the "best and Whitest." Because the mathematics experiences of Black children were largely ignored and pushed aside within the larger discussions of *Sputnik,* one could speculate that these children had virtually no access to "New Math" curricula and pedagogy that resulted from activities during this time. The "back to basics" movement was not a movement for Black children because they were already at basic from a pedagogical standpoint. The emphasis on testing and competencies through the 1990s and 2000s suggests again a pedagogical style that focused on basic skills and the acquisition of knowledge. It is plausible to conclude no real reforms have occurred for Black children and that the efforts asserted have been primarily for the purpose of addressing America's national interests as opposed to the realities of these children's lives.

It is interesting to note that change may be on the horizon. This change addresses questions four and five, focusing on variables researchers use in research focusing on Black children. When reviewing the literature on the *amount* and *type* of research that has occurred focusing on the experiences of Black children in mathematics education, our review shows that there has been a steady increase in research focusing on variables to understand the complexities of Black children. This does not suggest that relative to the larger knowledge base in mathematics education that research in this area has proliferated; in fact, our review suggests more research is necessary. While not exhaustive, our searches have found that that there is considerable current research focused on documenting Black children's failure and how Black children achieve relative to White children. In contrast, there are a growing number of Black mathematics education researchers publishing work on issues of racial and mathematics identity, socialization, student success, social justice, culturally relevant teaching, opportunities to learn, and critical mathematics education. The implications of research by this growing number of researchers will inform policies that impact the experiences of Black children and influence positive shifts in the broader field of mathematics education to consider variables beyond those traditionally used to investigate the experiences and outcomes of Black children. The challenge for these researchers is to convince funding agencies, such as the National Science Foundation and the U.S. Department of Education, that studying how Black children learn and do mathematics is a worthwhile endeavor. Then these researchers will have greater opportunities to interface with the broader mathematics education community to support what we already intuitively know: Black children are inherently creative, talented, and brilliant in mathematics.

REFERENCES

Alridge, D. P. (1999). Conceptualizing a Du Boisian philosophy of education: Towards a model for African American education. *Educational Theory, 49*(3), 359–379.

Alridge, D. P. (2008). *The educational thought of W. E. B. Du Bois: An intellectual history.* New York, NY: Teachers College Press.

Anderson, J. (1988). *The education of Blacks in the South, 1860–1935.* Chapel Hill, NC: University of North Carolina Press.

Apple, M. W. (1992). Do the standards go far enough? Power, policy, and practice in mathematics education. *Journal for Research in Mathematics Education, 23*(5), 412–431.

Berry, R. Q., III. (2005). Voices of success: Descriptive portraits of two successful African-American male middle school students. *Journal of African American Studies, 8*(4), 46–62.

Berry, R. Q., III. (2008). Access to upper-level mathematics: The stories of successful African American middle school boys. *Journal for Research in Mathematics Education, 39*(5), 464–488.

Burton, N. W. & Jones, L. V. (1982). Recent trends in achievement levels of Black and White youth. *Educational Researcher, 11,* 10–14.

Burrill, G. (2001). Mathematics education: The future and the past create a context for today's issues. In T. Loveless (Ed.), *The great curriculum debate: How should we teach reading and math?* (pp. 25–41). Washington, DC: Brookings Institution Press.

Coleman, J. S., Campbell, E. Q., Hobson, C. J., McPartland, J., Mood, A. M., Weinfeld, R. D., & York, R. L. (1966). *Equality of education: Summary report.* Washington, DC: U.S. Government Printing Office.

Department of Education, National Center for Education Statistics. (2004). *The condition of education 2004* (NCES 2004–077). Washington, DC: U.S. Government Printing Office.

Devault, M. V. & Weaver, J. F. (1970). Forces and issues related to curriculum and instruction, K–6. In A. F. Coxford & P. S. Jones (Eds.), *A history of mathematics education in the United States and Canada* (pp. 92–152). Washington, DC: National Council of Teachers of Mathematics.

Doughty, J. J. (1978). Diminishing the opportunities for resegregation. *Theory into Practice, 17*(2), 166–171.

Du Bois, W. E. B. (1915, July). Education. *The Crisis: A record of the darker races,* 132–133. Retrieved January 31, 2012, from http://www.library.umass.edu/spcoll/digital/dubois/EdEducation.pdf

Ellis, M. & Berry, R. Q. (2005). The paradigm shift in mathematics education: Explanations and implications of reforming conceptions of teaching and learning. *The Mathematics Educator, 15*(1), 7–17.

English, L. D. & Halford, G. S. (1995). *Mathematics education: Models and processes.* Mahwah, NJ: Lawrence Erlbaum Associates.

Feagin, J. R. & Barnett, B. M. (2004). Success and failure: How systemic racism trumped the Brown v. Board of Education decision. *University of Illinois Law Review, 2004,* 1099–1130.

Foster, M. (1997). *Black teachers on teaching.* New York, NY: The New Press.

Gould, S. J. (1996). *The mismeasure of man.* New York, NY: W. W. Norton.

Gutiérrez, R. (2008). A "gap gazing" fetish in mathematics education? Problematizing research on the achievement gap. *Journal for Research in Mathematics Education, 39*(4), 357–364.

Henriques, J., Hollway, W., Urwin, C., Venn, C., & Walkerdine, V. (1998). Constructing the subject. In J. Henriques, W. Hollway, C. Urwin, C. Venn & V. Walkerdine (Eds.), *Changing the subject: Psychology, social regulation and subjectivity* (pp. 91–118). London: Routledge.

Hill, P. L., Bell, B. W., Harris, T., Harris, W. J., Miller, R. B., O'Neale, S. A., & Porter, H.A. (Eds.). (1998). *Call & response: The Riverside anthology of the African American literary tradition.* New York, NY: Houghton Mifflin.

Hrabowski, F. A. III., Maton, K. I., & Greif, G. L. (1998). *Beating the odds: Raising academically successful African-American males.* New York, NY: Oxford University Press.

Hughes, L. W., Gordon, W. M., & Hillman, L. W. (1980). *Desegregating America's schools.* New York, NY: Longman.

Hughes, S. A. (2006). *Black hands in the biscuits not in the classroom.* New York, NY: Peter Lang Publishing

Jett, C. C. (2010). "Many are called but few are chosen": the role of spirituality and religion in the educational outcomes of "chosen" African American male mathematics majors. *Journal of Negro Education, 79,* 324–334.

Jetter, A. (February 21, 1993). "Mississippi Learning." *The New York Times Magazine,* 28–32; 50–51, 64, 72.

Joe, E. M. & Davis, J. E. (2009). Parental influence, school readiness and early academic achievement of African American boys. *Journal of Negro Education, 78*(3), 260–276.

Johnson, M. L. (1984). Blacks in mathematics: A status report. *Journal for Research in Mathematics Education, 15*(22), 145–153.

Jones, J. H. & Hancock, C. R. (2005). Brown v. Board of Education at 50: Where are we now? *The Negro Educational Review, 56*(1), 91–98.

Kelly, S. (2009). The Black-White gap in mathematics coursetaking. *Sociology of Education, 82,* 47–69.

Klein, D. (2003). *A brief history of American K–12 mathematics education in the 20th century.* Retrieved January 31, 2012, from http://www.csun.edu/~vcmth00m/AHistory.html

Levitan, S. A. & Taggart, R. (1976). The great society did succeed. *Political Science Quarterly, 91*(4), 601–618.

Lewis, C. W., James, M., Hancock, S., & Hill-Jackson, V. (2008). Framing African American students' success and failure in urban settings: A typology for change. *Urban Education, 43*(2), 127–153. doi:10.1177/0042085907312315

Logan, R. W. (1954). *The Negro in American life and thought: The nadir 1877–1901.* New York, NY: Dial Press.

Malloy, C. & Jones, M. G. (1998). An investigation of African American students' mathematical problem solving. *Journal for Research in Mathematical Education, 29,* 143–163.

Martin, D. B. (2000). *Mathematics success and failure among African American youth: The roles of sociohistorical context, community forces, school influence, and individual agency.* Mahwah, NJ: Lawrence Erlbaum Associates.

Martin, D. B. (2003). Hidden assumptions and unaddressed questions in mathematics for all rhetoric. *The Mathematics Educator, 13*(2), 7–21.

Martin, D. B. (2007). Beyond missionaries or cannibals: Who should teach mathematics to African American children? *The High School Journal, 91*(1), 6–28.

Martin, D. B. (2008). E(race)ing race from a national conversation on mathematics teaching and learning: The national mathematics advisory panel as white institutional space. *The Montana Mathematics Enthusiast, 5*(2&3), 387–398.

Martin, W. H. (1964). Unique contribution of Negro educators. In V. A. Clift, A. W. Anderson, & H. G. Hullfish (Eds.), *Negro education in America: Its adequacy, problems, and needs* (pp. 60–92). New York, NY: Harper & Row Publishers.

Mayo, J. B. (2007). Quiet warriors: Black teachers' memories of integration in two Virginia localities. *Multicultural Perspectives, 9*(2), 17–25.

Mickelson, R. A. (2001) Subverting *Swann*: First- and Second-Generation Segregation in the Charlotte-Mecklenburg Schools. *American Education Research Journal, 38,* 215–252.

Mintz, S. (2007). The Great Society and the drive for Black equality. *Digital History.* Retrieved January 31, 2012 from http://www.digitalhistory.uh.edu/database/article_display.cfm?HHID=372

National Center for Education Statistics. (2012). Retrieved October 3, 2012 from http://nces.ed.gov/pubs2012/2012026/tables/table_12b.asp

National Commission on Excellence in Education. (1983). *A nation at risk: The imperative for educational reform.* Washington, DC: U.S. Government Printing Office.

National Council of Teachers of Mathematics. (1980). *An agenda for action: Recommendations for school mathematics of the 1980s.* Reston, VA: Author.

National Council of Teachers of Mathematics. (1989). *Curriculum and evaluation standards for school mathematics.* Reston, VA: Author.

National Council of Teachers of Mathematics. (2000). *Principles and standards for school mathematics.* Reston, VA: Author.

National Education Goals Panel. (1990). Retrieved October 3, 2012 from http://www.ncrel.org/sdrs/areas/issues/envrnmnt/go/go4negp.htm

National Science Foundation. (2003). *The science and engineering workforce: Realizing America's potential.* arlington, VA: NSF.

Noble, R. (2011). Mathematics self-efficacy and African American male students: An examination of models of success. *Journal of African American Males in Education, 2*(2), 188–213.

Oakes, J. (1990). *Multiplying inequalities: The effects of race, social class, and tracking on opportunities to learn mathematics and science.* Santa Monica, CA: RAND.

Perry, T. (2003). Up from the parched earth: Toward a theory of African-American achievement. In T. Perry, C. Steele, & A. Hilliard (Eds.), *Young, gifted and*

Black: Promoting high achievement among African-American students (pp. 1–108). Boston, MA: Beacon Press.

Pincham, L. B. (2005). A league of willing workers: The impact of Northern philanthropy, Virginia Estelle Randolph and the Jeanes teachers in early twentieth-century Virginia. *The Journal of Negro Education, 74*(2), 112–123.

Planty, M., Provasnik, S., & Daniel, B. (2007). *High School Coursetaking: Findings from The Condition of Education 2007* (NCES 2007-065). Washington, DC: U.S. Department of Education, National Center for Education Statistics.

Porter, A. C., Kirst, M. W., Osthoff, E. J., Smithson, J. L., & Schneider, S. A. (1993). *Reform up close: An analysis of high school mathematics and science classrooms* (Final report to the National Science Foundation on Grant No. SAP-8953446 to the Consortium for Policy Research in Education). Madison, WI: University of Wisconsin-Madison, Consortium for Policy Research in Education.

Raizen, S. A., McLeod, D. B., & Rowe, M. B. (1997). The changing conceptions of reform. In S. A. Raizen & E. D. Britton (Eds.), *Bold ventures: Volume I: Patterns among U.S. innovations in science and mathematics* (pp. 97–130) Boston, MA: Kluwer Academic Pub.

Resnick, D. P. (1980). Minimum competency testing historically considered. *Review of research in education, 8,* 3–29.

Smith, S. L. (1950). *Builders of goodwill: The story of the state agents of Negro education in the south, 1910 to 1950.* Nashville, TN: Tennessee Book Company.

Snipes, V. & Waters, R. (2005). The mathematics education of African Americans in North Carolina from the Brown Decision to No Child Left Behind. *The Negro Educational Review, 56,*(2 & 3), 107–126.

Sowell, T. (1974). Black excellence: The case of Dunbar High School. *Public Interest 35,* 1–21.

Standish, H. A. (2006). *A case study of the voices of African American teachers in two Texas communities before and after desegregation, 1954 to 1975.* Unpublished doctoral dissertation, Texas A&M University, College Station, TX.

Stiff, L. V. & Harvey, W. B. (1988). On the education of Black children in mathematics. *Journal of Black Studies, 19,* 190–203.

Stinson, D. W. (2010). Negotiating the "white male math myth": African American male students and success in school mathematics. *Journal for Research in Mathematics Education,* 41. Retrieved from http://www.nctm.org/eresources/article_summary.asp?URI=JRME2010-06-2a&from=B

Tate, W. F. (1994) From inner city to ivory tower: Does my voice matter in the academy? *Urban Education, 29*(3), 245–269.

Tate, W. F. (1995). Returning to the root: A culturally relevant approach to mathematics pedagogy. *Theory into Practice, 34,* 166–173.

Tate, W. F. (1997). Race-ethnicity, SES, gender, and language proficiency trends in mathematics achievement: An update. *Journal for Research in Mathematics Education, 28*(6), 652–679.

Tate, W. F. (2000). Summary: Some final thoughts on changing the faces of mathematics. In W. G. Secada (Ed.), *Changing the faces of mathematics: Perspectives on African Americans* (pp. 201–207). Reston, VA: NCTM.

Tate, W. F. (2005). Ethics, engineering and the challenge of racial reform in education. *Race, ethnicity and education, 8*(1), 121–127.

Thayer, V. T. (1928). *The passing of the recitation.* Boston, MA: D. C. Heath and Company.

Thompson, L. & Lewis, B. (2005, April/May). Shooting for the stars: A case study of the mathematics achievement and career attainment of an African American male high school student. *The High School Journal, 88*(4), 6–18.

Thorndike, E. L. (1923). *The psychology of arithmetic.* New York, NY: The Macmillan Company.

United States Congress Office of Technology Assessment. (1992). *Testing in American schools: Asking the right questions, OTA-SET-519.* Washington, DC: U.S. Government Printing Office.

Walker, V. S. (2000). Valued segregated schools for African American children in the South, 1935–1969: A review of common themes and characteristics. *Review of Educational Research, 70*(3), 253–285.

Walker, E. N. (2006). Urban high school students' academic communities and their effects on mathematics success. *American Educational Research Journal, 43*(1), 43–73.

Walmsley, A. E. (2003). *The history of the "new mathematics" movement and its relationship with current mathematical reform.* Lanham, MD: University Press of America.

Washington, B. T. (1901). *Up from slavery.* New York, NY: Doubleday, Page & Company.

Willoughby, S. (2000). Perspectives on mathematics education. In M. Burke & F. Curcio (Eds.), *Learning mathematics for a new century* (pp. 1–15). Reston, VA: NCTM.

Wong, K. & Nicotera, A. (2004). Brown v. Board of Education and Coleman Report: Social science research and the debate on educational equality. *Peabody Journal of Education, 79*(2), 122–135.

Woodson, C. G. (1933/1999). *The mis-education of the Negro.* Washington, DC: Associated Publishers. (Original work published 1933)

CHAPTER 3

THE MATHEMATICAL LIVES OF BLACK CHILDREN

A Sociocultural-Historical Rendering of Black Brilliance

Maisie L. Gholson

INTRODUCTION

Math was my favorite. Math was my favorite. I love figures. Math was my favorite because I don't know. I guess it was something that you learned just by practice. My daddy only went to school for three years. He only had third grade, and he could count money better than anyone I knew. And I learned that being able to figure and count was a privilege.

Mrs. Ruby Gant, 83-year old, great-grandmother, wife to Henry Gant Sr. for 64 years, and self-proclaimed math person

In the above quote from Mrs. Ruby Gant, the memory of her father radiates pride in his ability to be able to *count* money well. Counting is a skill not highly regarded when judged by contemporary standards; yet, in the 1930s, this skill was essential to a Black person's financial survival, given their sub-

The Brilliance of Black Children in Mathematics:
Beyond the Numbers and Toward New Discourse, pages 55–76.

ordinated socioeconomic position under Jim Crow ordinances and laws. It was common for mathematical practices—literacy tests, poll taxes, unjust sharecropping laws and contracts—to be used toward the disenfranchisement of the Black community. Further contributing to this disenfranchisement was the fact that there were limited means of remunerating Black folks when they were cheated or deceived. The technological tools that surround us today, like electronic cash registers, mediate purchases and hide processes of counting and arithmetic, which can insulate consumers from deceptive practices. However, in the pre-civil rights era, these mathematical processes of counting and arithmetic were ever-present in the day-to-day transactions of Black men and women. Their social realities demanded mathematical competence. The ability to count money well was not only a means to protect one's self financially but equally important, as noted by Mrs. Gant, "being able to figure and count was a privilege."

Mrs. Gant's lessons learned about numeracy were not confined to her childhood. The practice of counting money and conducting mental calculations, as modeled by her father, was manifested in her shopping as an adult. In an interview that I conducted with Mrs. Gant, she recounted in great detail her memory of a White grocery store cashier who did not know how to give the appropriate change from the register. Contributing to that experience was the fact that Mrs. Gant, at that point in her life was the mother of twelve children and needed to purchase food on a train brakeman's salary. So this was not a trivial dispute for her; there were mouths to feed.

After engaging in a bit of verbal debate with the cashier, a manager was called to the checkout line to resolve the controversy. In an agitated voice, Mrs. Gant recalled her frustration, "But I told the manager, 'You know what? You wouldn't give me a job like this.'" She parenthetically noted in a more subdued tone, "'Cause that was during the time that [Black people] wouldn't get a job like that." Then, she continued, "But he hired a girl that couldn't even count!"

The narrative of this African American great-grandmother, born in 1929, the beginning of the Great Depression, and raised in a segregated community in south Texas, offers us a glimpse into historical moments that necessarily implicate important relationships between race and mathematics. These relationships between race and mathematics have been empirically established for African American children, adolescents, young adults and parents (Martin 2000, 2009). Accordingly, the learning of and engagement in mathematics is rightly conceptualized as *racialized forms of experience* (Martin, 2006).

It is important to note that the racial-mathematical moments in Mrs. Gant's life do not *really* begin in the grocery store line, but during her upbringing with a father who could count well and had to count well to sur-

vive the highly racialized context in which he existed. As part of her racial-mathematical socialization, Mrs. Gant's father relayed messages about
self-advocacy and mathematical power in a time that demanded a particular
mathematical competence. This sense of mathematical self-advocacy manifested itself not only small moments like the grocery store incident but also
throughout larger sociohistorical moments like the Black power movement
of the late 1960s.

The story of one's life can often force the confrontation of a complex
irony for many other lives. For example, the development of a positive and
strong sense of self, both racially and mathematically, was realizable by an
African American girl growing up with the overt racial oppression in the
pre-civil rights South. On the other hand, positive racial and mathematics
identities are deemed elusive, and often exceptional, for urban-dwelling
African American children who live in a post-civil rights, and supposedly
post-racial, era. How do we make sense of this schism?

In this chapter, I argue that the differences between Mrs. Gant and the
contemporary African American child are related to the processes of socialization, which have shifted over time. I provide a comprehensive historical
overview of important socialization processes that have encompassed the
mathematical lives of Black children and adolescents from the Jim Crow
era to the present day. My overview moves beyond static depictions of race
and highlights what race means in given sociopolitical contexts and conditions. I claim that these evolving meanings and conditions, and how they
have been negotiated, have also impacted racial identity development,
mathematical identity development, racial socialization, and mathematical
socialization for Black children. As these processes of socialization are put
into motion, African Americans throughout history have emerged as brilliant in their ability to use mathematics to survive in the world. In my view,
this brilliance remains an under-theorized aspect of the mathematical lives
of African American children.

SOCIALIZATION

I was a teacher's pet. My mother had to work when I was in school, and I
didn't have but one sister, and my sister was a lot older than I was. And the
boys didn't know how to do hair. Mrs. Swan would bring her comb and brush
to school, and she would do my hair. Not only me, she combed a lot of the
other girls' hair if they came to school without their hair combed, 'cause one
thing she taught us was how to be neat. That's part of an education—how to
be neat and how to be clean. That was one thing—she would comb my hair.
She did a lot for me. But you know what? She would whoop me too, if I got
bad, too. I had to learn.

In this section, two seemingly divergent recollections of Mrs. Gant are used to investigate the concept of socialization—hair combing and learning arithmetic facts. Socialization is a process in which cultural practices, meanings, and values are appropriated. For example, hair combing as experienced by Mrs. Gant and millions of other African American girls is a site for socialization. Similarly, the processes involved in learning how to solve a two-step linear equation and the recitation of one's multiplication facts are examples of mathematics socialization. Parents engaging their children in discussions about the potential for racial discrimination in their daily lives is an example of racial socialization. In general, socialization is a process by which *cultural tools*, like comb, brush, grease, barrettes, coefficient, variable, equation, number, and racial awareness, are used to mediate the respective activities. Socialization is also a process by which *cultural meanings and values*, like beauty, affection, mathematical equality and multiplicity, and discrimination are made relevant to the participating actors within the activity.

Using the practice of hair combing as an exemplar, the processes of socialization can be described further. First, it is important to note that hair combing, as a shared moment of beautification and hygiene maintenance, is not disconnected from the adult-child dyad of being, for example, female and African American in U.S. society. Thus, hair combing is the site of both cultural and gender socialization. That is, from a child's perspective, sitting between an adult's legs and getting one's hair pulled taut and braided is simultaneously a process of being and becoming African American, being an African American girl, and becoming a beautiful African American woman.

Further, socialization is always occurring across time, place, or situation. Socializing agents include parents, siblings, teachers, peers, curriculum, and media. Socializing agents and spaces will often be treated together and referred to as a *sphere of socialization*. This raises a critical point about the nature of socialization: discourses and practices that are formed during socialization can be propagated across various spheres of socialization as people co-participate with others in and across activities. African American children negotiate such practices and discourses as they move through various spheres of socialization.

In the text that opened this section, we learn from Mrs. Gant that her favorite teacher in grammar school, Mrs. Swan, would often comb her hair before school. This practice of combing hair in the schoolhouse seems markedly out of place when examining the general lack of care exercised in contemporary public schools (Noddings, 2005). Perhaps, this practice is so striking because there is nothing more intimate between a woman and a girl than the combing of hair, particularly in the African American community (Lewis, 1999). The talk, the physical closeness, and the heightened sense of maternal sensitivity to the cues given by the child that occur during hair combing create a key site for the development of intimate relationships. In mother-daughter dyads (who are the most common participants in this practice in an African American context), ideas of womanhood, beauty, hygiene, and affection are jointly created and appropriated. In fact, we learn from Mrs. Gant that Mrs. Swan's conception of a complete education extended beyond reading, writing, and arithmetic to personal wellbeing. This teaching was done with love and care, but also by being firm, as implied by her belief in corporal admonishment. In fond remembrance, Mrs. Gant says, "I will always remember her [Mrs. Swan]. I will *always* remember her. I was only 12 when she died but I will always remember her."

Later, when speaking specifically of the mathematics practices, Mrs. Gant stated:

> First, we started with doing numbers 1 through 10, and then ended up started by doing twos, fives, tens like that—you know. And when we got to the third grade, we start learning the times tables. We had to learn the times tables first through 12. We had to know them off by heart before we ever did any work like arithmetic, like multiplying. Mostly, from first and second grades, we only did adding and subtracting and learning the figures and counting. *You had to know how to count* [Spoken slowly with emphasis]. And, then, we had to know the basics. The basics in math were: If you were adding seven and eight, then you know the answer would have 15 or five. If you were adding seven and eight—say you were adding seven, eight, and two. Well, you know the seven and eight was gon' be a five and then add the two and that would make a seven. And you would carry. We would always carry numbers.

While these two memories are seemingly divergent, it is critical to acknowledge that the ritual of hair combing, again a particularly affectionate practice, was part of the same space for Mrs. Gant as learning addition, subtraction, and many other formal mathematical practices. The intersection of spaces of mathematics learning and intimate relational spaces was perhaps less visible thirty years ago and virtually non-existent today. Nevertheless, for Mrs. Gant, the site of socialization for becoming an African American girl and a mathematics doer were one in the same.

RACIAL SOCIALIZATION

> I remember the day I became colored.
> —Zora Neal Hurston, *"How It Feels to Be Colored Me"*

South Carolina is my home state and I am the aunt, granddaughter, and sister of Baptist ministers. Service was as essential part of my upbringing as eating and sleeping and going to school. The church was a hub of Black children's social existence, and caring Black adults were buffers against the segregated and hostile outside world that told us we weren't important. But our parents said it wasn't so, our teachers said it wasn't so, and our preachers said it wasn't

so. The message of my racially segregated childhood was clear: let no man or woman look down on you, and look down on no man or woman.
—Marian Wright Edelman, *The Measure of Our Success*

Racial Messages

Racial messages are endemic within U.S. discourse. Many of these messages have become hegemonic symbols and virtually intractable. Contemporary racial messages seem remarkably subtle and abstract (and, thus, more difficult to mediate) in contrast to racial messages from the Jim Crow era—"Whites Only"— that Mrs. Gant experienced as a small girl in South Texas.

There is a litany of contemporary examples: Consider that witches wear black, whereas princesses wear white. More globally, goodness and virtue are presented as light or white, but evil is presented as darkness or black (Murray & Mandara, 2002). Further, all popular superheroes are racially depicted as White (except for the one commercially popular Black superhero, Spawn, who is sent to hell and resurrected by the devil). Only recently have crayons and markers been expanded to include other skin tones beyond the once universal *fair* white skin (i.e., flesh tone color). Children's coloring practices are discussed by Holmes' (1995) ethnographic study in which Black children often colored themselves peach or pink even though they knew this was not their actual skin color. A dark skinned little girl in Holmes' study was quoted as saying, "I am Black on the outside, but my heart is peach." Further, the media perpetuates these symbols and serves an active role as a socializing agent. Many scholars have argued that television viewing is detrimental to Black children's self-concept, due to the lack of positive representations of Blacks, which serves to minimize their existence or frame it in a negative light.

Very few images exist in popular television culture that depict African Americans as competent mathematics doers with the noticeable exception of Dwayne Wayne, a central character in *The Cosby Show* spinoff, *A Different World*, who was a mathematics major at a historically Black college. Walker (2010) reveals the impact of such images when quoting a student, who was influenced by this television program: "I had this glorified image of going to an HBCU [historically black college and university] and majoring in math. Part of me said, 'I guess that will be okay,' because I saw it. [I said] 'Well, I saw it on TV so it's not crazy.'"

Understanding Racial Development and Socialization

Racial socialization can be simply considered the process of managing racial messages and material in situational contexts. Cognitivists believe that very young children move through several stages of racial development, which affect their ability to manage racial messages. These stages of development include: *racial classification* (i.e., distinguish racial difference), *awareness* (i.e., difference has concrete meaning), *constancy/stability* (i.e., race is permanent), and, then, *preference* (i.e., manifests in personal identification and attitudes). Supporting this thesis, Murray and Mandara (2002) relay a story of a three-year old Black girl, who considered her Black mother White because her skin color was light brown. She knew that there was a difference but could not reconcile her mother being Black as her skin was *light*. In this case, the three-year old understood that there was a racial difference, but this difference did not hold any meaning. Racial awareness culminates in racial identification. By age six or seven children have developed a sense of racial group membership (Harris & Graham, 2007). Slaughter-Defoe, Johnson, and Spencer (2009) state, "Before early adolescence, they [African American children] do not have a coherent concept of race that they can reflexively apply to themselves" (p. 802).

Stage theories of racial development, however, do not imply that very young children do not engage in shockingly complex racial arguments. Van Ausdale and Feagin (2001), in their gripping first chapter of *The First R: How Children Learn Race and Racism*, describe a three-year old child, who they call Carla, using *racial material* to navigate her social environment. Van Ausdale and Feagin (2001) describe Carla moving her sleeping mat across the room during naptime, justifying this move by calling another child a racial slur, and rationalizing that move by asserting that "Niggers are stinky" (p. 1). The authors note, "This shows a level of forethought. She has considered what a *nigger* is, to whom the appellation applies, and why such a label is useful in explaining her behavior to an adult" (Van Ausdale & Feagin, 2001, p. 2). This example indicates that at a very young age children are able to create and leverage racial messages to their desired end.

The seminal work that laid the foundation in children's racial socialization is the Clark and Clark doll studies conducted in the late 1930s and early 1940s. In these studies, Black children tended to connect positive traits with White dolls and negative traits with Black dolls. Methodological and interpretative critiques have been offered of the doll studies, namely, that group identity and personal identity are distinguishable. Thus, negative connections to Black dolls do not necessarily convey internalized racial hatred, but an acknowledgement of social positioning of one's racial group in a larger context.

Nevertheless, these findings were used as evidence of feelings of inferiority, poor self-concept, and low self-esteem of African American children in the *Brown v. Board of Education* desegregation case (Irons, 2002). Specifically, the plaintiffs argued that school segregation was socio-emotionally detrimental to Black children and, thus, public schools should be integrated. In other words, predominantly Black schools were responsible for socializing

Black children into negative self-conceptions. Shortly thereafter, in 1965, the Moynihan report, *The Negro American Family: The Case for National Action*, also implicated the Black family as a negative socialization agent for Black children and adolescents. It was institutionally memorialized via public policy documents and legal precedence: Black families and schools were considered detrimental to the socialization of the African American child.

SPHERES OF SOCIALIZATION ACROSS TIME

Black Parents and the Home as Socializing Agents

There are many spheres of socialization of the African American child. Potentially the most researched sphere of socialization is that of the parent and home environment. With respect to cultural socialization,[1] Hughes et al. (2006), for example, focused on parental practices that promote cultural customs, traditions, and ethnic pride directly or indirectly. Several examples are shared by Hughes et al. (2006), including "talking about important historical and cultural figures, exposing children to culturally relevant books, artifacts, music, and stories; celebrating cultural holidays; eating ethnic foods; and encouraging children to use their family's native language" (p. 749). Other scholars, like Boykin (1983), from a survivalist cultural perspective[2] have attempted to summarize dimensions of expressions that signify "African Americanness," such as: spirituality, harmony, movement, verve, affect, communalism, expressive individualism, orality, and social time perspective. Processes of socialization are believed to propagate these expressions in the African American home to the child. To the extent that such dimensions are indicative of a single African American cultural experience, divisions of African Americans along lines of social class almost certainly revise these different expressions of culture.

With respect to racial socialization, Hughes et al. (2006) noted that these practices primarily fall into three categories: *preparation for bias, promotion of mistrust, egalitarianism* and *silence about race* (also referred to as *mainstream socialization* by Boykin and Toms, 1985). Examining each in turn, preparation for bias relates to messages for coping with discrimination. These messages are thought to be passed down through collective memory of historical oppression. Promotion of mistrust, as it sounds, describes practices that encourage Black children to guard themselves from other racial groups or obstacles to success. Practices related to preparation for bias are distinguished from promotion of mistrust in the literature by the messages or advice given to effectively manage any experienced bias. Finally, many Black parents report conveying messages of egalitarianism or are completely silent about racial matters. For example, hard work, virtue, self-acceptance, and equality are values that many Black parents promote in lieu of overt racial messages, while others adhere to a colorblind or race-neutral philosophy. Parents' intuitive understandings of their child's development often lead them to give egalitarian messages to very young children and more complex messages related to mistrust or bias for adolescents navigating more complex racial experiences.

Daniel and Daniel (1999) noted that parent's engagement in racial socialization practices is often initiated by what they call *hot stove* encounters, that is unexpected moments of racial discrimination that hurt their child. They noted that socialization messages are often encoded in the theme of survival and usually communicated through narrative, in addition to proverbial messages of encouragement.

Brown and Lesane-Brown (2006) described the generational differences in the race socialization messages of Black parents across historical time. They examined racial socialization messages in three periods—pre-*Brown v. Board of Education* (prior to 1957), protest (1957 to 1968), and post-protest (1969 to 1980). The message that *Whites are prejudiced* was consistently transmitted across generations to Black children. The pre-*Brown* generation disproportionately recalled receiving messages of *deference to and fear of Whites*, in addition to *colorblind* messages. These findings about pre-*Brown* generation Black parents is corroborated by Rittenhouse (2006) in her book *Growing Up Jim Crow*, where middle class notions of child-rearing among Black parents were rooted in "a broader strategy of racial advancement through respectability and self help" (p. 83). The post-protest generation disproportionately recalled receiving messages of individual pride, racial group pride, and no messages.

Schools as Socializing Agents

Another site for socialization outside of the home is school. There is a great deal of variance in the values, meanings, tools, and practices between schools and communities, but there are some commonalities that persist even over history, which characterize the culture of U.S. schools. The ethos of U.S. schools promotes: individualism, competition, acquisition (i.e., receiving a diploma or grade promotion), obedience, production of work, and evaluations/rewards based on quantification (i.e., grades and grade point averages), and so on.

Socialization processes in schools are also related to the distribution of resources and opportunities to learn. For example, Black children are often enrolled in special education or lower tracks, while too few Black children are being counseled into gifted and talented programs (Carter & Goodwin, 1994). Such assignments are consequential to the Black child's socialization because the values and meanings constructed within these spaces have different histories and trajectories. Contemporary trends in course enrollment, expulsion, and suspension seem to suggest that, although Black children are technically integrated into the U.S. educational system, they are practically being socialized into vocational spaces or pushed out of school altogether.

Beyond the socioeconomic basis of socialization is the racial composition of schools (cf. Chavous, 2005). It is believed that most African American children attended all-Black schools until circa 1954 per the ruling of *Brown v. Board of Education*. However, ten years later (i.e., 1964), 98% and 70% of African American children were still in all-Black schools in the South and North, respectively (Darling-Hammond, 2010). The number of African American children attending all-Black schools declined through the 1960s, but stagnated between 1972 and 1986. During the 1990s into the 2000s, U.S. schools re-segregated such that in 2000, 72% of African American children attended predominantly minority and approximately 40% of African American children attended hyper-segregated schools(i.e. minority enrollment of 90 to 100%) (Darling-Hammond, 2010). Given the highly segregated nature of schools today, questions exist regarding the merits of the 1960 school desegregation movement.

Black Teachers as Socializing Agents

It is known that desegregation has had a particularly destructive and acute impact on the Black teaching force, perhaps the key socializing agents within all Black schools (Milner & Howard, 2004). Jones (1994) reports (as cited in Harris & Graham, 2007) that 6,000 Black teachers lost their jobs and another 25,584 Black teachers were displaced by White teachers during school desegregation. Many of these men and women served as anchors within their community and employed similar practices of racial socialization as Black children's parents. The continuity of racial socialization between home and school was certainly disrupted in the shift to integrate. Fairclough (2007) noted that many Black teachers held the White power brokers, who controlled courts and school boards, accountable for the poor integration of Black students into White schools.

> Black students entered majority-White schools feeling resentful that their old schools had been closed to facilitate integration. Their new schools offered nothing to inspire them, nothing with which they could identify, nothing to evoke loyalty, affection, and pride. They had lost their school, mascot, trophies, teams, magazines, and songs—all their school traditions, in fact. Moreover, many of the traditions of their new schools offended them. They could not warm to the schools named after Confederate generals and school mascots called "Rebel." And they felt deliberately excluded from the prestigious social positions—cheerleader, homecoming court—that high school students prized. (p. 397)

According to Fairclough (2007), Black teachers also complained that White teaching practices were too didactic, and White teachers' expectations of Black children were too low. Nearly forty-five years later, similar debates exist regarding school climate and White teacher expectations and their effects on Black student socialization. Contemporary portraits of Black teachers harken back images of pre-*Brown* teaching practices (Johnson, Nyamekye, Chazan, & Rosenthal, in press). Johnson et al. (in press) described Floyd Lee, a young, African American, male teacher of algebra, who employed the socialization practice of giving speeches that are themed to explicit rule setting, coaching and solidarity, and caring. These themes aligned neatly with what Milner and Howard (2004) described as Black teachers' position as "surrogate parent figures,…disciplinarians, counselors, role models and overall advocates for their academic, social, cultural, emotional, and moral development" (p. 286).

Other studies related to White teachers and Black children also indicated that White teachers have different expectations of Black students; and Black children perceive this difference as a lowered expectation (DeCuir-Gunby, 2009). Many interventions have been implemented by schools to support predominantly White teaching staffs in an all-Black context. One example of such intervention is described by Horn (2011), who revealed the socialization practices of the Knowledge Is Power Program charter school system, better known as KIPP. Many, if not most, of the KIPP charter schools serve large Black and Latino populations. Children engage in three weeks of summer school, where they learn the expectations and practices for the next grade level. Newly enrolled students learn SLANT, an acronym for "sit up straight, look and listen, ask, answer questions, nod to show understanding, [and] track the speaker." There is a "KIPP way" for seemingly rudimentary functions like walking, getting off the bus, sitting in the cafeteria, and going to the bathroom. Infractions to these expectations are usually met with immediate corrective action and may result in ostracism from the group in school and out of school. Children monitor each other and are expected to report the breaking of infractions, which is often referred to by some Black adolescents and adults as *snitching*—a cultural taboo for some.

Cultural artifacts in KIPP schools include a paycheck (e.g., a behavior card) that can be used to "purchase" items from the KIPP store and uniforms that include slogans, such as "Work Hard, Be Nice" and "No Shortcuts, No Excuses." Many of these practices are derivative of psychologist Martin Seligman, a proponent of learned optimism. Horn (2011) argued that such practices result in a submissive detachment "achieved by unrelenting and constant surveillance, harsh and sure verbal castigation, public humiliation and labeling, manipulative reinforcement, and ostracizing isolation" (p. 97). The worry that Horn expressed is that Black children subjected to such practices develop anger and resentment, which becomes internalized. Black children learn not to question organizational structures, which have historically existed as obstacles to their ancestors' lives. Thus, Black children assume that all consequences are solely the result of their personal choices and actions. Such interventions could be considered *deracialization* tactics such as: 1) give the child "complete" agency, but within a short menu of choices; 2) erase the history of racism in the United States; and 3) minimize the cultural capital of the Black community.[3] It is worth noting that such pedagogical tactics stand in direct opposition to the historical legacy of Black teachers' socialization practices.

Curriculum as a Socializing Agent

Much of the discussion above relates to the socialization in schools and with teachers as lived, but it is also critical to discuss the socialization processes of school curriculum as memorialized in text. Sleeter and Gant (1997) conducted race-specific analyses of a variety of disciplines, including mathematics. They describe a set of eight mathematics textbooks with respect to racial depictions of Blacks in the following way:

> Therefore, images of Blacks emanate mainly from pictures. Most books depict Blacks in a variety of everyday roles such as walking the dog, playing, or camping. However, the most common single role in which Blacks appear, with the exception of student, is that of athlete. Drawings of males playing, basketball, football, or baseball usually includes Blacks. Several texts include a few famous people, usually mathematicians; the only famous Blacks shown are not mathematicians but rather athletes such as Jackie Robinson. (Sleeter & Grant, 1997, p. 291)

Sleeter and Grant (1997) noted that "textbooks participate in social control when they render socially constructed relations among groups as natural" (p. 296).

A recent and rather sensationalistic example of curricular socialization involves four teachers from Georgia, who passed out a mathematics worksheet in their third-grade class. The worksheet was supposedly created to incorporate social studies learning objectives. A word problem on the worksheet stated: "Each tree has 56 oranges. If eight slaves pick them equally, then how much would each slave pick?" Another problem read, "If Frederick got two beatings each day, how many beatings did he get in one week?" What sense is an African American child to make of reading such mathematics word problems? What are the racial messages embedded in such a mathematics activity?

Peers as Socializing Agents

With respect to peers, Datnow and Cooper (1997) provided a thorough discussion of the role of peer culture as a context for socialization, citing some of the seminal works in the field related to Black children (Fine, 1991; Fordham & Ogbu, 1986; Labov, 1982; MacLeod, 1987). These studies suggest that African American adolescents' culture is formed in opposition to White, middle-class structures and, thus, the foundation of their culture does not relate to academic success. This finding was challenged by Mehan, Villanueva, Hubbard, and Lintz (1996), who followed a detracked group of African American and Latino adolescents. They found that African American and Latino adolescents formed peer groups and identities that mutually supported their culture as African Americans and Latinos, respectively, as well as their academic success.

Historically, Walker (2006) asserted that there is a hidden legacy among African Americans in the social support of education. A review of historical studies "reveal[ed] that, as communities of learners, Black and Latino/a students were driven to excel by each other as well as by other students (including their own siblings) and adults (both parents and supportive teachers and school administrators)" (p. 44). This assertion was supported by her study of high-achieving African American and Latino adolescents in mathematics classes, where these students maintained fluid relationships among peer groups (families, and school communities) that supported their academic success.

RACIAL-MATHEMATICAL SOCIALIZATION

Mathematics Socialization

When considering mathematics socialization, Gresalfi, Martin, Hand, and Greeno (2009) are right to remind us that mathematical competence (and practice) is largely a local social construction based on the idiosyncratic nature of a mathematics classroom. The local practices, taken day in and day out, over an educational career begin to define one's perception of the meaning and value of mathematics and one's place within the field of mathematics. This is what is meant by *mathematics socialization* (Martin, 2000).

Achievement Gap Rhetoric Shaping Mathematics Socialization

Ironically, to describe with accuracy the mathematics socialization of Black children, some attention must be paid to achievement gap discourses and the specter of a *racial hierarchy of mathematics* (Martin, 2006, 2009) because these conceptions strongly influenced Black children's mathematical experiences and constructions of competence. Schools, districts, and teachers, often fearful of sanctions and choked by community pressures, tend to employ a "by any means necessary" philosophy to improve test scores, close achievement gaps, and meet annual yearly progress (AYP), as mandated by *No Child Left Behind Act* of 2001. Davis and Martin (2008) noted

that the Black mathematics education experience is plagued with remedial strategies, "inundated with practice materials that include worksheets
and in-class practice tests devoted to states' assessments" and marred by
"repetition, drill, right-answering thinking that often focuses on memorization and rote learning, out-of-context mathematical computations, and
test-taking strategies" (p. 20). All of the above is in the service of boosting
performance on high-stakes standardized assessments and closing achievement gaps.

Davis and Martin's (2008) discussion supported several aspects of Lubienski's (2002) analyses of 1990, 1996, and 2000 National Assessment of
Educational Progress (NAEP) mathematics data. After controlling for socioeconomic status, Lubienski (2002) reported that more Black students
than White reported that there was only one way to solve a problem and
that learning mathematics is mostly about memorization. With respect to
instruction, Black children tended to have less access to calculators, which
indicates a continued focus on hand-computation, but equal access to
computers, which were used primarily by Black children for the purposes
of drill and practice. Black children were also assessed more frequently
with multiple-choice assessments. Given the findings of Davis and Martin
(2008), in addition to Lubienski (2002), didactic teaching is the most likely
method used in mathematics classrooms that consist of Black children and
adolescents.

When we examine the local, everyday practices of Black children under the pressure of testing regimes perpetuated by closing so-called "racial
achievement gaps," the mathematics that seems to be most valued in Black
children's schooling relates to finding correct answers through memorization or test-taking tactics.

Racial-Mathematical Identity

To account for socializing factors beyond the mathematics classroom,
like the achievement gap rhetoric, Martin (2000) describes a framework
that broadens the conception of mathematics identity, which includes: "(a)
ability to perform in mathematical contexts, (b) the instrumental importance of mathematical knowledge, (c) constraints and opportunities in
mathematical contexts, and (d) the resulting motivations and strategies
used to obtain mathematics knowledge" (p. 19). Given these dimensions,
Martin extracted the broader meanings and circumstances behind the being and doing of mathematics.

For example, in an interview conducted with Harold, a 55 year-old Black
man, Martin noted that socioeconomic context for Black people in the job
market mediated his decisions in mathematics class and his identity. Harold
did not pursue mathematics beyond "multiplication and division of fractions and decimals," because he knew that he was going to be a laborer

and "I wasn't going to be doing any math" (Martin, 2000, p. 42). This one example speaks to the power of context outside of the classroom in socializing Black student's mathematics identities.

In a recent study, English-Clarke, Slaughter-Defoe, and Martin (2012) studied the racial-mathematical socialization of ninth and tenth grade African American students from a Northeastern metropolitan area. Relationships between racial identity measures, racial-mathematical beliefs, and racial-mathematical messages were explored. Examples of racial-mathematical beliefs are: "People of my race are typically good at math" or "In my math classes, students of other races have gotten higher grades than Black students" (English-Clarke et al., 2012, pp. 71–73). These beliefs, among others, were considered with respect to scores on the *Multidimensional Inventory of Black Identity-Teen* (i.e., MIBI-T) and interview data regarding parents' socialization messages.

English-Clarke et al. (2012) found that about one third of the adolescents interviewed recalled receiving messages of discrimination in mathematical settings from parents or others. Further, racial-mathematical beliefs aligned more closely with racial identity constructs (e.g., racial ideology) for tenth graders versus ninth graders, which suggested that racial identity has a greater impact on Black adolescents' beliefs in mathematics class as they mature. English-Clarke et al. (2012) discussed another key finding:

> It is noteworthy that youth with positive views of African-Americans, as well as those who perceive that others have positive views of African-Americans, tended to believe that African-Americans were typically good at math and used math regularly. This suggests that general self-perceptions and beliefs about others' perceptions of one's own racial group carry over into mathematics. If an African-American girl perceives African-Americans generally in a negative light, she is likely to also perceive African-Americans' math use and performance negatively, which may result in her putting forth less effort in math class, reducing her mathematical aspirations, and developing a negative mathematical identity. (p. 74)

This study provides valuable insight into the *interrelationships* of race and mathematics. This finding shows that a high regard for Black people is connected to a positive perception of Black people doing math, which is also supported by Mrs. Gant. In response to the question of a racial achievement gap, Mrs. Gant expressed her feelings—stalwart against the notion of poor mathematics achievement by African American children. After leaning back in her chair and with her arms folded, she began emphatically:

> I don't believe it. I really don't believe it [racial achievement gap]. Because now it's understandable, it's, it's a, when they say Black students ah, they are taking a ratio, they are taking a ratio.

She leaned forward and began speaking more slowly and deliberately to make her mathematical argument:

Okay, so, ah you go into a school, and you take say, you got twenty students in a school and they are going to take this test, and it's only, out of the twenty that are taking this test you have only four that is black the other 16 is white, so you are gonna say that okay the black students have a score like this amount like this, they are figuring on the score what the whole, the whole average of the whole student body is, they don't have an average of just those four that is taking the test, they have the average of the whole test.

Here, Mrs. Gant argued that a Black student failing as one of four students in a class would have a disproportionate impact in the passing rate in comparison to a White student failing who is one of sixteen students in the class. Her body language softened a bit as she went on in her response:

Now, it's true it might be a lot of Black students who don't figure good, but then I would say there are a lot Black students who do. So I don't believe in their, you know...their percentage that they are basing that test on. I don't believe it. You know what I mean?

Mrs. Gant disavowed the inferiority of Blacks in mathematics using a mathematical argument. Black students' inability to do mathematics is inconceivable to Mrs. Gant. Her refusal could be read in her words, "I really don't believe it," and her body language, where she leaned back and folded her arms. Mrs. Gant conceded little when she stated, "Now, it's true it might be a lot of Black students who don't figure good, but then I would say there are a lot Black students who do."

RENDERING BRILLIANCE THROUGH SOCIALIZATION

Still lingering is the question of how this African American grandmother—a child of Jim Crow—could realize such a robust racial-mathematical identity? In an effort to reconcile the question of Mrs. Gant, recall her spheres of socialization beginning in the 1930s. First, Mrs. Gant's schooling experiences were characterized by care and commitment, as captured in her memory of hair combing. Her teacher was a woman for whom to this day she holds deep affection, but also a woman who was her mathematics instructor for five years.

Secondly, her mathematics learning was characterized by memorization of mathematics facts, mental calculation, drill and repetition, and execution of procedures using hand-me-down books from the White school children. This didactic pedagogy was characteristic of that time period (Jones, 1970). Therefore, the culture of mathematics for Mrs. Gant, a Black girl in the 1930s, seemed to be the same culture of mathematics experienced by most White children at that time (that is, at least with respect to learning

objectives) in the United States. In this sense, local and global constructions of mathematics competence were relatively closely aligned.

With respect to her parents and siblings I learned that after grammar
school, Mrs. Gant left for San Antonio, Texas, with her mother's encouragement to live with extended family to complete her education. Mrs. Gant
described the circumstances as it pertained to matriculating to high school:

> Because in Bloomington the schools only went to the ninth grade and my
> mother wanted me to finish the twelfth grade and I had cousins that lived in
> San Antonio, so I went to San Antonio to go to school, which was a privilege;
> that was privilege. My oldest brother never went to high school, but all the rest
> of my brothers and sisters went to school.

In various ways, the above quote from Mrs. Gant's life story counters typical
constructions of Blacks in the South with respect to education. It clearly
articulates the following qualities: 1) parental influence and insistence on
education, 2) the investment of education via moving to a different city,
and 3) a large Black family that was high school educated. Thus, her family
socialization comprised a supportive network that valued education.

Although Mrs. Gant did not share a great deal about her peers, there is
a general sense of her spheres of socialization—not the least of which includes the ominous shadow of a segregated community maintained by Jim
Crow and the railroad tracks.

> The only thing that I can say about the community was—in Bloomington,
> Bloomington was the sort of town that if you lived on one side of the railroad
> tracks you were Black and if you lived on the other side of the railroad tracks
> you were White. So nobody, so no matter how much money you had or what
> you accomplished, if you were Black you had to live on your side of the rail-
> road track. Going to school, we always went to Black school, and White kids
> went to White school.

Furthermore, Mrs. Gant is an example of a brilliant mathematics student. She recalls being exempt from mathematics tests because other students would copy her mathematics paper.

> Well, when I had Algebra I and II, when it was test time I was exempt, because
> the teacher used to give us a test and he would go out of the room, you know.
> And, when we would turn our papers in, and when I turned my paper in, the
> other kids would go and copy, you know what I mean? So, he stopped me
> from having a test, I didn't have to have a test in math. I never did in math.

Mrs. Gant did not go on to college or become a mathematician. She
attended a junior college after marriage and giving birth to her twelfth
child, and received certification in bookkeeping. Nonetheless, throughout her mathematical career as a student she exuded brilliance, which was

certainly a function of her own interests and persistence in juggling messages containing meanings, values, and so on from a variety of spheres of socialization. Moreover, Mrs. Gant's brilliance was a function of socialization forces. Certain spheres of socialization made Mrs. Gant's brilliance less visible—the segregated White community across the tracks. Other spheres of socialization provided opportunities for excellence, showmanship, and confidence—Mrs. Swan's classroom.

Walker (2009) draws a similar conclusion about providing opportunities for the display of brilliance in her study of Black mathematicians:

> Although black mathematicians since Banneker and Fuller were born into an era where they have had access to more opportunity, they did not always have access to the full opportunity of their white peers as many have so eloquently shared here. What seems to have helped their cause in addition to their talent, is *a series of serendipitous events*: First the establishment of an extensive network of historically black colleges and universities following the era of slavery...Second, the fact that a number of predominantly white institutions in the Northern region of the country—both for undergraduate and graduate students—admitted talented black students. Third, for this group of mathematicians, the investment by the federal government in science, mathematics, and technology education was critical. All of the mathematicians in this group in some way were touched by national initiatives, either because expertise was needed during World War II, or following the launch of Sputnik in 1957. (p. 73)

Walker's example and the extended example of Mrs. Gant reveal that brilliance is a function of forces, people, spaces, and events. Thus, it is more productive to view the brilliance of Black children, not as an attribute via achievement scores, rather as a communal rendering in which the child, parents, teachers, peers, schools, and communities are responsible. In this case, it is not the child who possesses or who does not possess brilliance, rather it is the community, the school, the parents, the teachers, and the peers who collectively illuminate or eclipse a child's brilliance. This implicates a broader network in the promotion of the Black child's brilliance.

CONCLUDING REMARKS

I would be remiss (and I am far too proud) to neglect to mention that Mrs. Gant is also *my* grandmother. Thus, I have a particular investment in this once Black child's mathematical life. Perhaps, Mrs. Gant's brilliance is visible to me, because I am her granddaughter (i.e., brilliance is easily rendered if one is close enough). I would offer that every African American child has more than moments, periods, and stories of brilliance. The challenge of research and teaching the African American child in mathematics is then to get close enough to forces and people such that brilliance is rendered visible. This challenge is far from being a platitude—it is an empirical stance that at the very least: 1) seeks out the racial material and messages from parents, teachers, peers, and curricula; 2) inspects the larger school structures for racial composition and climates, as well as mathematics opportunities globally (e.g., course-taking patterns) and locally (e.g., meaningful mathematics learning); and 3) acknowledges the sociopolitical climate of the community within history (e.g., Jim Crow). Such an empirical stance can make the difference to the African American child's brilliance in mathematics being rendered visible or invisible.

NOTES

1. Scholars have attempted to disentangle cultural socialization and racial socialization. For example, Quintana et al. (2006) draws a distinction between socializing children to Africentric cultural items, like toys, books, and clothing versus socializing children to handle racial messages and situations. These processes are often intimately related and potentially indistinguishable.

2. It is important to note that there is contestation among scholars studying African American people as to the various cultural ways of doing and being. Perspectives often differ according to theories of the cultural continuity to Africa, given the atrocities of the trans-Atlantic slave trade. Two divergent theories are: the *catastrophic* view that suggests African American culture was made anew in the United States, and the *survivalist* view that suggests that African (specifically, West African) cultural practices did indeed survive the horrors of slavery and are evidenced in African American culture today (although not in their native form) (cf. Hale-Benson, 1982).

3. Recent work by Byrd and Chavous (2011) pushes back on such interventions and shows that students with a strong sense of racial pride are better served by schools with fair and respectful racial climates. The conclusion of Byrd and Chavous is that changing the racial climate within schools is likely to be more effective (and I would argue humane) than attempting to change Black youth's perceptions of self.

REFERENCES

Boykin, A. W. (1983). The academic performance of Afro-American children. In J. Spence (Ed.), *Achievement and achievement motives* (pp. 321–371). San Francisco, CA: Freeman.

Boykin, A. W. & Toms, F. (1985). Black child socialization: A conceptual framework. In H. A. M. McAdoo, J. (Ed.), *Black children: Social, educational, and parental environments* (pp. 33–51). Thousand Oaks, CA: Sage Publications, Inc.

Brown, T. N. & Lesane-Brown, C. L. (2006). Racial socialization messages across historical time. *Social Psychology Quaterly, 69*(2), 201–213.

Byrd, C. & Chavous, T. (2011). Racial identity, school racial climate, and school intrinsic motivation among African American youth: The importance of person-context congruence. *Journal of Research on Adolescence, 21*(4), 849–860.

Carter, R. & Goodwin, A. L. (1994). Racial identity and education. *Review of Research in Education, 20*, 291–336.

Chavous, T. M. (2005). An intergroup contact-theory framework for evaluating racial climate on predominantly White college campuses. *American Journal of Community Psychology, 36*, 239–257.

Daniel, J. & Daniel, J. (1999). African-American childrearing: The context of a hot stove. In T. Socha & R. Diggs (Eds.), *Communication, race, and family: Exploring communication in Black, White, and biracial families* (pp. 27–48). Mahwah, NJ: Lawrence Erlbaum Associates.

Darling-Hammond, L. (2010). *The flat world and education.* New York, NY: Teachers College Press.

Datnow, A. & Cooper, R. (1997). Peer networks of African American students in independent schools: Affirming academic success and racial identity. *The Journal of Negro Education, 66*(1), 56–72.

Davis, J. & Martin, D. (2008). Racism, assessment, and instructional practices: Implications for mathematics teachers of African American students. *Journal of Urban Mathematics Education, 1*(1), 10–34.

DeCuir-Gunby, J. T. (2009). A review of the racial identity development of African American adolescents: The role of education. *Review of Educational Research, 79*(1), 103–124.

Edelman, M. W. (1992). *The measure of our success: A letter to my children and yours.* New York, NY: HarperCollins Publishers.

English-Clarke, T., Slaughter-Defoe, D., & Martin, D. B. (2012). 'What does race have to do with math?' Relationships between racial-mathematical socialization, mathematical identity, and racial identity. *Race and child development Contrib. Hum Dev. Basel, Krager, 25*, 57–79.

Fairclough, A. (2007). *A class of their own.* Cambridge, MA: Harvard University Press.

Fine, M. (1991) *Framing dropouts.* Albany, NY: State University of New York Press.

Fordham, S. & Ogbu, J. (1986) Black students' success: Coping with the burden of "acting White." *Urban Review, 18(3)*, 176–206.

Gresalfi, M., Martin, T., Hand, V., & Greeno, J. (2009). Constructing competence: An analysis of student participation in the activity systems of mathematics classrooms. *Educational Studies in Mathematics, 70*(1), 49–70.

Hale-Benson, J. E. (1982). *Black children: Their roots, culture, and learning styles.* Baltimore, MA: The John Hopkins University Press.

Harris, Y. & Graham, J. (2007). *The African American child: Development and challenges.* New York, NY: Springer Publishing Company.

Holmes, R. M. (1995). *How young children perceive race.* Thousand Oaks, CA: Sage.

Horn, J. (2011). Corporatism, KIPP, and cultural eugenics. In P. Kovacs (Ed.), *The Gates Foundation and the future of U.S. "public" schools* (pp. 80–103). New York, NY: Routledge.

Hughes, D., Smith, E., Stevenson, H., Rodriguez, J., Johnson, D., & Spicer, P. (2006). Parents' ethnic-racial socialization practices: A review of research and directions for future study. *Developmental Psychology, 42*(5), 747–770.

Irons, P. (2002). *Jim Crow's children: The broken promise of the Brown decision.* New York, NY: Penguin Press.

Johnson, W., Nyamekye, F., Chazan, D., & Rosenthal, B. (2013). Teaching with speeches: A Black teacher who uses the mathematics classroom to prepare students for life. *Teachers College Record, 115*(2).

Jones, P. S. (Ed.) (1970). *A history of mathematics education in the United States and Canada.* Reston, VA: National Council of Teachers of Mathematics.

Labov, W. (1982). Competing value systems in inner city schools. In P. Gilmore & A. Glathorn (Eds.), *Children in and out of school: Ethnography and education* (pp. 148–171). Washington, DC: Center for Applied Linguistics.

Lewis, M. (1999). Hair combing interactions: A new paradigm for research with African-American mothers. *American Journal of Orthopsychiatry, 69*(4), 504–514.

Lubienski, S. T. (2002). A closer look at Black-White mathematics gaps: Intersections for race and SES in NAEP achievement and instructional practices data. *The Journal of Negro Education, 71*(4), 269–287.

MacLeod, J. (1987) *Ain't no makin it.* Boulder, CO: Westview Press.

Martin, D. (2000). *Mathematics success and failure among African-American youth.* Mahwah, NY: Lawrence Erlbaum Associates.

Martin, D. (2006). Mathematics learning and participation as racialized forms of experience: African American parents speak on the struggle for mathematics literacy. *Mathematical Thinking and Learning, 8*(3), 197–229.

Martin, D. B. (2009). *Mathematics teaching, learning, and liberation in the lives of Black children.* London: Routledge

Mehan, H., Villaneuva, I., Hubbard, L., & Lintz, A. (1996). *Constructing school success: The consequence of untracking low-achieving students.* New York, NY: Cambridge University Press.

Milner, H. R. & Howard, T. C. (2004). Black teachers, Black students, Black communities, and Brown: Perspectives and insights from experts. *The Journal of Negro Education, 73*(3), 285–297.

Murray, C. & Mandara, J. (2002). Racial identity development in African American children: Cognitive and experiential antecedents. In H. McAdoo (Ed.), *Black Children: Social Educational, and Parental Environments* (2nd ed., pp. 73–96). Thousand Oaks, CA: Sage Publications.

Noddings, N. (2005). *The challenge to care in schools: An alternative approach to education.* New York, NY: Teachers College Press.

Quintana, S. M., Chao, R. K., Cross, W. E., Hughes, D., Gall, S., Aboud, F. E.,...Vietze, D. L. (2006). Race, ethnicity, and culture in child development: Contemporary research and future directions. *Child Development, 77*(5), 1129–1141.

Rittenhouse, J. (2006). *Growing up Jim Crow: How Black and White Southern children learned race.* Chapel Hill, NC: University of North Carolina Press.

Slaughter-Defoe, D., Johnson, D., & Spencer, M. B. (2009). Race and children's development. In R. Shweder (Ed.), *The child: An encyclopedic companion.* Chicago, IL: The University of Chicago Press.

Sleeter, C. & Grant, C. A. (1997). Race, class, gender, and disability in current text-books. In D. Flinders & S. Thorton (Eds.), *The curriculum studies reader.* New York, NY: Routledge, Inc.

Van Ausdale, D. & Feagin, J. (2001). *The first R: How children learn race and racism.* New York, NY: Rowman and Littlefield Publishers, Inc.

Walker, E. N. (2006). Urban high school students' academic communities and their effects on mathematics success. *American Educational Research Journal, 43*(1), 41–71.

Walker, E. N. (2009). "A border state": A historical exploration of the formative, educational, professional experiences of Black mathematicians in the United States. *The International Journal for the History of Mathematics Education, 4*(2), 53–78.

Walker, E. N. (June, 2010). *Reimaging mathematical spaces: Cultivating mathematics identity, learning, and achievement.* Presented at the Benjamin Banneker Association Conference, "Beyond the Numbers." Temple and Arcadia Universities, Philadelphia, PA.

SECTION II

POLICY AND BLACK CHILDREN'S
MATHEMATICS EDUCATION

CHAPTER 4

METHODS OF STUDYING BLACK STUDENTS' MATHEMATICAL ACHIEVEMENT

A Critical Analysis

Danté A. Tawfeeq and Paul W. Yu

INTRODUCTION

As more high-stakes mathematics tests are added to an already saturated pool of quantitative assessments in K–12 education, test scores of Black students will continue to be analyzed and compared to scores from other groups of students as evidence of growth, stagnation, or regression in terms of skills, abilities, and knowledge possessed. One reason that large-scale testing has gained such popularity is that it is being used as a major tool in an effort, as many educational policymakers and test developers claim, to ensure that all students are receiving adequate schooling and preparation for post-secondary education (Lindle, 2009). However, overemphasis on such outcomes only presupposes that the test scores provide the most

The Brilliance of Black Children in Mathematics:
Beyond the Numbers and Toward New Discourse, pages 79–94.
Copyright © 2013 by Information Age Publishing

meaningful data concerning the learning of Black students in urban low-performing schools. The results of particular assessments (e.g. National Assessment of Education Progress, Scholastic Aptitude Test, New York State Regents Exam, or College Board's Advance Placement Exams) may shed dim light on where Black males, in particular, are in terms of their mathematical development.

According to Lemann (1999), the creation of commercialized assessments that are supposed to facilitate access to higher education for African American and Latino/a students have had negative consequences for them. For example, high-stakes tests have reduced high school graduation rates in many large urban cities (Reardon, Arshan, Atteberry, & Kurlaender, 2010). Unfortunately, the use of test scores under the most popular current operational methods provides little information about the mechanisms influencing Black students' achievement or about the amount of learning that is taking place (Cole, 1991). Instead, data derived from test scores are used superficially to make predictions about students' future success or failure (Zwick & Sklar, 2005).

A few administrators of popular assessments are only recently backing away from the notion that their assessments are the only indicators of future academic success (American College Testing, 2005). Other administrators of these assessments seem to hold to the notion of assessment determinism so much so that assessments not only determine the future, they *control* the future as well because the impact of such assessments are used to deny or limit admission and access to higher education, especially for students of color. An example would be students denied a diploma based on scores on an exit high school exam. Yet a study by Vars and Bowen (1998) showed that a 100-point increase in the SAT combined score influence college GPA only by 0.1 when race, gender, and field of study are controlled. Thus, test scores do not have the predictive abilities they are purported to have in reality. This suggests that commercialized assessments are *not* the sole indicator of a student's success in post secondary education. One of the responsibilities of the psychometrician is to quantify the attributes of cognitive ability, student skill, and achievement so that these attributes become somewhat discernible. It should not be role of policymakers or administrators to determine the meaning of test scores.

The purpose of this chapter is to examine methods that can be applied to the study of Black students' mathematical achievement and brilliance. These methods can complement one another so as to establish quality control of inter-research methods, offer a critique that adds further structure to the findings of a previous study, or enhance the understanding of assessment outcomes relative to teaching and learning. The latter of these three scenarios is presented in this chapter.

EXPLANATORY ITEM RESPONSE MODELS (EIRM) AND
DIFFERENTIAL ITEM FUNCTION (DIF)

It is important to think about the causes of bias that may cause differences in test scores among different groups of students. Briggs (2008) employs a class of models called Explanatory Item Response Models (EIRM) (de Boeck & Wilson, 2004). The EIRM framework allows for the modeling of traditional analyses of group differences because it formulates these models as part of a broader generalized linear and non-linear mixed models framework. In addition, it allows for modeling of item-specific parameters. One common method of investigating items for fairness is called Differential Item Function (DIF). Hambleton, Swaminathan, and Rogers (1991) state that, "… an item shows DIF if individuals having the same ability, but from different groups, do not have the same probability of getting the item right" (p. 110). A comprehensive study of DIF and DIF-related procedures alternatively defines DIF by stating that DIF occurs when examinees from different groups with the same underlying ability level perform differently on an assessment item (Zumbo, 1999). In large-scale test development, the most common practice is to simply eliminate items with DIF. The reasons for DIF are seldom investigated. The EIRM framework also encompasses models that can predict DIF using item features. Few applications can be found in the literature in this area because it is a fairly new class of models. However, one study was successful in showing that certain features of items such as syntactic complexity tend to cause DIF (Cid, 2009). Using evidence from these sorts of analyses, DIF could possibly be eliminated through removing the features of items that cause DIF when creating the assessment.

To explore the sources of DIF, a brief review of the literature was conducted. Many of the studies that explore sources of DIF are specifically focused on explaining gender DIF. Several studies have explored whether items that include graphical representations favor non-reference groups (Kalaycioğlu & Berberoglu, 2011; Ryan & Chiu, 2001; Yildirim & Berberoglu, 2009). The results suggest that the inclusion of graphical representations in items may be an important factor in predicting DIF. For instance it was found that an item with graphical representation favors males verses females (Kalaycioğlu & Berberoglu, 2011). Another study examined graphical representations as a source of gender DIF but found that compared to other sources of gender DIF, the inclusion of graphical representations produced a relatively small amount of DIF (Ryan & Chiu, 2001). Yildirim and Berberoglu (2009) examined the source of country-DIF in an international test and found that problems with graphical representations tend to favor U.S. students verses students from other countries.

During the 1980s and 1990s, there were about twenty DIF-related studies comparing White students to Latino/a, Asian, and Black students (Santelices & Wilson, 2010). Freedle's (2003) work, unlike others before him, emphasized the role of linguistics and cultural differences to describe the association between DIF and item difficulty (Santelices & Wilson, 2010). In light of Freedle's (2003) emphasis on the influence of linguistic and cultural differences on assessment outcomes and Holland's (2008) presentation of *reverse causation* and *common causation* as ways of positioning the source of effect, we briefly engage qualitative inquiry methods that can help synthesize Freedle's and Holland's ideas in such a way that can help inform our understanding of Black students' performance on assessments. The discussion of *causation* of students' performance on assessments could be very variable laden. When we consider the number of multiple choice and free response questions on the AP calculus exam, there are many points of departure to consider when trying to understand how and why DIF occurs.

MIXED-METHODS OF INQUIRY

We believe that a broad use of methods, both quantitative and qualitative, may provide a rich platform that can enhance the quality, fairness, and validity of mathematics assessments that Black students engage. In fact, we suggest that mathematics education researchers and mathematics teacher educators should actively pursue a unique set of research standards and procedures that will provide, in essence, quality control in the design and evaluation of mathematical assessments, particularly those that have historically and repeatedly assessed Black students as being deficient.

We posit that mixed-method studies can produce a rich analysis of gaps that emerge from assessments. For example, a qualitative study on students who took the SAT could add structure to Freedle's (2003) findings and advance his argument by identifying variables, at the student level, that might *cause* this effect of low test scores. This *cause* may have no bearing on the ability of Black students to achieve mastery of mathematical concepts but implies that the test construction process disregards student level variables and inevitably ensures the continuation of achievement gaps on the assessment. Because the nebulous concept of *causation* is replete in the psychometric literature, perhaps emerging researchers of mathematics education should be encouraged to create new research paradigms regarding the *cause* of differences in test scores between socially constructed racial groups. While the study of causation can yield seemingly paradoxical conclusions, it is important to have a sense of the multiplicity of factors that possibly impact a researcher's interpretation of the outcomes of a studies related to Black students (Holland, 2008).

In the remainder of this chapter, we present a critical analysis of methods to study Black students' mathematics achievement; specifically, we will examine the following:

1. A brief report of high-achieving Black and Latino/a students' movement through three New York State Regents mathematics exams (2008–2011) and pre-calculus (2011–2012) in preparation for AP calculus, and
2. Black males' 1997–2011 performance on the AP calculus (AB) exam.

This discussion is presented to appeal to a variety of stakeholders, including mathematics teacher educators, educational policymakers, and practitioners. In this way, we will illustrate how mixed-methods inquiry can enrich our understanding of the mathematical achievement of Black students and impact the policy of curricula development and assessment.

PRELUDE TO THE STORY

First, we provide a snapshot of our current work within an urban high school in the metropolitan New York area. This high school, which consisted of predominantly Black and Latino/a students, was attempting to transition from a traditional didactic instructional model to inquiry-based learning (IBL) models (Tawfeeq & Yu, in press). In this project, we investigated professional development and classroom instructional strategies to help increase the number of students taking pre-calculus and subsequently AP calculus while still promoting IBL. The goal was to enhance the educational experiences of the Black and Latino/a students by increasing their opportunities to learn pre-calculus and calculus from an IBL perspective. While this chapter does not present the final findings of this research, it is a reflection on a mixed-methodology study that illuminates the brilliance of Black students within a context of a school culture deeply affected by standardized testing, in particular, the New York State Regents Exam program. While Leonard, McKee, and Williams (this volume) briefly engage the topic of AP calculus in terms of historical access, we analyze the process and assessment of these students' engagement of pre-calculus and calculus via inquiry-based learning (IBL).

This population of high-achieving eighth-grade mathematics students moved from a homogenous grouping of a first-year algebra course in middle school (AY 2008–2009) to three years of a heterogeneous grouping in their high school courses. In addition to the high-achieving cohort that took eighth-grade algebra, the heterogeneous grouping also included low- and middle-achieving students from differing grade levels across multiple course sections of geometry (AY 2009–2010), algebra II trigonometry (AY 2010–2011), and pre-calculus (AY 2011–2012). While this study is ongoing, the preliminary findings from this mixed-methods inquiry have illuminated characteristics that give insight into how high-achieving students who start out in homogeneous grouping as academically talented middle school children engage preparatory AP calculus courses with heterogeneous mixing in their high school mathematics program.

Second, we situate the identity of the Black student[1] as something other than the byproduct of the assessment industry and rather as an intellectual being of mettle, means, and fidelity towards intellectualism in spite of assessment outcomes. Because of the increased number of Black males taking the AP calculus exam and the excessive scrutinization of Black males' test outcomes, we have chosen to focus on the recorded experience of these students. We examine Black males' national AP calculus AB and BC scores from 1997–2011.[2] This is done so that the reader can place what is presented in this chapter in a national context. Further motivation for this focus comes from the increased attention placed on Black students' access to AP calculus courses (Oakes, Joseph, & Muir, 2001; Werkama & Case, 2005) and Black males' historical achievement gaps relative to the Black female and other racial and ethnic groups. According to *the Schott 50 State Report on Public Education and Black Males* (Schott Foundation for Public Education, 2010), Black males are not progressing academically as well as other groups and, therefore, require unique attention. We do not believe that focusing on the Black male ignores the plight of the Black female, or other underrepresented minorities in STEM education for that matter. We are simply of the opinion that social and institutional structures in America, as well as testing outcomes, have placed many Black males in precarious conditions. However, in spite of these structures, we also believe that these conditions can be rectified toward the better even in an educational context supported by standardized and national testing. Later in the chapter, we will generalize the issues illuminated to include underrepresented minorities as we engage particular instances of teaching and learning in urban high schools that serve significant minority populations.

A STORY OF BRILLIANCE:
THE STUDENTS OF METRO HIGH SCHOOL

In this section we will provide an overview and reflect on our project[3] to improve the enrollment and preparation of Black students (Latino/a students were also included) at Metro High School (Pseudonym) in pre-calculus coursework through a professional development project with their teachers. The focus of the professional development project was the incorporation of instruction that used student-centered IBL. Werkema and Case (2005) studied the transitioning of calculus into the mathematics curriculum in a Boston inner-city school. Werkema and Case narrated the shift in attitudes and experiences of teachers and students relative to the implementation of calculus. However, the difference between the Werkema and Case study and our project is that our study engages the implementation of professional development, curriculum design, and student involvement so as to enhance student learning and access to AP calculus.

In 2008, when Metro High School was first transitioning to an inquiry-based learning approach, fewer than 30% of its graduating seniors enrolled in a four-year college. The issue was that although borderline metropolitan schools were contributing to higher academic outcomes in general, Black and Latino/a students underperformed and continued to lag behind their White counterparts in critical academic areas, such as mathematics, science and technology. Moreover, these critical areas have been identified by educational research as key indicators for college eligibility and completion. An important part of these students' and teachers' academic stories is that their school and school district have had an unusually high rate of administrative turnover. The school has had four lead principals, and the district has had three different superintendents in the past five years. Also, because of the changing demographics of the school from majority Black to about 47% Black and 53% Latino/a, the school has had to close its doors on several occasions because of student fights between the Black and Latino/a students. In this time period, the administration was eager to move in a new direction and sought partners to help them move in a meaningful direction.

In a broader context, Metro High School and its students were required by law to participate in the New York State Regents Exam. In mathematics, this high-stakes testing program utilized criterion-based reference exams in integrated algebra (algebra I), geometry, and algebra II trigonometry. These tests are administered under uniform conditions in which each exam is given at a specific date and time with strictly held proctoring guidelines. For students, successful completion of the Regents exams is required for graduation from high school, regardless of the grade they received in their respective course. Students could receive an A in the integrated algebra course, yet fail the corresponding Regent's exam, thus making them ineligible to graduate.[4] On one hand, the schools were required to participate in state-wide high-stakes testing, while on the other hand, they were in a school district with a constantly shifting administrative structure.

This project targeted Black and Latino/a students and their teachers in this high-needs school by increasing their opportunities to learn pre-calculus and calculus. The project's goal was to increase the number of students enrolled in pre-calculus, high school calculus, or AP calculus in this low-resourced high school in New York City. The researchers also provided curriculum intervention in order to positively impact the students in the learning of mathematics with the goal of increasing their conceptual understanding of mathematical topics, and preparing them to become STEM majors in college. The inquiry-based learning approach was designed to move students beyond the learning of mathematics for the sake of doing mathematics to the learning of mathematics for the sake of application and conceptualizations of quantitative ideas. The researchers wanted the teachers and students in this study to distinguish rigor from regiment. Traditionally, the teacher encouraged rote procedures, where students' were relegated to working on skill-based exercises, three- to four-step exercises, and computational manipulation.

Many of the mathematics classes were dominated by the monologue of the teacher; the direction of the conversation was towards the students. The *Directional Bi-directional Discourse Model* and Table 4.1 provided the framework used in the professional development program and depicted a shift of dialogue in the mathematics classroom, a shift based on the role that the students and teachers have in the conversation (Tawfeeq & Yu, in press):

1. Teacher to student (teacher starting and moving the discussion with the student, which is the lowest level discourse);
2. Teacher to student (teacher starting the discussion with student, and that student responsible maintaining a fraction of discussion);
3. Student to student (students conversing about mathematical ideas); and
4. Student to teacher (student starting conversation with teacher, both equally responsible for keeping the conversation going, which is the highest level of discourse).

In Table 4.1., moving from the left to the right of the Five-Step Inquiry Domain Continuum, the content provider (teacher) and students encounter the same variables addressed in the *Directional Bi-Directional Discourse Model*. Unlike the Bonnstetter's (1998) Continuum, in the Five-Step Inquiry

TABLE 4.1. Five-Step Inquiry Domain Continuum That Develops Inquiry Maturity

Modality	Didactic	Structured Inquiry	Guided Inquiry	Student Initiated	Open Inquiry
Topic	Pre-Calculus/ AP Calculus Curricula	Pre-Calculus/ AP Calculus Curricula	Teacher/ Student	Student	Student
Question	Generated from Competencies	Generated by Teachers	Teacher/ Student	Student/ Teacher	Student
Materials	Textbooks	Textbooks	Teacher	Teacher/ Student	Student/ Teacher
Procedures/ Design	Teacher	Teacher	Teacher/ Student	Student/ Teacher	Student
Results	Teacher	Teacher/ Student	Teacher/ Student	Student	Student
Conclusion	Teacher	Teacher/ Student	Teacher/ Student	Student/ Teacher	Student

Continuum, the student moves towards intellectual autonomy faster, and the teacher-student positions are juxtaposed during the "fourth step" of the continuum (Yu & Tawfeeq, 2011). Also, the student is fully integrated in the process of his or her own learning by the fourth step as he or she begins to guide the instructor. The researchers encouraged the teachers to ask open-ended questions. The researchers provided a more structural and rigorous approach to asking open-ended questions. For example, they would encourage teachers to require their students to compare and contrast other students' responses. The researchers also encouraged students to verbally critique the questions and answers of other students and the teacher. The researchers wanted the students to become accustomed to such interactions and understand them to be necessary.

One component of this program was longitudinal, which involved the monitoring of a cohort of 49 mathematically high-achieving students starting in their freshman year and moving through their senior year. In fall 2009, this cohort of Black and Latino/a students who entered Metro High School were prepared to take the New York State Regents course in geometry. These students represented about 19% of the incoming freshmen class that year. During the previous year, these students were placed in two sections of the New York State Regents course of integrated algebra as eighth graders in the Metro Middle School. While this course is generally taken by ninth graders throughout the state of New York, these students' strong mathematical performance as seventh graders positioned them to take this Regents course as eighth graders in academic year 2008–2009.

During that academic year, two sections of integrated algebra offered at the middle school were completed by these high-achieving students. The instruction of the classes was rigorous and moved at an accelerated pace in comparison to other courses provided at the middle school. Upon successfully passing this class and the Regents exam, students received a 3 or 4 level rank, with 4 being the highest rank possible. This rank was assigned based on how high students scored on the Regents integrated algebra exam as eighth graders. During the time this population took this exam, a score of 85 and better received a ranking of a 4, while a score of 62–84 received a ranking of 3.

Upon entering ninth grade, these students took the Regents geometry course during the academic year of 2009–2010. This course is generally comprised of tenth and eleventh graders. These tenth and eleventh graders were those who did not take a Regents mathematics course as eighth graders and were not considered high achievers. In fact, from what we gathered from our four years of work with Metro High School, significant effort was geared toward mathematics instruction that supported remediation and the retention of basic skills for the majority of the student population. As with many schools in the metropolitan New York area that are similar to Metro

High School, a great deal of mathematical instruction that occurs could be described as skills- and remediation-based. As a result, many high-achieving students in Metro High School seldom, if ever, receive rigorous instruction that promotes conceptual understanding and development. Furthermore, while not having the opportunity to receive rigorous mathematical instruction is an impediment for higher-achieving students, we believe that it negatively impacts all students.

In the case of our experimental group, high-achieving eighth-grade participants who had taken integrated algebra were placed in a heterogeneous geometry course that was comprised of ninth, tenth, and eleventh graders of variable ability in academic year 2009–2010 (Year 1). This heterogeneous group remained intact for algebra II trigonometry in academic year 2010–2011 (Year 2). During academic year 2011–2012 (Year 3), 25 of the 49 students from the experimental group enrolled in pre-calculus.

Based on baseline data from June 2009, the experimental group's deviation of scores was less than the control group on the first Regents exam, reflecting a low spread across this particular group. However, the experimental group's standard deviation exceeded the control group's standard deviation on the June 2011 Regents exam, suggesting a wider spread of mathematical growth. Over the three-year period, the mean Regents scores of the experiment group regressed somewhat more than the control group (see Table 4.2). While the mean scores of the experimental group followed a statistical trend similar to that of the control group, the corresponding de-

TABLE 4.2. Metro High School's Scoring Trend from 2009-2011

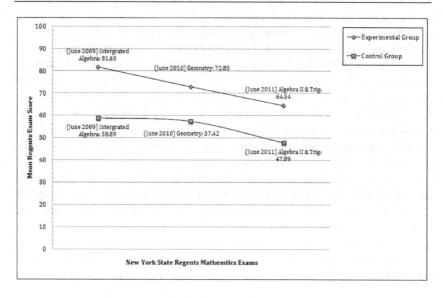

TABLE 4.3. Metro High School Standard Deviation Trend 2009–2011

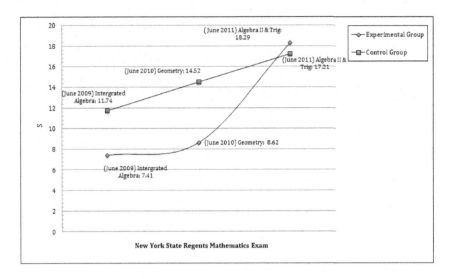

viation scores over that same time period had significantly increased for the experimental group, as can be seen in Table 4.3. Because we are still analyzing baseline data and currently incorporating more sophisticated statistical analyses, it would be premature to offer a final assessment of the outcomes. However, we pinpoint the persistence of the experimental group to continue through the curriculum, in spite of the fact that the instruction at Metro High School was geared toward the lower- to middle-achieving students.

ANOTHER PART OF THE STORY: BLACK MALES AND AP CALCULUS (AB) SCORES

In light of the cohort of students discussed in the previous section, we now turn our attention to the broader context of the Advanced Placement (AP) calculus program. Historically, from 1997–2011, the largest racial and gender population to take the AP calculus (AB) exam is White males. The percent increase for White males taking the exam from 1997 to 2011 is approximately 102%, which is 7.28% annually. Black male takers of the same exam during this time period increased by 261.5%, which is 18.76% annually (See Table 4.4).

However, because the overall number of Black males taking this AP (AB) exam is low, if the number of Black males taking the exam this year increases by 100, that would add about 6 percentage points to the total percentage increase over 15 years. This percentage increase would be significant if the population of Black males taking the AP calculus (AB) exam were statisti-

TABLE 4.4. Number of Black Males Taking the Advance Placement Calculus (AB & BC) Exam: 1997–2011

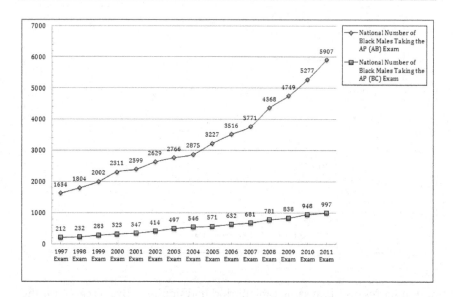

cally equivalent to their representation in the national school population. Of particular interest is that while the mean score of Black males taking this exam over this 14-year period (See Tables 4.5 & 4.6) is consistently in the 2's for the AB exam, the population of test takers among this group is increasing, which suggests that there is certainly a stream of information about the AP exam that is getting to this population. In spite of the test scores, Black males are increasingly taking advantage of this opportunity. This suggests a great deal about these Black males' commitment to academic advancement and their focus on post-secondary education. Even though many of these students received a 2 on the AP Calculus (AB) exam, they benefited from the opportunity to engage a rigorous course just prior to college enrollment.

Furthermore, while the increased numbers of Black male test takers is a positive, researchers need to turn their attention to studies that investigate issues that will help Black males increase their score from 2's and 3's to 4's and 5's. There are some institutions of higher learning that no longer allow a score of 3 and even 4 to receive credit for calculus I. So beyond the fact that there is an increased frequency of Black males taking the AP calculus exam, the issue of Black males' performance per test item is of great importance.

TABLE 4.5. National Mean Score of Black Males Taking the Advance Placement Calculus (AB & BC) Exam: 1997–2011

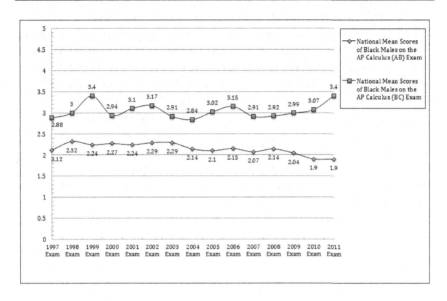

Table 4.6. The Number of Black Males That Scored 1, 2, 3, 4, or 5 on the AP Calculus (AB) Exam: 1997–2011

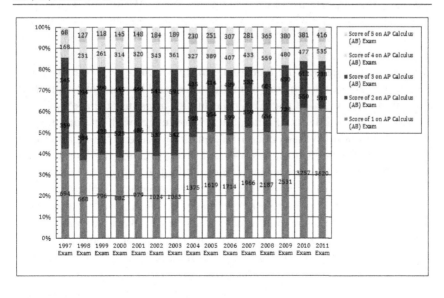

CONCLUSION

In this chapter, we presented our analyses of issues related to Black males and AP calculus and provided a preliminary report of a mixed-method inquiry project that promoted the advanced learning of mathematics among a group of students in an urban high school over a four-year period. When the project began in the fall of 2008, four students were taking AP calculus. According to the instructor of the course at the time, these students were not prepared for AP calculus. Since then, the researchers have worked to increase the number of students in pre-calculus and calculus. Because of the efforts in prior years, the number of students taking pre-calculus has increased school wide. The preliminary baseline data provided by this approach has shown that, despite heterogeneous grouping and teachers' tendency to promote less rigorous and skill-based instruction, a population of high-achieving Black students have persisted throughout the mathematics curriculum to become 75% of the population in a pre-calculus course. Further, there is a possibility that 30 Black and Latino/a students will be eligible to take AP calculus in fall 2012. Black males will comprise 25% of that number. While some may consider this a small stride, since 2008 the number of students enrolled in AP calculus has more than doubled. This is a great stride in the positive direction in spite of the trends reflected on the Regents Exam.

In the beginning of this chapter, we spoke about the overemphasis of quantitative assessments when attempting to measure and understand Black students' brilliance and achievement. Nowhere in this chapter do we disregard the importance of quantitative methods. We simply believe that the results of qualitative studies, such as the two described above, should inform psychometricians and test designers as they construct items for assessments. What we find most important about qualitative methods of inquiry is that the learner, the Black male in this instance, can come to know a great deal about his learning of mathematics and how to better interpret assessment outcomes. In the past, quantitative assessments have not fairly measured Black students' mathematical knowledge or their potential for college attainment. We also encourage qualitative researchers, particularly those who investigate issues of equity and access in mathematics education for Black students, to expand their methodological approaches to include mixed methods. Researchers who have swayed towards qualitative methods should also engage in quantitative perspectives in order to strengthen arguments regarding the mathematical success of Black students. Black children's brilliance and success in mathematics is not an anomaly or an outlier. We believe mixed methods designs will produce the studies that will showcase the brilliance we know is already there.

NOTES

1. While we use the term Black student and, in particular, focus on the Black male student in major parts of this chapter, this does not deter from the general application of this work towards the Latino/a population in urban school systems.
2. The tables and graphs were created by one of the authors of this chapter. The author disaggregated score data from the College Board website. This information is accessible to the public online.
3. This project was supported by the Educational Advancement Foundation.
4. Only the integrated algebra Regents exam is required for a general high school diploma. Successful completion of geometry and algebra II with trigonometry Regent exams are required to receive an Advanced Regent's diploma.

REFERENCES

American College Testing. (2005). *Crisis at the core: Preparing all students for college and work access.* Iowa City, IA: ACT, Inc.

Bonnstetter, R. (1998). Inquiry: Learning from the past with an eye on the future. *Electronic Journal of Science Education, 3*(1). Retrieved from http://wolfweb.unr.edu/homepage/jcannon/ejse/bonnstetter.html

Briggs, D. C. (2008). Using explanatory item response models to analyze group differences in science achievement. *Applied Measurement in Education, 21*(2), 89–118.

Cid, J. (2009). *Using explanatory item response models to examine the impact of linguistic features of reading comprehension test on English language learner.* Unpublished doctoral dissertation, James Madison University, Harrisonburg, VA. Available from ProQuest Dissertations & Theses database (AAT 3366261).

Cole, B. P. (1991). *College admissions & coaching.* In A. G. Hillard III (Ed.), *Testing African American students: Special re-issue of The Negro Educational Review* (pp. 97–110). Morristown, NJ: Aaron Press.

de Boeck, P. & Wilson, M. (Eds.). (2004). *Explanatory item response models: A generalized linear and nonlinear approach.* New York, NY: Springer.

Freedle, R. (2003). Correcting the SAT's ethnic and social-class bias: A method for re-estimating SAT scores. *Harvard Educational Review, 73*(1), 1–43.

Hambleton, R. K., Swaminathan, H., & Rogers, H. J. (1991). *Fundamentals of item response theory.* Newbury Park, CA: Sage.

Holland, P. W. (2008). Causation and race. In T. Zuberi & E. Bonilla-Silva, *White logic, white methods: Racism and methodology,* (pp. 93–109). Lanham, MD: Rowman & Littlefield Publishers.

Kalaycioğlu, D. B. & Berberoglu, G. (2011). Differential item functioning analysis of the science and mathematics items in the university entrance examinations in Turkey. *Journal of Psychoeducational Assessment, 29*(5), 467–478.

Lemann, N. (1999). *The big test: The secret history of the American meritocracy.* New York, NY: Farrar, Straus, and Giroux.

Lindle, J. C. (2009). Assessment policy and politics. In G. Sykes, B. L. Schneider, D. N. Oakes, J., Joseph, R., & Muir, K. (2001). Access and achievement in mathematics and science. In J. A. Banks & C. A. McGee Banks (Eds.), *Handbook of research on multicultural education* (pp. 69–90). San Francisco, CA: Jossey-Bass.

Plank, & T. G. Ford, *Handbook of education policy research* (pp. 319–327). New York, NY: Routledge.

Reardon, S. F., Arshan, N., Atteberry, A., & Kurlaender, M. (2010). Effects of failing a high school exit exam on course-taking, achievement, persistence, and graduation. *Educational Evaluation and Policy Analysis, 32*(4), 498–520.

Ryan, K. E. & Chiu, S. (2001). An examination of item context effects, DIF, and gender DIF. *Applied Measurement in Education, 14*, 73–90.

Santelices, M. & Wilson, M. (2010). Unfair treatment? The case of Freedle, the SAT, and the standardization approach to differential item functioning. *Harvard Educational Review, 80*(1), 106–133.

Schott Foundation for Public Education. (2010). *Schott state report on black males and education.* Retrieved from http://www.schottfoundation.org/funds/black-male-initiative.

Tawfeeq, D. A. & Yu, P. W. (in press). Developing a socio-cultural pragmatic mathematics methods course. *Journal of Urban Mathematics Education.*

Vars, F. & Bowen, W. (1998). Scholastic aptitude, test scores, race, and academic performance in selective colleges and universities. In C. Jencks & M. Phillips (Eds.), *The Black-White test score gap* (pp. 457–479). Washington, DC: The Brookings Institute.

Werkema, R. D. & Case, R. (2005). Calculus as a catalyst: The transformation of an inner-city high school in Boston. *Urban Education, 40*(5), 497–520.

Yildirim, H. H. & Berberoglu, G. (2009). Judgmental and statistical DIF analyses of the PISA-2003 mathematics literacy items. *International Journal of Testing, 9,* 108–121.

Yu, P. W. & Tawfeeq, D. A. (2011). Can a kite be a triangle? Bidirectional discourse and student inquiry in a middle school interactive geometric lesson. *New England Journal of Mathematics, 43,* 7–20.

Zumbo, B. D. (1999). *A handbook on the theory and methods of Differential Item Functioning (DIF): Logistic regression modeling as a unitary framework for binary and likert-type (ordinal) item scores.* Ottawa, ON: Directorate of Human Resources Research and Evaluation, Department of National Defense.

Zwick, R. & Sklar, J. (2005). Predicting college grades and degree completion using high school grades and SAT scores: The role of student ethnicity and first language. *American Educational Research Journal, 42*(3), 439–464.

CHAPTER 5

NOT "WAITING FOR SUPERMAN"

Policy Implications for Black[1] Children Attending Public Schools

Jacqueline Leonard, Malaika McKee, and York M. Williams

INTRODUCTION

Waiting for Superman (Guggenheim, 2010) was a highly anticipated film that debuted in the fall of 2010. Davis Guggenheim, whose most recent documentary was the academy award winning, *An Inconvenient Truth*, applied the same directorial panache to the tough issue of public school inequity. *Waiting for Superman* addressed the failings of public education in America, and it proposed enticing alternatives of charter schools and other entrepreneurial market-based strategies for educational reform. According to Guggenheim, public schools are bureaucratic behemoths where the average citizen is disempowered by the *system*; the policy practices of union-backed teachers and staff breed mediocrity at best and academic failure at worst.

The media hullaballoo about the film was equally matched by the impassioned views of its director. An unprecedented publicity machine sup-

The Brilliance of Black Children in Mathematics:
Beyond the Numbers and Toward New Discourse, pages 95–120.

ported the showing of the documentary. The children cast in the film were featured on the Oprah Winfrey Show and visited the White House to meet with President Obama (Ravitch, 2010).American Federation of Teachers president, Randi Weingarten, was featured on many of the popular television programs to defend her support of the collective bargaining portrayed in the film. Town halls were convened across the country with teachers' unions and PTAs debating the merits of collective bargaining based on information depicted in the film.

Waiting for Superman's cinematic power derived from the director's position as an outsider. His narration is peppered with statistical facts and data to give credence to his position that public schools are *failure factories*. This indictment of the system of public education gave Guggenheim the authority to inform the viewer that even though he was part of a grassroots organization to create a charter school, he did not consider a charter school for his own children. Instead, he claimed he was *lucky* to be able to send his children to a private school.

Though Guggenheim's description of his school choice experience as *luck* was not the issue he probably wanted to highlight, it is the very issue that disempowers the message of the film and hence warrants critique. The lottery process is a cinematic metaphor by which Guggenheim builds a larger narrative to explain how parents can mitigate personal hope for their children to have a better future. Luck is continually repeated as an operant factor for school success—the students are lucky when their names are chosen in the lottery, privileged children are lucky to live in neighborhoods where their property tax dollars support excellent schools; the country is awaiting an improbable arrival of a character created by DC comics to save Americans from a poor and ineffective public school system. Through the stories of the five children and their families who hope to gain access to high-quality education, it is luck, or the lack thereof, that impacts their ability to profit from a failing school system.

We empathize with these families as we learn the children in the film have faced unlucky adversities that are difficult to leave at the schoolhouse door. Daisy's father was unemployed. Bianca's mother tried unsuccessfully to sustain tuition at a private school. Francisco's mother, while educated and gainfully employed, was a single parent who struggled to keep him interested in school. Anthony's father died of a drug overdose. As a rhetorical finish, the denouement of the film features each family courageously engaging the lottery system to enroll their child into a selective charter school. They wait for a lottery-superman; they wait for luck.

Ineffective schooling is an outcome of poor policy decisions, which are often made by those who are disconnected from the lives of poor children. Though luck has some influence on events in a metaphysical sense, research urges us to question the randomness of birth with a real conver-

sation and analysis of social policy and political wrangling. We had anticipated that a documentary of such magnitude as *Superman* would give its
viewers the tools to do just that and become equipped to make a substantive
difference in their communities. However, Guggenheim's misleading data
unfortunately disempowers; without clear analysis of the facts, the viewers
of the film and the parents are left with no guidance about how to change
their predicament. They are simply left knowing that they are at the mercy
of *luck* or, as award winning New York education innovator and Harvard
Graduate School of Education alumnus Geoffery Canada laments in the
film, "No one is coming with enough power to save us." Geoffrey Canada
made this comment in response to a conversation about why he had to take
an entrepreneurial approach to start a charter school.

Herein lies the profound irony of the film; Geoffrey Canada, whose
schools are best known for their cozy ties to Wall Street funding, does
not describe to the audience how to make relationships with hedge fund
managers (Otterman, 2010) to change their school funding woes. Guggenheim's *luck* is more of a factor of privileges fueled by a system of great
financial gains from being an award-winning director that advantages him
as an affluent White male (Anyon, 1981). In this matter, Guggenheim is his
own Superman in that he is able to rescue himself from the failures of the
system he indicts. Canada is able to call Superman at JP Morgan.

Why are these ironies so critical to understanding how policy decisions
impact Black children in public schools? Powerful documentaries like *Waiting for Superman* take on a historical truth of their own. They attempt to inform viewers by arming them with information through powerful narratives
of human emotion sprinkled with *facts* to give credence to the human story.
But without critical analysis, they can end up reifying the very themes of
inequality they attempt to reveal through misinformation and misleading
statistics. Oftentimes, directors hide behind artistic license as a means of
justifying their approach. However, in a volatile policy environment where
the sound bite often triumphs over fact, this kind of approach warrants
a check of academic integrity. Moreover, we contend that the underlying
meme of luck obfuscates the more substantive conversation of unpacking
the statistical complexities of poor public policy decisions. Just as the heroic
figure Superman's power came from within, we contend that communities
under siege from poor public policy decisions regarding their schools must
realize their potential for self-determination. Power begins with understanding how the system works and using mathematics to demand change
in public policy. Future viewers of this admittedly powerful film need to be
given the tools to engage their outrage rather than directorial finesse that
only tickles their pathos.

In this chapter, we have chosen to illuminate the misleading data and in-
consistencies in the film to uncover what is not shown and place it in a larg-

er conversation about the U.S. educational system and how it systemically works to constrain access to high-quality education. This chapter discusses the importance of mathematics literacy as a tool to demystify data in the film and enable families to become informed citizens and use mathematics for empowerment. We encourage other educators to use this film and our critique as a tool to work alongside families and communities to challenge inequities, so they are not waiting for Superman and waiting on luck.

THE NEED TO UNDERSTAND TRUE POPULATION

One of the central tenets of basic statistics is understanding the notion of true population. When taking a sample for a given survey, is it truly reflective of the population? Public schools are not strangers to the vagaries of selective sampling. As the subject of constant critique, Guggenheim used the meme of failure factories to selectively sample public schools and provided a negative portrayal while cherry picking certain charter schools as beacons of success. This illusion is highlighted by Guggenheim's decision to depict only failure of public schools and not success—success is only seen at the charter schools he advocates to the viewer. Not one example of a high-achieving public school (although there are many) was mentioned in the film (Ravitch, 2010).

The *2002 reauthorization of the Elementary and Secondary Education Act, popularly known as No Child Left Behind* legislation (2001), as an education policy inadvertently supports the failure factory perspective. Scores of public schools, which have been identified as low performing, have been closed or placed into turnaround status in several urban cities, including Philadelphia (Leonard, Taylor, Sanford-De Shields & Spearman, 2004), New York City (Millner, 2011), and Chicago (Lipman, 2003). Charter schools or for-profit schools often reopen in place of these public schools. The public is told about charter school systems like the Knowledge is Power Program (KIPP), which boasts of moving students from the 32nd to 60th percentile in reading and from the 40th to 80th percentile in mathematics (Guggeheim, 2010).

More often than not, these schools have limited space to meet the needs of communities where public schools have closed (Barr, Sadovnik, & Visconti, 2006; Leonard, 2002), and as a result of scarce supply, the demand for entry turns what should have been a public education into a commoditized item. Moreover, *"the merits of a marketplace model for public education have been among the most prominent themes in education policy discussions over the last two decades. The No Child Left Behind Act (NCLB), has accelerated the trend toward private, for-profit activities in public education"* (Molinar & Garcia , 2007, p. 11). Commoditized public education takes two forms. The first is actual takeover of public schools and school systems by for profit companies for profit gain. Examples of this include Edison Schools, Agora Cyber Academy, or

K12, a company sponsored by William Bennett, the former Secretary of Education. The other form of commoditization is more subtle and particularly acute in underserved communities. Families are seduced into competing for too few spaces in charter schools. The illusion of scarcity makes school entry a commodity instead of a basic right, and parents end up fighting for lottery space as opposed to changing the true source of inequality, which is unequal funding.

While the cited performance outcomes of charter schools are commendable, the more important question is whether underrepresented students, who are predominant in charter schools, are engaged in the academic rigor needed to attain a college education. Guggenheim admits that only one in five charter schools succeeds in terms of performance outcomes, even though he presents charter schools as the best educational choice for urban families. He selectively chooses the population of charter schools to portray his argument, which is not only a mathematical error but is disingenuous to his audience.

PUBLIC POLICY: LIES, DAMNED LIES, AND STATISTICS

Education policy "defines what knowledge, values, and behavior are legitimate and shapes public meaning about groups of students and communities and purposes of education" (Lipman, 2003, p. 334). In the current climate, education policy has done little to improve educational outcomes for Black children living in poverty (Ladson-Billings, 2006). A mounting education debt, similar to the U.S. budget deficit, has steadily accumulated in our nation's underserved schools since separate and *un*equal schools were legitimized for Black students (Ladson-Billings, 2006).

The enormous education debt has opened a rancorous debate about accountability. Under the guise of accountability, what should be a democratic conversation about better schooling instead becomes,

> … a product of surveillance and containment…which produc[es] inequality
> and racial oppression and the reproduction of a highly stratified, economi-
> cally polarized labor force…. In this sense, school accountability and mecha-
> nisms that promote competition for resources and rewards contribute to a
> larger ideological shift that erodes our capacity to act together for the com-
> mon good." (Lipman, 2003, p. 331)

Black and Latino/a children's educational experiences, which are differentiated by race, ethnicity, and class, are regulated and controlled by education policies that dictate the kind of education they receive (Leonard, 2008; Lipman, 2003). When academic success is determined solely by performance on high-stakes tests, schools are forced to go into remedial and basic education mode, which guarantees that the students in such programs, who are predominantly Black, Latino/a, and immigrant, receive substandard education in comparison to their White counterparts to whom they are unfairly compared (Buckley, 2010; Lipman, 2003; Spencer, 2009). School-based evidence in the Midwest (Buckley, 2010; Lipman, 2003) and Southern California (Spencer, 2009) revealed that modifying mathematics curriculum in the name of school reform has resulted in glorified tracking and curriculum deprivation that do not lead to mathematical fluency (Buckley, 2010). Instead, low achievement is perpetuated, which ensures that the recipients of such an educational system remain in a cycle of poverty and low-paid labor. Thus, the public is lured into believing that failure to pass high-stakes tests—instead of race— is the impetus for placing schools on probation, turning them over to state control, shutting them down, and reopening them as charter schools.

UNWILLINGNESS TO ENGAGE THEORY

"That's too academic!" or "She's in an ivory tower." are familiar refrains when attempts are made to engage intellectual space to explain social problems. In the seminal work, *Anti-Intellectualism in America*, Richard Hofstadter (1962) argues that these statements are endemic to American culture:

> Intellectuals in the twentieth century thus have found themselves engaged in incompatible efforts: they have tried to be good and believing citizens of a democratic society and at the same time tried to resist the vulgarization of culture which that society constantly produces. It is rare for an American intellectual to confront candidly the unresolvable conflict between the elite character of his own class and his democratic aspirations. (p. 407-408)

Similarly, *Waiting for Superman* offers no theoretical underpinnings to explain why social inequality exists except to say that they are only due to observational conclusions that are all logical extensions of each other: poverty, joblessness, being a person of color.

Critical Race Theory (CRT) (Delgado, 1995) is a theoretical construct that offers a more nuanced understanding to the intersections of race, social class, and gender and can be used as a powerful theoretical construct against the clamor of anti-intellectualism to deconstruct the racialized data presented in the film. Even more importantly, it is a theoretical perspective that legitimates the narratives of families portrayed in the documentary as essential for intellectual exploration.

In the field of education, Gloria Ladson-Billings and William Tate proposed to extract CRT from its historical roots in legal studies to "… examine the role of race and racism in education" (Dixson & Rousseau, 2005, p. 8). Traced to the legal work of Derrick Bell and Alan Freeman in the mid 1970s (Delgado, 1995), CRT holds the premise that racism is normative in the U.S. (Dixson & Rousseau, 2005; Ladson-Billings, 1998). Rather than referring to biological or genetic traits, scholars within the CRT tradition suggest that race is both an *ideological* and a *social* construction that operates to protect the interests of those who are privileged and in power (Essed, 2002). Thus, race becomes a common way of "referring to and disguising forces, events, classes, and expressions of social decay" and economic stratification (Morrison, 1992, p. 63). Similarly, racism is both a *structure* and a *process* (Essed, 2002). As a structure, one racial group oppresses one or more subordinated groups. As a process, the domination of one racial group over another results in the subjugation of others to lower status based on race. Racism is reproduced when rules, laws, and regulations are used to maintain the privileges and power of the dominant group (Essed, 2002; Martin, 2009a; Matias, in press). In *Waiting for Superman*, the systemization of race as a socially reproducible inequity embedded in the law is left to the wayside. Race is an artifact of students' condition and not a cause of their misfortune.

To understand how CRT helps uncover the purpose of race in systemic education inequity requires the use of the students' *counternarratives* in the film to be analyzed in juxtaposition to the *masternarrative* of the dominant group—which in the U.S. are those who are legally and socially constructed as White (Matias, in press). If Guggenheim had used a CRT framework, he would have problematized how his protagonists, primarily Black Americans and people of color, are markedly different from other ethnic groups by way of collective history and condition (Martin, 2009a). Instead *Waiting for Superman* is essentially a master-narrative that implies that race is a causal variable when it comes to achievement and one in which Black children are viewed as inferior to White children (Martin, 2009b). The consequence of this master-narrative is most evident in large urban cities, where students are most likely to attend schools that are segregated by race. In these settings, Black and Latino/a children are overrepresented in urban school districts and exposed to less than adequate educational opportunities. On the contrary, White children are more likely to attend suburban, rural, and private or parochial schools (Tillman, 2004).

CRT uses counter-narratives as a method for qualitative analysis (Duncan, 2005; Solórzano & Yosso, 2002). The primary strategy is "one of unmasking and exposing racism in its various permutations" (Ladson-Billings, 1998, p. 11). The counter-narrative is used to analyze myths and presuppositions about race that marginalize and dehumanize Blacks and other people of color (Ladson-Billings, 1998). This method is used to deconstruct the myths about the education of Black children purported in *Waiting for Superman*. Thus, CRT's counter-storytelling form of analysis is a process that serves to *debunk* what is accepted as normative (Matias, in press).

CRT allows for examination of deeper questions around equity and race. Are Black students who attend predominantly Black schools given opportunities to advance and succeed in mathematics? Are systems in place for brilliant students to exceed what is presented in a given textbook from year to year? For example, what happens when fourth graders who learn fifth-grade curriculum are promoted to the next grade? Is the school or the district able to handle multiple levels and provide advanced curriculum? In poor and urban school districts, the answer is often a resounding no.

CRT allows scholars of color to examine how sociotemporal notions of race become normalized forms of oppression and inequality and is particularly useful to unveil "how society's laws, policies, and practices routinely continue to converge in subjugating Black children" (Martin, 2009a, p. 30). Critical race theory (including LatCrit; TribCrit; AsianCrit) names the injustices that are endemic in U.S. society and allows scholars to raise concerns about racial issues and engage in social justice activities to promote anti-racist practices. In this regard, education is touted as the great equalizer.

Cream always rises to the top, which purports neutrality, colorblindness, and meritocracy operate to determine one's success. But race is not a neutral construct. Justice is not colorblind. And according to President Obama, meritocracy—"pull yourself up by your own bootstraps, even if you don't have any boots"[2]—does not always translate into success.

CHARTER SCHOOLS AS A PANACEA

Education reform, which has been going on for decades, is a moving target that has become the pursuit of "something that resembles equity but never reach[es] it" (Kozol, 2003). Rather than establishing equitable funding so that all public schools provide high-quality education and promote student success, education policy is predicated on the notion of providing students with an *adequate* education. In the current decade, school reform has become synonymous with charter schools. While charter schools offer families a free alternative to traditional public school education, this, too, may well be another form of hegemony that erodes educational opportunities for students of color.

The charter school concept, which began in Minnesota in the early 1990s, has now become the modus operandi of school reform. The term *charter school* is commonly understood to mean "an independent public school of choice, freed from rules but accountable for results" (Manno, Finn, & Vanourek, 2000, p. 737). Charter schools may be approved and funded by the state or local school district. In the year 2000, thirty-six states and the District of Columbia had approved more than 1,700 charters that impacted roughly 350,000 low- and middle-income students (Manno et al., 2000). At that time, many charter schools were sponsored by grassroots organizations dedicated to creating better schools (Leonard, 2002; Wells, Lopez, Scott, & Holme, 1999). For example, charter schools like *Harambee Institute of Science and Technology* (elementary school) and *Imhotep* (secondary school) in Philadelphia provide students with an Afrocentric curriculum, small student learning communities, and cultural rites of passage. With supports such as advanced placement courses and senior internships, 100% of Imhotep students were admitted to college for nine consecutive years (Barboza, 2011). Thus, charter schools became popular and some policymakers and educators believed they would create fair competition and improve educational outcomes (Berends, Goldring, Stein, & Cravens, 2010; Leonard, 2002).

On the surface, the emergence of charter schools in response to low-performing public schools seems reasonable. Increasing student achievement and college access among Black and Latino/a students is laudable. However, we must examine history in order to expose the *root causes* of underachieving public schools. Low-achieving schools are a product of the sharecropper schools[3] established for Black children during the Jim Crow era, which are still having a ripple effect today (Ladson-Billings, 2006; Moses & Cobb, 2001). *De facto* and *de jure* segregation has resulted in large urban school districts where Black student populations range from 50% Black in Trenton, New Jersey (Barr et al., 2006), to 80% Black in Prince George's County, Maryland (Leonard, Napp & Adeleke, 2009). In such districts, Black students are more likely to have substitute teachers, novice teachers, unqualified teachers (i.e. teaching out of subject area), and multiple teachers in a single year due to high turnover rates (Kozol, 2003; Leonard et al., 2009). Such conditions contribute to low achievement in urban schools.

Urban school districts like St. Louis, Kansas City, and Chicago continue to struggle with accreditation issues (Lipman, 2003; Shapiro, 2011). As a result, scores of public schools have been closed while charter schools have experienced growth in these and other cities (Shapiro, 2011). In 2011, the number of charter schools tripled to more than 5,500 schools nationwide (Millner, 2011). The National Association for the Advancement of Colored People (NAACP) filed a lawsuit to prevent the closing of more than 20 public schools and to challenge the opening of additional charter schools in New York City (Millner, 2011). Since funding for charter schools comes from the same revenue streams as traditional public schools, funding charter schools has a negative impact on traditional public schools, particularly turnaround[4] and underperforming public schools.

More often than not, charter schools are located in urban areas and are disproportionately located in poor and minority school districts (Arsen, Plank & Sykes, 1999; Barr et al., 2006). In 2001, 68% percent of charter school enrollment in New Jersey was African American (Barr et al., 2006). This figure was higher than the African American population in the resident districts, which was 50 percent. Thus, race is the underlying factor when it comes to charter school education. Critical race theory (CRT) begs the question, who benefits from the emergence of these charter schools? Edison and KIPP dominate the charter school market, sponsoring scores of schools across the nation. Who are the CEOs benefiting from tax dollars to educate predominately poor Black children? Are Black children in charter schools being used as canaries to test business models in the name of education reform? Can competition improve education for the poor when there are fewer dollars allocated for poor children in the first place?

Given few alternatives and hoping that their children will receive a better education, urban families have flocked to charter schools (Barr et al., 2006; Leonard, 2002; Resmovits, 2011). Charter schools are favorable among some urban families, like the families showcased in the *Waiting for Superman* documentary, because many of the schools are small and often students are given individual assistance (Good & Braden, 2000; Leonard, 2002). However, the small size of the charter school itself limits opportunities to educational access. Seventy percent of charter schools have a waiting list for admission (Barr et al., 2006). As a result, disparities within and among school districts may actually increase because of the emergence of charter schools (Berends et al., 2010; Leonard, 2002; Wells et al., 1999). Families are caught in the middle of the political fray (Ravitch, 2010; Resmovits, 2011), struggling for access to quality education for their children in charter schools while the neighborhood public school is shut down without being given substantial support to succeed.

The controversy regarding charter school success is ongoing (Barr et al., 2006; Berends et al., 2010). Ten years ago there was virtually no research to show that charter schools were working (Garcia & Garcia, 1996; Good & Braden, 2000). In 2006, the U.S. Department of Education found that fourth-grade students in traditional public schools outperformed similar students attending charter schools in reading and mathematics (Barr et al., 2006). However, mathematics achievement, unlike reading, is more closely linked to school than home variables (Barr et al., 2006; Lubienski & Lubienski, 2006). Studies that compare mathematics scores by school type instead of Black-White test gaps shed important findings about mathematics achievement in different types of learning environments.

A study comparing the mathematics scores of fourth-grade students attending charter and traditional public schools in Trenton, New Jersey, showed mixed results. Data revealed pockets of excellence and mediocrity in both charter and traditional schools (Barr et al., 2006). In a study that used a large data set obtained from the Northwest Evaluation Center (NWEA), Berends et al. (2010) found charter schools had no effect on students' achievement gains in mathematics. Moreover, they found a negative association with teacher innovations, suggesting innovation for innovation's sake should not be the sole focus of reform regardless of school type (Berends et al., 2010). Lubienski and Lubienski (2006) conducted a broad study that analyzed 2003 fourth-grade mathematics scores using the National Association for Educational Progress (NAEP) database and found public schools significantly out-performed Catholic schools. When charters and non-charters were compared, charter schools scored significantly lower than non-charter schools. Among private schools, Lutheran schools had the highest scores, and conservative Christian schools had the lowest scores. These data concur with earlier findings that public schools perform better than charter schools in mathematics.

THE CASE FOR MATHEMATICS LITERACY

Waiting for Superman underscores the importance of mathematics literacy as the key to ameliorate the conditions of low academic achievement. However, the snippets of mathematics we see in the film are gratuitous or misleading statistically.

Gratuitous mathematical literacy is illustrated in the case of Anthony, an African American male in the fifth grade who attended school in Washington, DC. Anthony is given mediocre mathematics problems to solve. At the beginning of the film, Anthony was asked to solve a problem similar to (actual problem was inaudible): *You have 4 cookies and ate 2; what percentage of cookies did you eat?* Anthony reasoned that he needed to *cross-multiply* to solve the problem. Apparently, he used mental math to set up a proportion similar to *2/4 = x/100* to answer 50%. Anthony showed flexible thinking when he solved a mediocre mathematics problem in an unconventional way. As children continue to develop mathematics literacy and engage in classroom discourse, they often become more confident and resilient (Berry, 2008; McGee, this volume). While Anthony could have simply stated half of four is two or 50%, he showed his ingenuity by using a more difficult, but more insightful, strategy to solve the problem.

Black children are among the most resilient despite their living conditions and life circumstances (Martin, 2009a). Guggenheim could have taken this opportunity to link the issues of intellectual resilience with life resilience. Indeed, Anthony showed resilience by persevering to stay in school although he was held back a grade for absenteeism after his father died. Only a fifth grader, Anthony spoke like a sage when he explained that he wanted to receive a quality education so he could be a role model when he had his own children.

The second mathematics problem discussed in the film was presented during an actual mathematics lesson. The topic was division of fractions, which is commendable because most teachers avoid teaching this topic in elementary school. The mathematics teacher had four equal pie shapes drawn on the chalkboard divided into thirds to represent 4 divided by 2/3. Rather than presenting the children with a measurement interpretation for fraction division (How many 2/3rds are there in 4?), the teacher asked what is 12 divided by 2 (the answer to the reciprocal of the problem (4 times 3 divided by 2). Anthony answered correctly by saying 6. Research has shown teaching fractions, especially division of fractions, is a very difficult concept for teachers to teach and students to learn (Newton, 2008). What is most remarkable about these two scenes is Anthony's ability to find ways to cultivate his own intellectual flexibility in spite of being a product of a failing school. This issue of resilience on behalf of children of color is the more compelling narrative.

Given Anthony's inquisitiveness and underlying brilliance, the audience is left wondering if he would be admitted to a charter school. A prevailing theme of the film is that access to high-achieving schools is a matter of *luck*. Ironically, the only student to gain access to a charter school during the actual lottery shown near the end of the film was a White student who already had access to a state of the art suburban school. While probable, this outcome was less likely since more children of color were in the pool. This implies that a lottery system may not be the fairest way to select students. Nevertheless, Anthony, who was originally waitlisted, was selected to attend the SEED Charter School. Were the odds in his favor or was Anthony's admission to the SEED boarding school the result of pressure to admit at least one student of color into a charter school for the sake of the film?

In addition to the vignettes described above, mathematics presented in the movie was in the form of statistical data. For example, the film states that only three out of 100 students in public schools have taken the classes necessary for college entrance. However, little was said about the availability of college preparatory classes for students to take. In 2008, Chester High School in Pennsylvania had no advanced placement courses (AP) while students in nearby Wallingford had 25 AP courses (Leonard et al., 2009). The film also depicted *bad* teachers as those who teach only 50% of the curriculum while *good* teachers are described as those who teach three times as much or 150%. Yet, 150% does not have real meaning in this context because 100%, by virtue of the meaning of percent, implies *all*.

Data on per pupil expenditures (PPE) were also presented in the film. Guggenheim claimed that per pupil expenditures—$4,300 in 1971 compared to $9,000 in 2009—have doubled while test scores have remained flat for the past 40 years. This suggests that more money is not the answer to alleviate the problem of underachievement in education (Ravitch, 2010). However, these data are misleading. Accounting for inflation, $4,300 in 1971 dollars had the same spending power as $22,778 in 2009. Thus, it would take 2.5 times more than the $9,000 PPE average mentioned by Guggenheim in the film for students in 2009 to enjoy a comparable level of PPE as students did in 1971.

The film also presented data on high dropout rates in urban high schools (NCES, 2010). More than 50% of urban students are dropping out of high school. However, as Professor Marc L. Hill (2011) pointed out, metal detectors are used to prevent the *wrong* students from entering a school, but no one is stopping potentially good students from leaving. A study of Black and Latino students in a federal job program cited mathematics difficulties as the primary reason for dropping out (Viadero, 2005). If this is true, then more mathematics rather than less mathematics, advanced levels rather than proficient levels, and relevant contextualized problems rather than disconnected decontextualized problems should be used to motivate and engage students.

Finally, the data on teacher credentials and tenure presented in the film are skewed. Guggenheim failed to provide the public with a balanced view of teachers and the teaching profession. The film compared teaching to other professions, like medicine and law, and claimed that only one out 2,500 teachers loses the teaching credential. However, this information is also misleading. Having taught in the state of Maryland, Leonard (first author of this chapter) can attest that Maryland does not offer a life certificate. Teachers in Maryland (and many other states) have to complete advanced study or a significant number of professional development courses within a specified time period in order to renew their credential. In 1984, Texas House Bill 72 required all Texas public school educators to take and pass the Texas Examination of Current Administrators and Teachers (TECAT) as a condition of continued certification. Leonard, along with every teacher and administrator in Texas, took an examination in reading and writing. Several teachers retired and others lost their jobs as a result of not passing the TECAT. Many states, including Pennsylvania, require teacher candidates to pass PRAXIS exams before they can receive a teaching certificate. Regardless of possessing a teaching credential, nearly half of all teachers quit during their first five years in the profession (Ingersoll & Smith, 2003; Ravitch, 2010). Clearly, education reform, in terms of increasing teacher standards, has had a negative impact on teacher recruitment and retention.

Waiting for Superman also presented the issue of tenure in an uneven and very negative light. Guggenheim blamed poor educational outcomes on teachers' unions and tenure rules. Washington, DC, chancellor, Michelle Rhee, was actually shown in the act of firing a school principal in the film. While she herself did not have the qualifications of most school superintendents, she closed 23 schools and fired 30 principals and 100 teachers during her tenure. Yet, the film purported that it is nearly impossible to fire a teacher (let alone a principal) because of unions and tenure rules. While we are told that ineffective teachers with tenure were sent for retraining in New York City, no retraining was offered for teachers in the District of Columbia. Thus, the sweeping claim that *bad* teachers are protected by unions and tenure is contradictory.

Furthermore, collective bargaining and the tenure process are under siege. Several states including Ohio, Wisconsin, and Massachusetts have passed bills limiting collective bargaining among teachers, and legislators in Florida, Tennessee, Indiana, Colorado, and Nevada have eliminated tenure altogether (Moore, 2012). While such actions appear to be tough on accountability, students of color will experience even more instability under such legislation as more teachers and principals leave the field. A study of teacher dissatisfaction estimates that 12% of all teachers leave the field each year while only 25% of those cases are due to retirement (Ingersoll, 2002; Moore, 2012).

Guggenheim's use of data in *Waiting for Superman* masks structural and policy issues and ignores the educational well-being of children of color. The latest report from the Children's Defense Fund shows 40% of Black children lived in poverty in 2009 (Lang, 2011). In 2011, Blacks experienced a 16% unemployment rate compared to 8% for Whites (Lang, 2011). While some may attribute the high unemployment rate for Blacks to high school dropout rates, Blacks also have the highest unemployment rate once they attain degrees, even in fields like engineering (NSF, 2011). Thus, the 16% unemployment statistic reflects a disproportionate workforce, especially in science, mathematics, engineering and technology (STEM). Martin (2009a) contends that even if Blacks were more represented in the STEM pipeline, these workers would only be assimilated into the STEM workforce to the degree that they do not upset the status and security of White workers. Regardless of their educational level, twice as many Blacks are unemployed compared to Whites. In order to have equal access to high-tech jobs and economic empowerment, students must have access to high-quality teachers and innovative STEM education, not for workforce readiness alone but to unveil the creativity and brilliance Black children possess (Bracey, this volume), which can be nurtured in public or charter schools.

Mathematics literacy and familiarity with basic statistics allows one to interpret and understand data and disentangle fact from fiction (Gutstein, 2006). Critical mathematics literacy can be used to debunk myths about Black children and charter schools. Critical mathematics literacy can be used to critique educational policy that uses charter school reform as the means to *save* Black children. When the oppressed tell their own story and name their reality, they have the power to change it (Ladson-Billings, 1998).

HOW DOES THE FILM TEACH CITIZENS TO BE THE CHANGE THEY SEEK?

Waiting for Superman falls short of engaging underserved families in informed citizenship. Instead Black and Latino families appear to be pawns in the film (Ravitch, 2010). In light of these shortcomings, we discuss two other films—*Lean on Me* and *Stand and Deliver*—as counter-narratives to show Black and Latino/a students' potential and brilliance in mathematics. These two films are used as springboards to discuss their limitations as well as their potential to show how striving for mathematical excellence provides opportunities for Black and Latino/a children to exhibit mathematical brilliance.

Lean on ME (Mathematical Excellence) Not Minimal Basic Skills

Avildsen's (1989) film, *Lean of Me*, was based on the true story of Joe Clark, a principal in Paterson, New Jersey, who took the initiative to keep drug pushers out of Eastside High School and improve student performance on the state's standardized test. The title of the film, taken from a hit song performed by Bill Withers, suggests that weaker members of the community should lean on stronger members of the community to get through difficult times. The essence of the film is captured in a rap recorded by Big Daddy Kane (1989). The lyrics are as follows:

> Lean on me! (2X)
> "Big Daddy Kane, a teacher…" (2X)
> "Big Daddy…Big Daddy…"
> "Goes a little somethin' like this!"
> "Can I get a yes" (4X)
> "Just lean on me" [Big Daddy Kane]
> "Can I get a yes?" No "Why?" I ain't happy
> 'Cause I came back for these troubles to grab me
> And they've gotten a hold on me
> My oh my, things ain't what they used to be
> Looks to me like Eastside got problems
> Now it's time to see what we can do to solve them
> Change for the better, letter
> New rule be enforced in school so we can get rid of
> all the bad apples in the bunch
> Clean up the wall and cook better lunch
> Put the students in class where they should be
> to stop roamin' the halls, and oh could we
> PLEASE, eliminate the drug situation
> Stop smokin' and get an education
> Point blank, let's change the whole scenery.
> You need help? Well hey, lean on me.
> {exc}"Just lean on me."
> {exc}"LEAN ON ME!" (2X)

Our interpretation of these lyrics is that Big Daddy Kane is the teacher or messenger who wants to highlight salient portions of the film. Kane drew attention to the film, first of all, by naming it at the beginning and end of the rap. He identified the problem at Eastside by stating "things ain't what they used to be" to emphasize that getting an education has changed. The drug culture infiltrated the schools, changing them from safe havens to places where drug dealers ran rampant. While demographic variables such as race, SES, and family structure are often blamed for poor outcomes (National Center for Education Statistics, 2011), few admit that America has lost the war on drugs, which has had a devastating impact on communities, schools, and families.

In the rap, Big Daddy Kane mentioned "change for the better, letter...." It was a letter that authorized Clark to establish a new rule to clean up Eastside. His philosophy was simple: get rid of the *bad apples* (riff-raff), clean up the school, serve better food in the lunch room, keep students in class, and stop drug dealers from infiltrating the school. To get rid of the bad apples, he expelled students he believed were incorrigible but did not provide them with some form of alternative education. As a result, 300 of the 3,000 students (10%) enrolled at Eastside were pushed out of school and most likely became part of the state's dropout statistics. The overcrowded conditions of the school required strong leadership. However, Clark's tough love drew fire and criticism from some parents, the school board, and the district superintendent.

Such is the major critique of *Lean on Me*. While the film portrayed Clark (role played by actor Morgan Freeman) as a super hero, his methods were rather unorthodox to some. He blamed teachers for failing students instead of the system that propelled students to high school status without learning to read or write. The film depicted Joe Clark intimidating and to some degree bullying teachers. He chided the very teachers who were committed to the students and their families.

In contrast, a shining beacon of excellence in the film was the performance of the school song as well as Negro spirituals under the direction of Mrs. Powers. Eastside students came together when they were required to learn the school song, which produced pride and school spirit. Visual and performing arts are just as important to a child's growth and development as the core curriculum (Dewey, 1934; Goodlad, 1992; Leonard, 2009). Further, students became more engaged in their school work when Mr. Clark developed personal relationships with the them, listened to them, and helped to orchestrate schedule changes. For example, one Latina, Louisa, was moved from the home economics class to the auto shop class because she was more interested in cars than cooking. Clark approved a schedule change stating, "Auto mechanics make $17.00 per hour." He also developed relationships with Sams and Keneesha, who would have gone to foster care had he not helped her mother find a job. Clark was also depicted allowing students like Lillian, who was a single parent, to use the office phone to make a phone call to check on her baby. Thus, in the film, Keneesha claimed, "Mr. Clark is like a father...." Students could depend on Joe Clark because he genuinely cared for them.

In the end, it was the relationships developed with the students, teachers, and parents that led to success on the state exam. Clark instituted peer coaching and Saturday school to raise minimum competency scores from 33% to 75% or more. While this goal was reached during his tenure as principal at Eastside High, minimum competency is not enough to gain access to college education and jobs that pay more than minimum wage.

Like Joe Clark, Geoffrey Canada is viewed as a hero, but he blames the plight of failing schools on teachers' unions, lobbyists, and *bad* teachers in public schools. Neither Clark nor Canada question the system that demands accountability without responsibility. *No Child Left Behind* (2001), demands accountability. The goal of reaching every child is laudable. The intent of NCLB was to ensure that educators teach *all* children, regardless of race, ethnicity, socioeconomic status or special needs. However, the use of standardized tests as the sole means to measure student learning has had stifling affects on teaching for excellence as well as student creativity. How does one measure excellence and creativity? Focusing on development of minimal skills to obtain proficiency means that excellence is not the goal. Instead of teaching for excellence, teachers are trapped into teaching the test and focusing on low-level skills (Leonard et al., 2009).

"Stand and Deliver" More Not Less Mathematics

Stand and Deliver (Menendez, 1988) is a mathematics-focused film that is based on a true story about Latino/a students at Garfield High School in East Los Angeles. Again, the idea of a failing school that is turned around by one person's heroic efforts is portrayed. This time, the storyline focuses on the heroic efforts of Jaime Escalante, who left his high-tech job to become a computer instructor at a high school that did not have any computers. He was assigned to teach a basic mathematics course but taught the students algebra instead. He used a step-by-step method of instruction along with culturally responsive teaching. One of the algebra problems presented in the movie was:

> Juan has five times and many girlfriends as Pedro. Carlos has one girlfriend less than Pedro. The total number of girlfriends between them is 20. How many girlfriends does each gigolo have?

This problem might be considered sexist and inappropriate by some. Yet, the students were highly engaged in the problem and several students attempted to solve the problem using variables. Armondo had difficulty representing all of the gigolos in terms of *X*. He stated, "Juan is *X*, and

Pedro is Y". Claudia solved part of the problem, stating, "Juan's girlfriends is $5X$". Raphael asked, "Can you get negative girlfriends?" Javier thought it was a "trick problem" claiming, "You can't solve it unless you know how many girlfriends they have in common." Ana showed her brilliance when she solved the problem the following way: X = Pedro's girlfriends; $5X$ = Juan's girlfriends; $X - 1$ = Carlos' girlfriends; $X + 5X + (X - 1) = 20$ so $X = 3$.

After teaching his students algebra in an unconventional but effective way, Escalante requested that he teach his students calculus the following year. This request was met with criticism by the department chair and laughter by the school principal. Undeterred, Escalante coached his students during the summer to ensure they were prepared. However, one critique of the story may be that Escalante worked the students too hard, and he did not understand how economics dictated some of the students' choices. For example, Francisco wanted to work as an auto mechanic, which was a good job. As we learned in *Lean on Me*, auto mechanics made $17.00 an hour. Francisco should have been allowed to make his own choice rather than being pressured. Yet, Escalante clearly understood the importance of mathematical rigor and hard work in order to instill mathematical excellence among his students. Escalante wanted his students to go to college so that they would have a greater range of career choices. A college education, in Mr. Escalante's opinion, would provide his students with greater economic empowerment.

Not only did all of the students in his pilot course achieve passing scores on the AP calculus exam, several were accepted into college with advanced course credit. What was even more compelling was the increase in the number of students who enrolled in the course and passed the AP calculus exam in subsequent years. This finding concurs with Tawfeeq and Yu's (this volume) finding that innovative teaching increased the number of students who enrolled in AP calculus at a New York City high school.

It was reported in *Stand and Deliver* that fewer than 2% of all high school students attempt to take the AP calculus exam. Taking the AP calculus course increased Garfield High School students' confidence and helped them to believe they were college material. Eighteen students took and passed the test in 1982. Table 5.1 shows the number of students passing the AP calculus exam at Garfield High School from 1982 through 1987. Additional studies should be conducted to determine if the use of culturally relevant pedagogy and a step-by-step approach is transportable to other groups of underrepresented students.

Jaime Escalante believed in his students and worked hard for them to achieve excellence. He, too, developed personal relationships with his students. He invited them to dinner in his home and took them on fieldtrips to inspire them to strive toward excellence. Further, Escalante interfaced with families, convincing Ana's parents that she should be in school rather

TABLE 1. Garfield High School AP Calculus Exam Data*

Year of Exam	Number of Students Passing Exam
1982	18
1983	31
1984	63
1985	77
1986	78
1987	87

*Source: Credits in film Stand and Deliver (1988)

than working in the family restaurant. Moreover, when his students were accused of cheating, he was an advocate, challenging the ETS monitors' interpretation of the results. Caring relationships (Gay, 2000), high expectations (Cooper, 2002), and strong parent and student relationships (Ogbu, 2003) helped to produce outstanding results.

A pivotal moment in this film is when Escalante provide Angel with three textbooks: one for his locker, one for home, and one for school. This act is one that should be modeled when necessary. Sometimes male students are ostracized by their peers for carrying books and appearing too studious (Ogbu, 2003; Martin, 2000). What is phenomenal about this film is that Latino/a students achieved at high levels in rigorous mathematics courses despite their socioeconomic status when they had a dedicated teacher and culturally responsive instruction.

Simply attaining mathematics proficiency, as touted in *Waiting for Superman,* does not change the array of mathematics courses available to Black and Latino/a children (e.g. lack of AP courses) in charter schools. Teachers must be able to deliver mathematics instruction in culturally appropriate ways and provide students with more challenging and rigorous problems rather than simply covering the material, as suggested in the documentary. Black and Latino/a students should be exposed to academically rigorous mathematics courses regardless of school type. Recent studies (Berry, 2008; McGee, this volume) showcasing Black students' success in algebra and advanced mathematics courses shatter myths about Black students not being the *right* kind of mathematics students and promote the concept of brilliance.

COMMUNITY EMPOWERMENT

A historical examination of events show that Black community involvement has great potential to produce change. In 1868, a group of leaders in the African Methodist Episcopal (AME) Church formed a powerful legislative bloc to challenge the education system in Florida (Mizell, 2010). Black members of AME and Baptist congregations ran for office, and record numbers of Blacks voted to elect 18 Blacks to the state constitutional convention. As a result, a statewide system of free common schools was created in Florida in 1869. These Blacks viewed education as not simply a *civil* right or even a *human* right but a *divine* right (Mizell, 2010). Black leaders in churches and political organizations like the NAACP have organized to address educational issues in this era as well. An educational forum was held at an African Methodist Episcopal Church in Denver, Colorado, in May, 2011. Professors McKee (second author of this chapter) and Mizell were co-facilitators. The movie *Waiting for Superman* was shown and deconstructed to empower parents and the community to act. More than 50 volunteers went into three schools named for members of the church community prior to the close of the academic school year. The volunteers witnessed the profound impact their visit had on the students, staff, and administrators at these three schools. Leonard currently serves as an educational liaison to continue these community efforts.

Parents who live in urban school communities and whose children attend traditional public schools and charter schools must collectively work to challenge the power dynamics of one-sided school reforms that purport change, but in fact, promote *more of the same* education reforms (Ladson-Billings, 2006; Ravitch, 2010). Culturally responsive school-home collaboration is a hallmark of school reform. Parents of children of color can begin to address school-based decisions such as curricula, placement, access to advanced and honors courses, graduation rates, and other concerns that the data reveal are a result of gaps in access to equitable educational resources for their children (Anyon, 2005; Noguera, 2003; Oakes & Rogers, 2006) . Additionally, parents must keep an active critical eye on both the mission that charters argue they can deliver, as well as the often times fledgling educational results these charters produce far after they have been granted licensure to start or operate a school (Williams, 2007).

The current education policy that allows public and charter schools to compete, with the consequence of closing low-performing schools, is a hegemonic form of survival of the fittest. Few citizens truly understand the dire consequences of the policy. Instead charter schools continue to expand to the demise of public schools. Yet, some would say "the game is rigged" because charter schools mirror the same educational inequities of the community at large as do the public schools (Barr et al., 2006). Until there is public indignation of Guggenheim's implication that "other people's children are of less inherent value than [his] own" (Kozol, 2003, p. 291), change in education policy will not occur.

Reflection on Guggenheim's documentary reveals that poor parents are pitted against each other to compete for spots in charter schools that may or may not succeed in educating their children. The Black community must demand an educational policy that supports more than highly-qualified teachers. The majority of Black children attend schools that are substandard in terms of facilities and resources. What is needed is equitable school funding (Kozol, 2003). Nevertheless, it should be clear that equitable funding is not a panacea. It takes the creativity, dedication, commitment, and cultural responsiveness of administrators and teachers like Joe Clark and Jaime Escalante, in addition to resources, to improve educational outcomes for students of color. Geoffrey Canada admits his naiveté, stating he believed he could fix all of the problems in education in three years after graduating from the Harvard School of Education, which was mathematically impossible given 300 years of inequity.

Waiting for Superman portrays Superman as the *Messiah* who is not coming to rescue Black children living in poverty. The film title presupposes that *salvation* must come from outside of the Black community. Can any good thing come from Nazareth (Morris, 2004)? Morris' study revealed that parents, teachers, and communities can work together to make urban public schools successful. In this regard, Rev. Dr. Martin Luther King prophetically realized the urgency of the moment. As Dr. King stated in 1963 in Detroit, "*Gradualism* is little more than *escapism* and do *nothingism*, which ends up in stand *stillism*."[5] Social justice is not a *noun* but a *verb* that challenges the Black community to *act* in this current climate on behalf of Black children. The Black community need not wait for Superman because it has historically *made bricks without straw*.[6] Likewise, the Black community should draw upon its history and strengths to bring about change in this decade on behalf of Black children.

Waiting for Superman suggests "it's difficult to hold two competing ideas at the same time." F. Scott Fitzgerald said, "The test of a first-rate intelligence is the ability to hold two opposed ideas in the mind at the same time, and still retain the ability to function." Can we hold two competing ideas about traditional public education and high-quality schools? Do we have the will to educate *Black* children (Hilliard, 1991)? Can we ensure both equity and justice exist simultaneously in terms of access and opportunity in our nation's public schools (Kozol, 2003; Leonard, 2009)? If so, then we are on a path that truly leads to liberation and ultimately education reform.

The first step in the liberation process is *waking up* (Harro, 2000). The Black community needs to wake up and challenge the two-tiered educational system (one for affluent and one for poor minority students), which perpetuates failing schools (Leonard, 2008; Ravitch, 2010; Tillman, 2004). The next step is taking action—that is pushing to change education policy toward Black children. Black political and community leaders should organize and promote legislation to improve and maintain better schools for children of color. Liberation involves naming the problem, critical transformation, monitoring change as it occurs, and maintaining the rights that have been won (Harro, 2000). Rather than simply calling racism out, as Freire (1982) suggests, we must engage "in the incessant struggle to regain humanity...so that reflection will become liberation" (p. 25).

Power concedes nothing without a demand.

—Frederick Douglass

NOTES

1. Black is used to describe members of the African diaspora including Africans, Caribbean Americans, African Americans, and Black Hispanics. The term African American is used only in the context of cited research.
2. Quote by Barack Obama during his acceptance speech at the Democratic National Convention on August 28, 2008.
3. Sharecropper schools were inferior schools for children of Black sharecroppers in the South.
4. A turnaround school may be described as a low-performing school that was closed and reopened with new administrators, teachers, and staff.
5. Speech at the Great March on Detroit by Martin Luther King on June 23, 1963.
6. Reference to Genesis story where the Hebrew people were told to make bricks without straw in Egypt implies demanding quality production with little or no investment of capital goods.

REFERENCES

Anyon, J. (1981). Social class and school knowledge. *Curriculum Inquiry, 11*(1), 3–42.

Anyon, J. (2005). *Radical possibilities: Public policy, urban education, and a new social movement.* New York, NY: Routledge.

Arsen, D., Plank, D. N., & Sykes, G. (1999, April). *The economic and social geography of choice is urban Michigan.* Paper presented at the annual meeting of the American Educational Research Association, Montreal, Canada.

Avildsen, J. G. (Director). (1989). *Lean on me* [Film]. Burbank, CA: Warner Brothers.

Barboza, S. (2011). *Waiting for culturally relevant education, not Superman.* Retrieved on October 11, 2011 from http://thyblackman.com/2011/02/15/waiting-for-culturally-relevant-education-not-superman/

Barr, J. M., Sadovnik, A. R., & Visconti, L. (2006). Charter schools and urban education improvement: A comparison of Newark's district and charter schools. *The Urban Review, 38*(4), 291–311.

Berends, M., Goldring, E., Stein, M., & Cravens, X. (2010). Instructional conditions in charter schools and students' mathematics achievement gains. *American Journal of Education, 116,* 303–335.

Berry, R. Q., III. (2008). Access to upper-level mathematics: The stories of successful African American middle school boys. *Journal for Research in Mathematics Education, 39*(5), 464–488.

Buckley, L. A. (2010). Unfulfilled hopes in education for equity: Redesigning the mathematics curriculum in a US high school. *Curriculum Studies, 42*(1), 51–78.

Cooper, P. (2002). Does race matter: A comparison of effective black and white teachers of African American students. In J. J. Irvine (Ed.), *In search of wholeness: African American teachers and their specific classroom practices* (pp. 47–63). New York, NY: Palgrave.

Delgado, R. (Ed.). (1995). *Critical race theory: The cutting edge.* Philadelphia, PA: Temple University Press.

Dewey, J. (1934). *Art as experience.* New York, NY: Minton, Balch & Company.

Dixson, A. D. & Rousseau, C. K. (2005). And we are still not saved: Critical race theory in education ten years later. *Race Ethnicity and Education, 8*(1), 7–27.

Duncan, G. A. (2005). Critical race ethnography in education: Narrative, inequality and the problem of epistemology. *Race Ethnicity and Education, 8*(1), 93–114.

Essed, P. (2002). Everyday racism: A new approach to the study of racism. In P. Essed & D. Goldberg (Eds.), *Race critical theories* (pp. 176–194). Malden, MA: Blackwell Publishers.

Freire, P. (1982). *Pedagogy of the Oppressed.* Harmondsworth, UK: Penguin.

Garcia, G. F. & Garcia, M. (1996). Charter schools—Another top-down innovation. *Educational Researcher, 25*(8), 34–36.

Gay, G. (2000). *Culturally responsive teaching: Theory, practice and research.* New York, NY: Teachers College Press.

Good, T. L. & Braden, J. S. (2000, June). Charter schools: Another reform failure or a worthwhile investment? *Phi Delta Kappan, 81*(10), 745–750.

Goodlad, J. I. (1992). Toward a place in the curriculum for the arts. In K. J. Rehage, B. Reimer, & R. A. Smith (Eds.), *The arts, education, and aesthetic knowing: The ninety-first yearbook of the National Society for the Study of Education* (pp. 192–212). Chicago, IL: National Society for the Study of Education.

Guggenheim, D. (Director). (2010). *Waiting for Superman* [Film]. Hollywood, CA: Paramount Pictures.

Gutstein, E. (2006). *Reading and writing the world with mathematics: Toward pedagogy for social justice.* New York, NY: Routledge.

Harro, B. (2000). The cycle of liberation. In M. Adams, W. J. Blumenfeld, R. Castaneda, H. W. Hackman, M. L. Peters, & X. Z niga (Eds.), *Readings for diversity and social justice,* (pp. 463–469). New York, NY: Routledge.

Hill, M. L. (2011, July). *Beats, rhymes and classroom life: Hip-hop pedagogy and the politics of identity.* Lecture presented at Arcadia University, Glenside, Pennsylvania.

Hilliard, A. G., III. (1991). Do we have the will to educate all children? *Educational Leadership, 49*(1), 31–36.

Hofstadter, R. (1962) *Anti-intellectualism in American life.* New York, NY. Vintage Books.

Ingersoll, R. M. (2002). The teacher shortage: A case of wrong diagnosis and wrong prescription. *NASSP Bulletin, 88,* 16–31.

Ingersoll, R. M. & Smith, T. M. (2003).The wrong solution to the teacher shortage. *Educational Leadership, 60*(8), 30.

Kane, B. D. (1989). *Lean on Me (Rap Summary).* New York: Cold Chillin' Records and Video, Inc.

Kozol, J. (2003). Savage inequalities: Children in America's Schools. In T. E. Ore (Ed.), *The social construction of difference and inequality: Race, class, gender, and sexuality* (pp. 290–296). New York, NY: McGraw Hill.

Ladson-Billings, G. (1998). Just what is critical race theory and what's it doing in a *nice* field like education? *Qualitative Studies in Education, 11*(1), 7–24.

Ladson-Billings, G. (2006). From the achievement gap to the education debt: Understanding achievement in U.S. schools (2006 Presidential Address). *Educational Researcher, 35*(7), 3–12.

Lang, C. (2011, August 28). Race, class, and Obama. *The Chronicle Review.* Retrieved on August 28, 2011 from http://chronicle.com/article/Race-ClassO-bama/128787/

Leonard, J. (2002). The case of the first-year charter school. *Urban Education, 37*(2), 219–240.

Leonard, J. (2008). *Culturally specific pedagogy in the mathematics classroom: Strategies for teachers and students.* New York, NY: Routledge.

Leonard, J. (2009). "Still not saved": The power of mathematics to liberate the oppressed. In D. B. Martin (Ed.), *Mathematics teaching, learning, and liberation in the lives of Black children,* (pp. 304–330). New York, NY: Routledge.

Leonard, J., Napp, C., & Adeleke, S. (2009). The complexities of culturally relevant pedagogy: A case study of two mathematics teachers and their ESOL students. *High School Journal, 93*(1), 3–22.

Leonard, J., Taylor, K. L., Sanford-De Shields, J., & Spearman, P. (2004). Professional development schools revisited: Reform, authentic partnerships and new visions. *Urban Education, 39*(5), 561–583.

Lipman, P. (2003). Chicago school policy: regulating Black and Latino youth in the global city. *Race Ethnicity and Education, 6*(4), 331–355.

Lubienski, C. & Lubienski, S. T. (2006). *Charter, private, public schools and academic achievement: New evidence from NAEP data.* New York, NY: National Center for the Study of Privatization of Education.

Manno, V. B., Finn, C. E., Jr., &Vanourek, G. (2000, June). Beyond the schoolhouse door: How charter schools are transforming U.S. education. *Phi Delta Kappan, 78*(1), 18–23.

Martin, D. B. (2000). *Mathematics success and failure among African American youth: The roles of sociohistorical context, community forces, school influence, and individual agency.* Mahwah, NJ: Lawrence Erlbaum.

Martin, D. B. (2009a). Little Black boys and little Black girls: How mathematics education and research treat them? In S. L. Swars, D. W. Stinson, & S. Lemons-Smith (Eds.), *Proceedings of the 31ˢᵗ Annual Meeting of the North American Chapter of the International Group for the Psychology of Mathematics Education* (pp. 22–41). Atlanta, GA: Georgia State University.

Martin, D. B. (2009b). Liberating the production of knowledge about African American children and mathematics. In D. B. Martin (Ed.), *Mathematics teaching, learning, and liberation in the lives of Black children* (pp. 3–38). New York, NY: Routledge.

Matias, C. (in press). Who you callin' White? A critical counterstory on colouring white identity. *Race Ethnicity and Education.*

Menendez, R. (Director). (1988) *Stand and deliver* [Film]. Burbank, CA: Warner Brothers.

Millner, D. (2011, July). Charter schools: Fad or future? *Jet, 120*(3/4), 12 & 14.

Mizell, L. (2010). "The holy cause of education": Lessons from the history of a freedom-loving people. In T. Perry, R. P. Moses, J. T. Wynne, E. Cortes, & L. Delpit (Eds.), *Creating a grassroots movement to transform public schools* (pp. xvi). Boston, MA: Beacon Press.

Molinar, A. & Garica, D. R. (2007) The expanding role of privatization in education: Implications for teacher education and development. *Teacher Education Quarterly*, Spring. 11–24.

Moore, C. M. (2012). The role of school environment in teacher dissatisfaction among U.S. public school teachers. *Sage Open.* DOI: 10.1177/215824401243888

Morris, J. (2004). Can anything good come from Nazareth? *American Educational Research Journal, 41*(1), 69–112.

Morrison, T. (1992). *Playing in the dark: Whiteness and the literacy imagination.* Cambridge, MA: Harvard University Press.

Moses, R. P. & Cobb, Jr., C. E. (2001). *Radical equations: Math literacy and civil rights.* Boston: Beacon Press.

National Center for Education Statistics. (2010). *Trends in high school dropout and completions rates in the United States, 1972–2008,* No. NCES 2011012. Retrieved on September 24, 2012 from http://nces.ed.gov/pubsearch/pubsinfo. asp?pubid=2011012

National Center for Education Statistics. (2011). *Table 48. Number of gifted and talented students in public elementary and secondary schools, by sex, race/ethnicity, and state: 2004 and 2006.* Retrieved on September 24, 2012 from http://nces. ed.gov/pubs2011/2011015.pdf

National Science Foundation. (2011). *Women, minorities, and persons with disabilities in science and engineering: 2011.* Arlington, VA: U.S. Government Printing Office. Retrieved April 8, 2011 from http://www.nsf.gov/statistics/wmpd/

Newton, K. J. (2008). An extensive analysis of preservice elementary teachers' knowledge of fractions. *American Educational Research Journal 45*(4), 1080–1110.

No child left behind: Reauthorization of the Elementary and Secondary Education Act of 2001. (2001). A report to the nation and the Secretary of Education, U.S. Department of Education, President Bush Initiative.

Noguera, P. A. (2003) Schools, prisons, and the social implications of punishment: Rethinking disciplinary practices. *Theory into Practice*, 42(4), 341–351.

Oakes, J. & Rogers, J. (2006). *Learning power: Organizing for education and justice.* New York, NY: Teachers College Press.

Ogbu, J. (2003). *Black American students in an affluent suburb: A study of academic disengagement.* Mahwah, NJ: Lawrence Erlbaum.

Otterman, S. (2010, October 12). Lauded Harlem Schools have their own problems. *The New York Times*. Retrieved from http://www.nytimes.com/2010/10/13/education/13harlem.html?_r=1&ref=geoffreycanada#

Ravitch, D. (2010). The myth of charter schools. *The New York Review of Books*. Retrieved on August 19, 2011 from http://www.nybooks.com/articles/archives/2010/nov/11/myth-charter-schools/?page=1

Resmovits, J. (2011, June 3). Charter schools challenged in St. Louis, California, Georgia. *Huffpost Education*. Retrieved on August 5, 2011 from http://www.huffingtonpost.com/2011/06/03/charter-school-challenges_n_871163.html

Shapiro, J. (2011, April 27). Missouri House votes to expand charter schools. *Columbia Missourian*. Retrieved on August 5, 2011 from http://www.combianmissourian.com/stories/2011/04/27/missouri-house-votes-expand-charter-schools/

Solórzano, D. G. & Yosso, T. J. (2002). Critical race methodology: Counter-storytelling as an analytical framework for educational research. *Qualitative Inquiry, 8*(1), 23–44.

Spencer, J. A. (2009). Identity at the crossroads: Understanding the practices and forces that shape African American success and struggle in mathematics. In D. B. Martin (Ed.), *Mathematics teaching, learning, and liberation in the lives of Black children,* (pp. 200–230). New York, NY: Routledge.

Tillman, L. C. (2004). (Un)intended consequences?: The impact of the *Brown v. Board of Education* decision on the employment status of black educators. *Education and Urban Society, 36*(3), 280–303.

Viadero, D. (2005). *Transition mathematics project.* Retrieved on August 12, 2005 from http://www.transitionmathproject.org/marketingarticle.asp

Wells, A. S., Lopez, A., Scott, J., & Holme, J. J. (1999). Charter schools as postmodern paradox: Rethinking social stratification in an age of deregulated school choice. *Harvard Educational Review, 69*(2), 172–204.

Williams, Y. (2007). *Charter school reform as rational choice: An analysis of African American parents' perceptions of a charter school located in a residentially segregated community.* Unpublished doctoral dissertation, Temple University, Philadelphia, PA.

SECTION III

LEARNING AND LEARNING ENVIRONMENTS

CHAPTER 6

ADVANCING A FRAMEWORK FOR CULTURALLY RELEVANT, COGNITIVELY DEMANDING MATHEMATICS TASKS

Lou Edward Matthews, Shelly M. Jones, and Yolanda A. Parker

INTRODUCTION

Over the last decade of working with prospective and practicing elementary, middle and secondary school teachers of diverse student communities in Connecticut, Illinois, South Carolina, Georgia, Texas, and Bermuda, our work as mathematics educators has been largely informed by two major tensions. The first of these consisted of assisting efforts in these communities to infuse standards-led mathematics teaching perspectives into local teaching practice. The second, even more daunting challenge lay in helping educators in these very same communities reframe mathematics instruction to become more relevant for Black children and work to impact achievement disparities for Black students that have long been associated with school mathematics.

The Brilliance of Black Children in Mathematics:
Beyond the Numbers and Toward New Discourse, pages 123–150.
Copyright © 2013 by Information Age Publishing
123

National standards-based reform heralded by the world's largest mathematics education body, the National Council of Teachers of Mathematics (NCTM, 2000), suggests that the teaching of mathematics shift away from teacher-centered, traditional approaches which view mathematics as a static set of isolated facts and procedures taught with little connection to concepts. The new paradigm being encouraged is a more dynamic view of mathematics involving the teaching of rich, related mathematical ideas, in classrooms where students are engaged in doing challenging mathematics tasks, utilizing a variety of approaches and communicating their constructed ideas. These shifts are echoed in many state standards, the Common Core State Standards Initiative (CCSSI), and international standards and have been continuing focuses of our individual but related work and involvement in the United States and Bermuda.

Addressing the first challenge in our capacity as mathematics educators has played out in teaching undergraduate and graduate courses for prospective and practicing teachers, as well as facilitating professional development experiences with practicing teachers. In this chapter we will report on work completed with practicing teachers as professional development in Bermuda and graduate level study in the United States. Undoubtedly each experience with practicing teachers has its own idiosyncrasies based on the local political, racial, and cultural issues that guide educational reform in their communities. What remains the same, however, is the need for mathematics educators and mathematics teachers of Black children in Bermuda and the United States to move toward excellence in teaching for these children. As Matthews (2009) states, "systemic efforts to impact the mathematics education of Black children in Bermuda have mirrored international efforts for mathematics reform" in the US and other countries (p. 65).

The crux of our experiences has been to promote mathematics reform through the use of student-centered teaching methods for teaching mathematics. These experiences are captured in a term Matthews uses, "Let go!", which describes how teachers recognize that common and traditional teaching roles place an overemphasis on teacher control behaviors such as telling important concepts versus allowing students to develop an understanding of these concepts through active engagement. This notion of letting go also means understanding that teachers must ask more questions that allow students to reveal their strategies in doing mathematics, and further reflect on and refine these strategies. This leads to another focus of our work, which has been to help teachers understand that much of this shift in teaching is more easily facilitated when mathematics tasks are designed to challenge students cognitively. In these cases, Matthews often asks students in his courses, "where is the struggle?" and often encourages them to create such a struggle in its absence.

Making Mathematics Meaningful

The second challenge of our work, and more germane to this chapter, is encapsulated in the title of an annual summer mathematics institute Matthews helped develop and facilitate in Bermuda between 2000 and 2005, *Making Mathematics Meaningful*. As the name suggests, the purpose of the institute was to assist teachers in reframing mathematics classroom experiences for Bermuda's predominantly-Black school population so that tasks are seen as meaningful and relevant for students and work to impact student underachievement and engagement. The institutes were organized to utilize the knowledge base drawn from exemplary teachers whose pedagogical practices aim to empower students academically while building strong cultural, social and political identities (Gutstein, Lipman, Hernandez, & de los Reyes, 1997; Ladson-Billings, 1994, 1995a, 1995b; Matthews, 2003; Sleeter, 1997; Tate, 1995). Persistent mathematics achievement disparities on standardized tests for racial subgroups (AERA, 2004; African-American Student Achievement Committee, 2001; Martin, 2009; SEF South Carolina Task Force and Advisory Committee, 2002) have meant that our work with local districts and schools has focused on maximizing achievement outcomes. Much of these efforts has been in helping teachers emphasize content and cultural connections that help students understand and apply mathematics concepts.

Literature suggests that there is indeed a space where mathematics reform goals and critical, cultural approaches to teaching can work together to impact student learning and citizenship possibilities (Enyedy & Mukhopadhyay, 2007; Gutstein et al., 1997; Leonard & Guha, 2002; Matthews, 2003; Tate, 2004). But doing so often requires that teachers re-engineer mathematics classroom content and social structures (tasks, interrelations, etc.) to support culturally relevant approaches to teaching. This is important because most recognize mathematics tasks given to students as the "currency" of mathematics classroom possibilities (Herbst, 2006).

In other work (Matthews, 2003, 2008) Matthews has written of how teachers struggled to see mathematics as a relevant, cultural discipline from which cultural and societal inquiry can emanate and flourish. In some episodes Matthews witnessed instructional choices of several of the teachers focused on strict adherence to textbook tasks and context with little or no potential to make connections to students' culture. On the other hand, according to Enyedy and Mukhopadhyay (2007), even when mathematics tasks are based around the context of students' lives, instruction may fail to maximize its potential to engage students. Students of color are often subjected to instructional strategies that emphasize authoritative, didactic, and/or whole group instruction, which may not be conducive to their learning styles (Gay, 2000).

The intention of this chapter is to present and analyze the kinds of tasks that overcome these kinds of difficulties by explicating a guiding framework for building mathematics tasks that are both cognitively demanding and culturally relevant. This work very intentionally delves into the world of practice for teachers. What the scholarship has done very well is to frame culturally relevant teaching theoretically; there is a documented need to help teachers actualize it in the enterprise of teaching (Enyedy & Mukhopadhyay, 2007; Matthews, 2003).

The basis of this framework will be drawn from two sources: a) a definition of higher level, cognitively demanding mathematics tasks (Stein, Smith, Henningsen, & Silver, 2000), and b) features of mathematics tasks extracted from the existing literature on culturally relevant teaching. In the paragraphs that follow, Stein et al.'s (2000) extant framework for cognitively demanding mathematics tasks and the components of culturally relevant pedagogy and its implications for developing mathematics tasks are explored. Finally a grafted framework for culturally relevant, cognitively demanding mathematics tasks is proposed and investigated.

Cognitively Demanding Mathematics Tasks

The suggested shifts in mathematics teaching carry implications that require teachers to be able to identify, create and even transform traditional mathematics tasks in order to engage students' thinking. These tasks should emphasize the learning of rich mathematics concepts, multiple representations and strategies, and the communication of one's reasoning when problem solving (NCTM, 2000). Stein et al. (2000) suggest that tasks with which mathematics lessons are framed are the "basis of opportunities" for learning mathematics (p. 11). In addition to focusing on covering a full spectrum of mathematical content, the National Assessment of Educational Progress (NAEP), or the Nation's Report Card, emphasizes cognitive demand as a key component of its mathematics tasks administered nationally. According to the National Assessment Governing Board (2006), this focus is captured in the dimension "mathematical complexity," which is elaborated as follows.

Mathematical complexity, the second dimension, attempts to focus on the cognitive demands of the assessment question. Mathematical complexity is categorized as low, moderate, or high, and each level of complexity includes aspects of knowing and doing mathematics, such as reasoning, performing procedures, understanding concepts, or solving problems. The levels of complexity form an ordered description of the demands an item may make on a student. Items at the low level of complexity, for example, may ask a student to recall a property. At the moderate level, an item may ask the student to make a connection between two properties; at the high level, an item may ask a student to analyze the assumptions made in a mathematical model. The complexity dimension builds on the dimensions of mathematical ability (conceptual understanding, procedural knowledge, and problem solving) and mathematical power (reasoning, connections, and communication) that were used in the mathematics framework for the 1996–2003 NAEP assessments.

Underscoring the different kinds of potential experiences that result from varying complexity, Stein et al. (2000) state:

> Tasks that require students to perform a memorized procedure in a routine manner lead to one type of opportunity for student thinking; tasks that demand engagement with concepts and that stimulate students to make purposeful connections to meaning or relevant mathematics ideas lead to a different set of opportunities for student thinking. (p. 11)

In distinguishing between *lower-level* and *higher-level* demands, Stein et al. (2000) defined several categories of common mathematics tasks. Lower-level tasks are classified as *memorization* and *procedures without connections to understanding, meaning or concepts.* An example of a *memorization* task using the framework would be: "What is the fractional equivalent for 75%?" Stein et al.(2000) argue that these kinds of tasks would still be considered low level because "...there is no connection to concepts or meaning required, and the focus is on producing the correct answer" (p.17). On the other hand a *procedures-without-connections* task might be, "What is 23% of 58?" In this case a student might merely be required to perform a procedure such as multiplying 23 by 58 and dividing by 100. In such case no conceptual understanding of percent is needed. While both of these tasks are important, they send the message that the teacher is focusing on speed and procedural skill; they also represent limited opportunities for students to understand important concepts.

At the other end of the spectrum, Stein et al. (2000) classify two types of higher level cognitively demanding tasks: *procedures with connections to understanding, meaning and concepts,* and *doing mathematics.* Tasks classified as *procedures with connections* require students to use procedures (or algorithms) in ways that build conceptual understanding of important mathematical ideas. An example of this is given:

> Students might be asked to use a 10 x 10 grid to illustrate how the fraction 3/5 represents the same quantity as the decimal 0.6 or 60%. Students would also be asked to record their results on a chart containing the decimal, fraction, percent, and pictorial representations, thereby allowing them to make connections among the various representations and attaching meaning to their work by referring to the pictorial representation of the quantity every step of the way. (Stein et al., 2000, p. 12)

The second category of higher level tasks, *doing mathematics*, is often non-algorithmic, unpredictable, and requires multiple ways of representing concepts. Open-ended word problems which might involve a series of steps and representations including symbolism, graphs and verbal explanations often fall into this category. Stein et al. (2000) further suggest that an essential feature of higher level mathematics tasks is that they require students to draw from "relevant knowledge and experiences and make appropriate use of them in working through the task" (p. 16). The idea of drawing from the experiences of students to build mathematical knowledge has been steadily promoted in mathematics reform messages (NCTM, 1989, 1995, 2000). It must also be understood that these experiences are undoubtedly cultural in nature.

In the last 15 years there has been a dedicated stream of mathematics and science education literature that has sought to frame mathematics and science teaching and learning around cultural approaches to pedagogy. Several mathematics educators (Enyedy & Mukhopadhyay, 2007; Gutstein et al., 1997; Leonard, 2008; Leonard & Guha, 2002; Matthews, 2003; Tate, 2004) have explicated critical features necessary for implementing culturally relevant pedagogical practices in mathematics. This literature base and its implications for creating mathematics tasks that draw from students' experiences more directly and explicitly are discussed in the sections that follow.

CULTURALLY RELEVANT TEACHING OF MATHEMATICS

Historically, connections between mathematics and what is being experienced in students' day-to-day lives have been overlooked (NCTM, 1989). As articulated by Ladson-Billings (1994, 1995a, 1995b, 1997), the purpose of culturally relevant teaching is to frame teaching more broadly in ways that help students to achieve academically, socially, politically and culturally. In practice, it is embodied in the work of teachers who foster empowering relationships with students and center teaching activity culturally to honor students' individual, community, and ethnic identities, while extending curriculum towards personal meaning, citizenship and engagement. Culturally relevant pedagogy supports three goals for students: 1) academic success, 2) cultural competence, and 3) the ability to critique the existing social order (Ladson-Billings, 1994).

According to Enyedy and Mukhopadhyay (2007), the notion of relevance in teaching mathematics consists of relevance of the content or context of the lesson, relevance to students' community/social/cultural selves, and relevance as the *process* of instruction. Their portrayal of a multidimensional focus for relevance was seen through the use of statistics tasks given to high school students in South Los Angeles. Using geographic information system (GIS) software, which allows users to visualize and interact with geographic maps, students were given opportunities to examine educational inequities in their communities that still persist, even after 50 years of desegregation. Students had the backdrop of a pending lawsuit that argued inequitable conditions and funding in many school districts around the state. One student group used GIS mapping to represent the relationship between the percentage of the Hispanic population with the percentage of fully certified teachers (less than 80%) in the same area.

Examining the nature of the culturally relevant teaching of mathematics in a Mexican American context, Gutstein et al. (1997) sought to better understand the connections between culturally relevant pedagogy and mathematics reform. They pose a model for culturally relevant mathematics teaching that holds that cultural knowledge can be seen as a deliberate, primary source of informal knowledge that can be used to build students' mathematical ideas, and that experiences should frame mathematics as a way of thinking critically about society, its structure and knowledge. Another element put forth by Gutstein et al.(1997) is that teachers must possess empowered, non-deficit views of their students, which would allow them to simultaneously empower students as they teach mathematics. An implication of this type of orientation might be supported by a mathematics task that allows students to ask questions of self and others—tasks that focus on the classroom community or the community at large.

A key feature in the work of middle school teachers in Gutstein et al. (1997) is seen in how student inquiry into cultural knowledge becomes central to understanding important mathematical ideas. They describe the following account of one teacher:

> Mr. Chamorro helps students understand the mathematical idea of scale by drawing a map of the United States and asking if that is the real size of the country. He helps students understand inexactness of measurement, another mathematical concept, by referring to their experiences being measured with and without shoes. For the concept of "as the crow flies" distance, he uses the visually graphic image of a flying crow stopping for red lights and making sharp turns, and he additionally uses their knowledge of their community. Finally, his example of the road from Mexico to El Paso, places known to students, uses students' informal knowledge of the geometry of roads as nonlinear objects and helps them visualize further the relationship of actual distance to the road distance. (Gutstein et al., 1997, pp. 723–724)

Matthews (2003) illuminated four unique scenarios or "complexities" arising when teachers draw from perspectives on culturally relevant teaching in the mathematics classroom. One complexity represented tasks whose goals appeared to build relationships or thinking about relationships, as well as the mathematics. Two of the complexities highlighted how teachers attempted to utilize instructional tasks that drew from the informal cultural thinking of Bermudian students as well as promoting critical thinking about society and knowledge itself. More problematically, a final complexity was seen in situations where a mathematics task merely represented information from which students might read cultural information. This final complexity described by Matthews represented a transmission of culture similar to that of lower-level *memorization* tasks, which represent a transmission of knowledge.

What kinds of tasks allow for the students to become empowered in their relationships with themselves, teachers and community? In this sense mathematics tasks might focus on issues of human living as inquiry. For example, a mathematics task in data analysis might deliberately focus on having students interpret data related to urban sprawl, crime, or marriage statistics—tasks that require students to think about their interrelated roles as citizens. In Matthews (2003) the following is an abbreviated account of a primary five (grade four) classroom in Bermuda:

> "Where do we use big numbers?," Ms. Tiffani began, while also informing him that they were continuing a discussion from a previous lesson. One student then responded, "Marketplace," referring to a popular local grocery store chain. "Town!" another student chimed in, referring to the central city of Hamilton. As usual, Ms. Tiffani continued to press her students to explain and justify how big numbers were being used in these places and as her students began to chime in, she redirected the burgeoning conversation to focus on the recent attacks on the World Trade Center in New York City. The conversation soon grew to include discussion about how the disastrous events of September 11, 2001 could impact on Bermuda's economy, spending, and how all of these things involved relatively large numbers. In addition to talking about how millions and billions might be associated with the tragic events, the students also shared their own views and ideas about how these events might affect Bermudian life. Throughout the discussion, which lasted about 20 minutes, Ms. Tiffani attempted to get her students to think critically about the notion of big numbers, but also make connections to the tragic September events. he came to call moments like these, *critical teaching moments...* (p. 71)

Tate's (1995) descriptive account of the pedagogy of Sandra Mason, a middle school teacher in Dallas, Texas, also provides insight into the kinds of mathematics tasks that facilitate culturally relevant teaching. According to Tate, Sandra's pedagogical tasks are designed to politically and socially empower her students through the *mathematizing* (the way students count, measure, classify, and infer mathematical meaning; D'Ambrosio, 1985) of community issues, which emphasizes collaborative communication and group work, learning through investigation, critical questioning, relevant rich problem solving, and social action. In one instance, students embarked on an effort to close or relocate 13 liquor stores within 1000 feet of their school.

> This required the students to think about mathematics as a way to model their reality. Real situations in the students' lives were transformed into mathematical representations...The actions of the students have resulted in

some change. The police have issued over 200 citations to liquor store own-ers, and two of the 13 liquor stores were closed down for major violations. (D'Ambrosio, 1985, pp. 170–171)

Leonard and Guha (2002) similarly describe several mathematics tasks crafted directly from mathematization of a familiar community institution, the church. Teachers and students took a walking tour of the children's neighborhood around the church, which served as the context for students developing word problems. This experience involved students' critical ex-amination of their neighborhood through a mathematical lens and a new level of critical consciousness about the effects of math in their daily lives. Leonard (2008) contends that while the literature shows a critical need to use cultural pedagogy, there are yet scarce examples of how culture can be connected to mathematics pedagogy.

Earlier we stated that higher level tasks are considered as the desired basis of opportunity for defining culturally relevant mathematics tasks. Fol-lowing is an explanation of why lower level tasks are problematic. A fea-ture of both *memorization* and *procedures-without-connections* tasks (Stein et al., 2000) is that, with both kinds of tasks, explicit connections are not made to students and their communities. These tasks, while serving some function cognitively, limit the opportunity for students to make connections with familiar contexts, and even work to affirm the notion of mathematics as a disconnected, static domain of knowledge. The first classification, *memoriza-tion,* could also be taken as limited opportunities to relate if the student is only required to receive cultural and community information through the task. This was the case in Matthews (2003) when it was noted that some at-tempts at incorporating culturally relevant teaching involved attempts to build cultural knowledge by transmission rather than by student engage-ment in cultural activity.

As Stein et al. (2000) note, "low-level tasks...can appear to be high-lev-el...when they contain superficial reform characteristics such as manipula-tive usage, real-world contexts, multiple steps, and diagrams" (p. 17). In a similar way, transmission tasks that contain cultural information that is deemed important, though well intentioned, may serve as gloss to increase student interest and engagement. It also serves to promote limiting no-tions of multicultural teaching as merely content-modification (Bartolome, 1994).

A FRAMEWORK FOR CULTURALLY RELEVANT, COGNITIVELY DEMANDING MATHEMATICS TASKS

In review of the features taken from accounts of practice involving cultur-ally relevant teaching, and utilizing higher level cognitive demand as a starting point, the authors put forth the following definition of *Culturally Relevant, Cognitively Demanding (CRCD) Mathematics Tasks*:

> Culturally relevant, cognitively demanding tasks should be mathematically rich and embedded in activities that provide opportunities for students to experience personal and social change. The context of the task may be drawn from students' cultural knowledge and their local communities. But, the use of context goes beyond content modification and explicitly requires students to inquire (at times problematically) about themselves, their communities, and the world about them. In doing so, the task features an empowerment (versus deficit or color-blind orientation) toward students' culture, drawing on connections to other subjects and issues. CRCD tasks ask students to engage in and overcome the discontinuity and divide between school, their own lives, community and society, explicitly through mathematical activity. The tasks are real-world focused, requiring students to make sense of the world, and explicitly to critique society—that is, make empowered decisions about themselves, communities and world.

One way of connecting higher level tasks with the perspectives of culturally relevant teaching may be to propose a refinement to the descriptions of the two categories presented by Stein et al. (2000) as a) *procedures with connections to concepts, meaning and understanding of mathematics, culture and community*, and b) *doing mathematics for the purpose of becoming empowered intellectually, culturally, politically and socially*. These two renamed classifications are given for the purpose of having teachers of mathematics rethink the extent to which selected mathematics tasks challenge students to ask relevant questions of themselves and the world around them. Figure 6.1 shows an extension of the Stein et al. framework to include features of culturally relevant mathematics tasks.

The second thing to consider in CRCD mathematics instruction is intent. Much of the reform language reference "experiences" and "informal thinking" as culture-free and in unproblematic terms. That is, those who look at student experiences need not venture too deeply into ideas of race, ethnicity, and class, or question problematic deficit orientations. This classification explicitly points to culture as an essential source for building student understanding. Thinking in this manner will undoubtedly challenge the reliance that practitioners have on the use of textbook contexts to frame the cognitive experiences of children.

Teachers, themselves, will have to become more knowledgeable and build relationships with students and communities. Current canons of knowledge on teaching mathematics do not adequately include these kinds of endeavors and still require developing mathematics teachers to think of the subject as engaging, but one where contexts are trivially contrived and its purposes are uncritically addressed. That is, we want students to practice good mathematics as long as it doesn't ask the kinds of critical questions

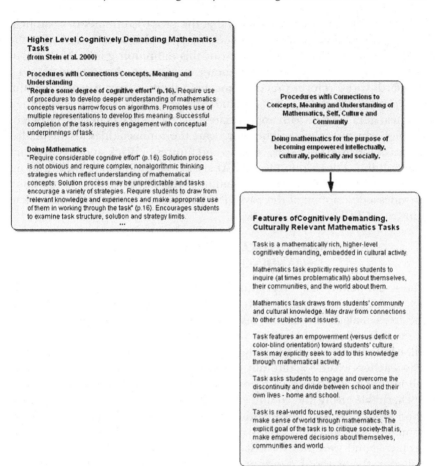

Higher Level Cognitively Demanding Mathematics Tasks
(from Stein et al. 2000)

Procedures with Connections Concepts, Meaning and Understanding
"Require some degree of cognitive effort" (p.16). Require use of procedures to develop deeper understanding of mathematics concepts versus narrow focus on algorithms. Promotes use of multiple representations to develop this meaning. Successful completion of the task requires engagement with conceptual underpinnings of task.

Doing Mathematics
"Require considerable cognitive effort" (p.16). Solution process is not obvious and require complex, nonalgorithmic thinking strategies which reflect understanding of mathematical concepts. Solution process may be unpredictable and tasks encourage a variety of strategies. Require students to draw from "relevant knowledge and experiences and make appropriate use of them in working through the task" (p.16). Encourages students to examine task structure, solution and strategy limits.
...

Procedures with Connections to Concepts, Meaning and Understanding of Mathematics, Self, Culture and Community

Doing mathematics for the purpose of becoming empowered intellectually, culturally, politically and socially.

Features of Cognitively Demanding, Culturally Relevant Mathematics Tasks

Task is a mathematically rich, higher-level cognitively demanding, embedded in cultural activity.

Mathematics task explicitly requires students to inquire (at times problematically) about themselves, their communities, and the world about them.

Mathematics task draws from students' community and cultural knowledge. May draw from connections to other subjects and issues.

Task features an empowerment (versus deficit or color-blind orientation) toward students' culture. Task may explicitly seek to add to this knowledge through mathematical activity.

Task asks students to engage and overcome the discontinuity and divide between school and their own lives - home and school.

Task is real-world focused, requiring students to make sense of world through mathematics. The explicit goal of the task is to critique society-that is, make empowered decisions about themselves, communities and world.

FIGURE 6.1.   Framework for Culturally Relevant, Cognitively Demanding Mathematics Tasks.

that might change students or the world around them. Culturally relevant pedagogues, however, have shown that this need not be the case. A framework for culturally relevant, cognitively demanding mathematics tasks does not provide a magic bullet but instead provides a tool to guide teachers in the selection of tasks that, coupled with the social and learning conditions in which the tasks are experienced, will provide cognitively demanding mathematics for Black children. This involves a process in which the users of the tool question their views of knowledge, self and others, and social relations—all of which are foundational for a culturally relevant pedagogy.

The key endeavor of this work is that the proposed framework can help to navigate the tensions of excellence and equity that were described in the introduction of this chapter. How can this definition guide the tensions stated in the introduction? That is, how can we assist educators in adopting standards-based reform practices and help them to create a more culturally relevant classroom for Black students? The guide works to strengthen how teachers struggle with appropriate engagement tasks by offering a lens through which teachers might interrogate the connectedness of the mathematics to the lived concerns, realities and futures of Black students. It seems that building a description for CRCD tasks from the existing classifications of Stein et al. (2000) serves as a solution to this dilemma. The next section provides a description of the ways in which teachers used this framework to construct CRCD mathematics tasks.

CREATING CULTURALLY RELEVANT COGNITIVELY DEMANDING MATHEMATICS TASKS

The assignment for eleven graduate students, who are also practicing teachers, was to construct culturally relevant tasks for middle grades students in the areas of algebra, data analysis, and geometry. The assignment read: You are to select five tasks that are "higher level, cognitively demanding" that will be modified into a culturally relevant middle grades mathematics task. The teachers wrote 55 culturally relevant mathematics tasks that seemed to fall into one of the following contextual categories: 1) use of technology, 2) charitable giving and volunteerism, 3) student part-time work, 4) getting good grades, 5) issues with voting, 6) visual and performing arts, and 7) community connections. Many of the tasks were modified from textbook problems and were not necessarily considered culturally relevant as presented in the textbook; however, when asked to modify the tasks using the *CRCD Mathematics Task Framework*, the teachers were able to create CRCD mathematics tasks. One of the most salient features of the tool teachers considered was one of empowering students "intellectually, socially, emotionally, and politically by using cultural referents to impart knowledge, skills, and attitudes" (Ladson-Billings, 1994, p. 20). The framework gave teachers the opportunity to consider the complexities of creating CRCD mathematics tasks.

It should be reiterated here that task creation is by far only the beginning. Culturally relevant pedagogy necessitates that teachers learn about students, their culture, and their backgrounds (Guberman, 1999; Ladson-Billings, 1997; Nieto, 2002). Ladson-Billings (1994) indicates that the teacher must be the driving force to creating a culturally relevant classroom. The contexts of the tasks alone will not necessarily make for a culturally relevant environment. It is the thinking behind the tasks and the actions during implementation that make them culturally relevant. Without the appropriate set up of the task and the accompanying discussion and connection to the students and/or their communities, the task, although created as culturally relevant, will lose its relevance. Gay (2000) states,

> However important they are, good intentions and awareness are not enough to bring about the changes needed in educational programs and procedures to prevent academic inequities among diverse students. Good will must be accompanied by pedagogical knowledge and skills as well as the courage to dismantle the status quo. (p. 13)

We share seven of the 55 CRCD mathematics tasks in Figure 6.2 as a starting point for teachers who wish to investigate use of the CRCD mathematics task framework for creating a more culturally relevant classroom. Culturally relevant teaching opens the classroom discourse to different ways of knowing and talking about mathematics (Leonard & Guha, 2002). Following the tasks are the teachers' rationales for why they believe the tasks are culturally relevant.

Task 1: *So You Think You Can Draw* is a problem that demands an appreciation for the informal mathematics knowledge that is used in the creation of street art. Street art is actually a very sophisticated use of mathematical ratios combined with artistic talent. Having students learn and appreciate the process by which murals are created gives a great degree of respect for the artists and the communities that have the art. This problem fosters critical thinking because it addresses using coordinate mapping applied to art. The problem supports and integrates the informal use of mathematics within their community. It builds on the student's cultural knowledge and appreciation of street art as a vehicle to make the community aware of social and political concerns of the artist. The appreciation of the sophistication of the construction of the art causes the student to inquire whether society at large has an appreciation of this culturally rich art form.

Task 2: *Buying Beads for Christmas Stockings* is a problem that invites the student to think about giving to those less fortunate. By mentioning the idea of foster children it makes students think about the responsibilities of parenting and the unfortunate circumstances when parents can't take care of their children. It is empowering because it speaks of our community responsibility to help take care of those who are less fortunate. The ramifications are subtle, but there are also complex issues behind sales tax, social programs, and state and county government. As students connect the amount of money they spend on tax, they may relate to the amount of services they think the government should provide. But, they must make the loop back to the fact that services are paid for by taxes on wages and goods: higher services = higher taxes.

The ultimate goal for the student who engages in Task 3, *Show Me The Money!* is to realize two things: the compounding power of money and how

Task 1. So You Think You Can Draw

 Your sister loves street art. You would like to recreate one of her favorite pieces for her birthday. You decide to create a poster board replica of this piece even though you're not an artist. Suddenly a deeper side of the image strikes you.

This is going to be easy! You notice the tip of his nose at (0,0), the bottom lip (0,-2)......Where is his right eye, ...the bottom of his chin, ...the large patch of grass? What is the domain and range? Explain your reasoning. Try creating a replica on poster board.

Artwork ©1994 Dave Kinsey (aka Büst) in Atlanta, GA. Photographer © 1994 Ted Mikalsen. Used with permission from www.graffiti.org

Task 2. Buying Beads for Christmas Stockings

Your Sunday school class is filling stockings for foster children this Christmas. You have been saving your money and decide to use it to buy dread beads for the girls' stockings. When you go shopping, you find out that the store adds a 5 percent sales tax to every purchase. If you did not have to pay tax, you could have bought three more beads for the same amount of money. How many beads did you buy?

Task 3. "Show Me The Money!"

Stephon Marbury has introduced a new line of basketball shoes that cost $15 a pair. The average price for Nike's Zoom LeBron IV basketball shoes is $150 a pair. If your parents bought you two new pairs of shoes every year since you were 4 years old, and bought the Stephon Marbury shoes instead of the Nikes, to date, how much money could they have saved for you to go to college?

If your parents invested that money at a compound interest rate of 7% per year, how much money could you have?

FIGURE 6.2. Selected Culturally Relevant Cognitively Demanding Mathematics Tasks

the marketing of overpriced basketball shoes, especially to African American males, can adversely affect their family's ability to save money for them to go to college.

Task 4: *Weighty Issues* can be considered culturally relevant to middle school students since it is based on real statistical data about students in and around their age group. Cultural factors, such as an increasingly sedentary lifestyle due to video games, television, Internet, and other modern conveniences, must be considered when determining if a constant growth rate can be predicted. It is culturally relevant because it "embeds student culture into the curriculum" (Leonard & Guha, 2002, p. 114). Through the mathematical analysis, students are empowered with the knowledge that they, their communities, or their peers face this potential health threat, and they can be encouraged to use that knowledge to educate and inform others.

Task 4. Weighty Issues

According to a 2006 report from the Centers for Disease Control, 33.6% of Americans between the ages of 2 and 19 are either overweight or on the brink of becoming so, up from 28.2% in 2000. That's about 25 million overweight kids. Assuming this trend continues at a constant rate, how many kids do you predict to be overweight in the year 2050? What about the year 2100? Is it reasonable to assume a constant growth for this trend? Why or why not?

Task 5. Curfew Controversy

The members of your community were asked to vote on a mandatory curfew of 11:00pm for minors under the age of 18 years old. The results were printed in the paper, but there was a misprint, one digit was incorrect.

Yes votes: 13,657   42%
No votes :186,491   58%
Find the correct number of votes if the misprint is in the "yes" total.
Find the correct number of votes if the misprint is in the "no" total.

   Based on the population of your community, describe what you would expect the results to be if minors were allowed to vote on this issue?

Task 6. Voting Dilemma

The newspaper reports that only 42% of legal voting age of Dunwoody will vote in the next election. If 13,779 people voted, what is the population of people of voting age in Dunwoody? A local community group has a goal of increasing voting by minorities by 10%. If racial minority groups make up 18.2% of the population of people of voting age, how many people must the group convince to vote to reach their goal? Explain how you came to your answer.

Task 7. Reduce, Reuse, Recycle

You can recycle five aluminum cans to make one new one. How many new cans would you be able to eventually make from 250 aluminum cans? (Remember to recycle the cans you make into more cans!) Based on your family's aluminum can usage, estimate how long it would take your family to recycle 250 aluminum cans. Why is recycling important to our environment?

FIGURE 6.2. (Continued)

   Task 5: *Curfew Controversy* can be considered culturally relevant because it not only deals with a community issue of curfews for minors, but also the importance of voting. The mathematical demands of the problem themselves are not essential to the cultural issues; rather the mathematical concepts are presented within the context of a real-world situation that may have an effect on the students themselves. Without this context, there is no motivation for students to solve simple percent calculations that have no effect on them. Within context, students are able to see how percentages are used to represent data from the results of voting. By using the issue of a curfew for minors, the students' own cultures are placed at the heart of the issue of voting. Leonard and Guha's (2002) article states: "Mathematics problems that tap the culture of the students have the potential to engage them" (p. 114). This problem does exactly that, and may incite discussion not only about the issue of a curfew, but about the importance of voting as well. This discussion can empower students as they realize the importance of the voting process in issues that relate to their daily lives. The problem could be changed to include any issue relating to students' lives or that may impact the community in which they live. The last question of the task requires students to consider their own community populations and consider how the amount of votes might change if they themselves were allowed to vote. Undoubtedly students would argue that minors would be more involved in the voting process if it involved an issue that would affect them. This would heighten their awareness about the importance of being politically aware of issues and promote the importance of participating in the political system. This task weaves the mathematics of percentages into the context of citizenship and the importance of voting, an idea that youth in America saw in action in the 2008 presidential election. Based on exit polls, 66% of the 23 million voters under the age of 30 voted for President Barack Obama (Keeter, Horowitz, & Tyson, 2008; The Center for Information & Research on Civic Learning & Engagement [CIRCLE] staff, 2008). From 2000 to 2008, there was a steady increase of youth voters, from roughly 41% to 53% of that population exercising their right to vote (CIRCLE staff, 2008).

Task 6: *Voting Dilemma* continues with the theme of voting and is an example of a culturally relevant mathematics task because it has a basis in the community and important civic issues. One goal of public schools is to help students to become responsible citizens. Tasks such as this one incorporate that goal into the mathematics classroom by asking students to think about issues that face their own community. This problem was set in a specific city because it is the community in which the students live. Setting the problem in the students' community exposes them to a local problem. In "Issues of Culture in Mathematics Teaching and Learning" Malloy and Malloy (1998) state that "culturally relevant teachers...seek excellence within the students' culture thereby enabling students to become more emancipated lifetime learners" (p. 253). This task achieves this by asking students to investigate a real problem within their own community. The figures used are all authentic and therefore the racial makeup of the community described in the task is true to the city in which this group of students lives. The students are also investigating two real problems within society as a whole—low voter turnout for the general population and low voter turnout for minority groups, which often leads to under-representation of minority groups in elected office. Also, since this task asks the students to investigate a real-world issue, it is teaching that mathematics is an important tool for all citizens.

Culturally relevant pedagogy promotes rethinking the relationship between the best interests of all students and the teachers, the curriculum, and society (Matthews, 2003). Task 7: *Reduce, Reuse, Recycle* supports all three of these components. It addresses the student-teacher relationship through the inevitable discussion on recycling that this problem stimulates, which would include questions like "How many students' families recycle?" "Does the teacher recycle?" and "Why is recycling important?" Secondly, this problem boosts the student-curriculum relationship because it allows opportunities for additional mathematical concepts to be addressed (surveying, tabulating, graphing,) as well as the science issues it addresses (environmental responsibility, ecology). Finally, the student can clearly see the relationship to society after discussing the effects that recycling has on his/her world.

CULTURALLY SPECIFIC PEDAGOGY

For this particular assignment, the teachers were not asked to create culturally specific tasks for Black children; however, for the purpose of this chapter, the seven tasks shared are the most culturally specific to Black children. Culturally specific pedagogy has been defined as: "intentional behavior by a teacher to use gestures, language, history, literature, and other cultural aspects of a particular race, ethnic or gender group to engage students belonging to that group in authentic student-centered learning" (Leonard, 2008, p. 9). Culturally specific pedagogy acknowledges the style, language, behavior, and tradition of the students' community (Cooper as cited in Leonard, 2008) and supports the use of curriculum materials that highlight a distinct ethnic background. "Historical facts and statistical data" can also be used to "develop culturally specific mathematics" lessons (Leonard, 2008, p. 52).

When considering the brilliance of Black children, we must be mindful in choosing the mathematical activities in which they will be engaged. These seven tasks are all connected to Black children's experiences and the issues they may face in their communities. Several tasks focus on social issues in the Black community such as the issue of diabetes within the *Weighty Issues* task and issues related to the importance of voter responsibility in the *Curfew Controversy* and *Voting Dilemma* tasks. The teacher could extend this discussion with statistics about voting in our communities, voter registration, and the election of Barack Obama, our nation's first Black president.

The *Reduce, Reuse, Recycle* task could be used as a springboard for discussions about the economy, the jobless rate in the Black community, and the future of green jobs. Students could research green jobs to investigate the training needed, the types of jobs available, and the cities and states where the most green jobs exist. They could also contact local politicians to ask for better recycling programs if needed. "Exemplary mathematics teaching is committed to organizing classrooms and instruction in ways that stress relevance, by fostering empowering relationships with students. The mathematics classroom is a place where larger community and societal goals are achieved" (Matthews, 2009, p. 80).

The *Buying Beads for Christmas Stockings* problem can be culturally specific to Black children if the connection is made to the history of the beads and to the culture of dreadlocks, also known as locs. In today's time, locs can be described as a natural hair style worn by many Black people where the hair is allowed to matt into locs. Locs have a long and rich history in Africa and the Western World, the Caribbean, and North and South America. It is also widely associated with the Rastafari movement in Jamaica. Beads are used to adorn the locs. The beads are made from a variety of materials including crocheted yarn, wood, and other materials.

In this section we've highlighted seven tasks that could be an effective starting point for a teacher using culturally specific pedagogy in the classroom. In the next section we begin a discussion of a group of graduate students who used the CRCD Rubric (see Figure 6.3) as a tool to examine a subset of the original 55 tasks.

CRITIQUES OF CULTURAL RELEVANCE: EXAMINING TASKS USING THE ASSESSMENT RUBRIC

Another group of graduate students ($n = 206$) were given a subset of ten of the algebra tasks (taken from the original 55 tasks) and asked to rate them as CRCD based on the assessment rubric in Figure 6.3 in order to practice using the rubric as a tool and to provide feedback. They only received the tasks and not the original graduate students' rationale for designing the tasks. The graduate students ($n = 11$) in this study were not asked, however, to create tasks that were culturally specific to Black students.

The graduate students were enrolled in an online MEd program and were typically K–12 teachers in a self-contained, mathematics, or science classroom, while some were instructional specialists for their school or district. Graduate students who analyzed the CRCD tasks discussed were enrolled in an algebra curriculum course. The students were asked to choose one task to evaluate and provide a rationale based on the assessment rubric and course readings. Their responses were coded for common themes and patterns, and the following categories emerged: Connections, Choices, Culture, Culturally Specific, and Math in the World around Them. The ten tasks they could choose from are included in Appendix A.

The graduate students chose one task and rated it as either culturally relevant, moderately culturally relevant, or not culturally relevant, based on the assessment rubric. For the tasks they didn't consider culturally relevant, many gave suggestions to modify the task or provided comments on what the task was lacking according to the rubric. A majority of the graduate students chose to critique *Earning Money* (n=51) or *Renting Video Games* (n=49). There were 23 who chose *Getting a Discount,* 16 who chose *Buying*

Description	Degree in Task Structure		
	high	Moderate	low
Mathematics task explicitly requires students to inquire (at time problematically) about themselves, their communities, and the world about them.			
May draw from connections to other subjects and issues.			
Mathematics task draws from students' community and cultural knowledge.			
Task may explicitly seek to add to this knowledge through mathematical activity.			
Task is mathematically rich and cognitively demanding, embedded in cultural activity.			
Tasks asks students to engage the discontinuity and divide between school and their own lives – home and school.			
Task is real-world focused, requiring students to make sense of world through mathematics.			
The explicit goal of the task is to critique society—that is, make empowered decisions about themselves, communities and world.			

FIGURE 6.3.   Assessment rubric of culturally relevant, cognitively demanding mathematics tasks

Pencils, 15 each who chose *So You Think You Can Draw* and *Weighty Issues,* 12 who chose *Driving Decisions,* eight who chose *Math-a-Thon,* six who chose *How Much Does a Locust Eat,* and one who chose *Buying Beads at Christmas.* In order for tasks to be culturally relevant, many of these graduate students pointed out where they lacked in several of the aforementioned categories. For instance, the graduate students noted that there needed to be connections between the K–12 student and the topic, between the student and the teacher, and between the topic and mathematics, which is why we have the category Connections. The K–12 students need to feel a connection to the topic either by prior knowledge or experience. Also, the teachers need to know students and their backgrounds, using that information to foster that connection. Some who felt the task was culturally relevant still offered suggestions for improvement. For instance, in response to *Weighty Issues,* Student 115 stated the following:

> This task is highly culturally relevant. I believe that the most culturally relevant aspects of the [rubric] are the first and last items....A task is the most

culturally relevant when it offers kids an opportunity to reflect on their culture, neighborhood, and personal lives. This task will reveal to students a very serious problem in the American lifestyle. It can be made even more culturally relevant if the problem is revisited with the data on obesity...broken down into subpopulations. This problem will likely make many students uncomfortable, particularly the overweight ones. That can be a scary fact for teachers, but a truly culturally relevant task will often cause healthy discomfort within our students. Tasks that require students to critique themselves and their culture are described by Ladson-Billings as community problem solving. The author explains that culturally relevant tasks force students to critique themselves and the community. Learning is most useful and meaningful when it improves our world (Ladson-Billings, 1995a).

Another student commented on the *Weighty Issues* tasks with:

According to the rubric and readings, yes, it is culturally relevant. These kids all know someone that is overweight if it is not themselves. I think that if you would take the question a step further and ask them to find other countries data on this issue and examine it closely [and] ask them to see if they can think of any correlation between the numbers and the cultures... this would make it even more relevant to the student. They can use their own knowledge about what creates this epidemic and expound upon it with their experiences. "Critical educators hold that a central purpose of education should be to prepare students for active participation in public life toward a more just and democratic society" (Gutstein et al., 1997, p. 714). Examining the possible causes of being overweight will help kids understand issues surrounding this problem and not judge people. (Student 202)

This task could push students to engage in community problem solving on the American lifestyle that leads to a large number of obese people in our population. Students may connect to the task if they or members of their family are overweight. Also the statistics in the problem as well as additional data from other countries not only connects the student to the concept but also to the mathematics of investigating trends and correlations. Though this is a sensitive topic, it could be beneficial for the students who may not realize the dangers of an unhealthy lifestyle. This could lead to a discussion about making changes in their eating and exercise habits and possibly improve their world around them if they are in a family that consists of several overweight members. The response by Student 202 also adds a component that could lead to addressing or curtailing bullying, which has become an increasing problem in schools. Children who are overweight are among those considered at risk of being bullied (U.S. Department of Health and Human Services, 2012). Approaching the task in a way that promotes tolerance and understanding, the teacher can create a safe environment for all students.

Additional critiques on connections pointed to gender relevance, cultural knowledge, and motivation. In response to *So You Think You Can Draw,* a student made the following comment:

> I believe this task is culturally relevant not only through the use of the rubric...but because many of my students are interested in art and drawing and I believe they would be really interested in recreating art similar to this. Street art is also very culturally relevant to a majority of my students because street art is more common where they live. Many of my students have either "tagged" something in their neighborhood or something in town, or know someone who has "tags" quite often. This street art takes it one step further and turns the "tag" into something they all can appreciate. Also, this street art looks like many of them. Since it looks like them, it relates to them even more. (Student 087)

Tagging is one of the most common forms of graffiti. A "tag" is the most basic writing of an artist's name, usually his or her personalized signature. Controversies that surround graffiti continue to create disagreement among city officials/law enforcement and writers/artists who wish to display and appreciate their work in public locations. Students see images of graffiti in their neighborhood, and to see that in a mathematics classroom makes a connection between home and school that they may have never seen before. This shows them that the teacher values where they come from and is showing them how to mathematize their out-of-school experiences.

The graduate students also brought up obstacles to making connections to the tasks as well. This occurred most often in the *Earning Money* task that referred to mowing lawns. Many commented that a female student may relate more to babysitting than mowing lawns. Also students in the inner city or an area that doesn't have lawns wouldn't relate to this task because of no prior experience with lawns. Perhaps students in a colder climate may relate to shoveling snow during a bad snowstorm instead of mowing lawns.

The graduate students felt that the K–12 students could relate to the topic of obesity either by knowing a family member struggling with health issues, their own issues, or the attention that better health receives in the media. With the *Weighty Issues* task some offered comments about lifestyle choices. For instance, Student 116 stated: "It concretizes an otherwise abstract but important issue that they face daily. Recognizing the potential hazards and seriousness of childhood obesity, the students themselves are empowered and may be inclined to make better lifestyle choices on the basis of their findings, which leads to the next theme for the responses."

Choices

According to the rubric, a CR task has as a goal for students to "critique society—that is, make empowered decisions about themselves, communities and world" (Figure 6.3). Of the tasks the students had as options, most of the comments about empowerment pertained to the control the students had as a result of the task. Either they felt the students weren't empowered because they didn't have control over the task, or they felt the task was structured in a context that gave students the decision making power to effect change in their lives, have control over economic and/or lifestyle choices or even over their future. For instance a student made the following comment about the *Earning Money* task:

> I feel it is empowering and can be considered critical mathematical thinking as mentioned by Gutstein, Lipman, Hernandez, and de los Reyes (1997) because a student can realize that it takes a lot of work to save $300 mowing yards and decide that mowing yards for a living is not something they want to do and will therefore graduate, go on to college and make something of themselves. (Student 109)

The response is aligned with the nationwide focus on producing students who are college and career ready. By the line of reasoning presented by Student 109, K–12 students would be motivated to go to college and increase their earning power based on the realistic picture of whether or not it is profitable to mow lawns. However, this reality check could also make a student think more like an entrepreneur and consider how to have a lucrative landscaping business in the future, which doesn't necessarily require a college degree. Thus, a task such as this could lead to a conversation about life after high school and/or college and what it means to have a "good job." According to The American Diploma Project (2004), these are jobs that "pay enough to support a family well above the poverty level, provide benefits, and offer clear pathways for career advancement through further education and training" (p. 105).

Culturally specific tasks that cause students to critique the world around them lead to better decision making and cause them to place more thought and meaning into the process. Many of these choices are based on prior knowledge and experiences, but they may also be based on their culture—family, ethnic, or age.

Culture

When making statements about culture, these graduate students made several references and insinuations to American culture as well as teenage culture or region of the country; there were rare specific statements about an ethnic group. For instance, there were six specific statements about African American culture and eight specific statements about Hispanic culture. Of these, the most common references were to familiarity with street art and health issues for both groups; thus students would have a connection to *So You Think You Can Draw* and *Weighty Issues,* respectively. For Hispanic culture there were references to familiarity with landscaping, thus students could have a connection to the lawn mowing aspect of the *Earning Money* task.

Many students rated the tasks as culturally relevant because they felt the task was related to teen/youth culture. Student 120, however, had trouble making the determination as indicated on the following response:

> I believe this problem is culturally relevant because video games are a huge part of our youth's culture. I struggle with this problem because I am trying to make sure it is culturally relevant, and not just connected to real-life situations. The article titled, "Creating Cultural Relevance in Teaching and Learning Mathematics" was helpful in describing how to let the students take charge of the task by making it their own, simply by taking pictures in their neighborhood. This problem isn't like that, but I can't get away from the fact that this problem does attract the students because of the fact that they love video games. An extension to this problem that would make it even more culturally relevant, would be to somehow get students to study sales data of their favorite games and decide which video game creating company to invest in. The sentence in the article, "mathematics may be used to empower people to make needed changes both politically and economically," (Leonard & Guha, 2006, p. 114) sticks out the most to me, and helps me believe this problem is culturally relevant.

The issue that Student 120 had with determining if this was a culturally relevant task means that there is a need for clarity when discussing *culture.* In the assignment no specific guidelines were given in order to find what attributes students would use to define culture. Most people, for whatever reason, try to stray away from ethnicity. In an effort to be inclusive and tolerant, some believe we have become a color-blind society, which does not benefit students. Teachers should have a specific focus on the culture of students that does not lead into perpetuating stereotypes, which is what some of the graduate students felt, but it lets the students know that the teacher is willing to make a connection to the students to help them understand their mathematics better. When the graduate students did mention culture in their critiques, they didn't go much further than saying that the task relates to the student's own cultural experiences, leaving it open for interpretation but not necessarily defining, again, what is meant by culture. Student 081 stated, "The students will have to reflect on their own cultural experiences and develop meaningful mathematical concepts that reflect their own experiences to solve the task." This is a vague statement, not giving any insight on the role the teacher will play in setting up a task that will allow this to occur for the students.

When asked to evaluate the cultural *relevance* of these tasks, as opposed to cultural *specificity*, the graduate students had a much broader sense of what culture is. These responses were related to community culture, community structure and connections between home and school. This includes the family, home/school connection, inner city, where the students live and making the connection among them. Since culture was generic or not specific to any ethnicity in the statements that were coded in this theme, we determined that we need to be deliberate in our instructions to students to create culturally specific tasks, which could have been the root of Student 120's dilemma—cultural relevance is too broad. If we want them to focus on Black children, then that is exactly what needs to be asked of them. The assessment rubric is a great start for students to think about cultural relevance in a broad sense; however, if the students in front of them are Black children, they will need to dig deep and find tasks with which these specific children can relate and engage.

Culturally Specific

Statements coded in this theme were ones that specified a race/ethnicity. The more elaborate responses were mostly given for an ethnicity other than African American. Though these graduate students are teachers from across the United States, a large number of them teach in an area where the student population is largely Hispanic. In response to the *Earning Money* task, another student made the following connection for K–12 students from the Hispanic culture:

> For my particular group of students, I believe this task would need some adjustments to be culturally relevant. Earning $300 in one week is a bit much to expect for fourth graders. Also, my students all come from the neighborhood surrounding our school. A majority of the homes are extremely modest and very few if any have lawns. It is doubtful that many of the families own a lawn mower, so lawn mowing for neighbors would not be a possibility. Trimming bushes is probably not age appropriate for 9 and 10 year olds. (There are few bushes in the neighborhood anyway.) However, I could change the problem to say that your tio or dad's boss has agreed to let you help with some of the smaller lawns in their lawn trimming business. They could earn $20 for each yard mowed, and for the bigger lawns $10 for picking up a yard for the crew. This would make the task a bit more cognitively demanding as well since it would be open ended with two independent variables. Another possibility for making it culturally relevant would be to say that your tia and abuela are going to Mexico and want you to pet sit—possibly for a lower amount of money per week, or $20 for your tia's two dogs and $10 per day for your abuela's bird. Maybe earning enough for school supplies and a particular pair of shoes would be easier for them to relate to than the $300 that the original task mentioned. (Student 205)

Several of these teachers were able to relate more with Hispanic culture than with African American culture, which is why without prompting, they saw a connection. There could be many reasons why teachers quickly make a connection to the Latino/Hispanic culture than any other. Perhaps they are Latino and recall their K–12 learning experiences and know first-hand strategies for enhancing that experience for their students. Also, there is increased attention on programs that cater to students who are English as a Second Language (ESL) and English Language Learners (ELL), especially in the southern states bordering Mexico. So when teachers think about modifying their instruction to meet the needs of culturally and linguistically diverse (CLD) students in their classroom, they focus more on the linguistically diverse Latino students. It can be more feasible to address language and vocabulary obstacles by changing words in a task as opposed to trying to create a scenario to which students may or may not be able to relate. So, to make a connection for African American or Black students, teachers need to be specifically asked to develop mathematics tasks specific to these students' culture.

Math in the World around Them

Comments similar to Student 120 earlier ("I struggle with this problem because I am trying to make sure it is culturally relevant, and not just connected to real-life situations") were common. Just because a task is real-world focused doesn't necessarily mean that it is culturally relevant *or* culturally specific, and this real-world connection may actually be a superficial reform characteristic (Stein et al., 2000). However, with these graduate students, if the task related to a certain age group and students could relate to the age appropriateness of the tasks, then they regarded the task as culturally relevant. These teachers are correct only if the culture we are speaking of is age. This means that if teachers feel that they are using culturally relevant tasks, they will be under a false impression that they are teaching culturally relevant mathematics when in fact they are not with many of these types of tasks.

The comments presented in these sections provide a sense of how the graduate students examined these tasks using the CRCD rubric. Of the themes that emerged, we discussed how culturally relevant tasks need to foster connections between the students and the topic, the teacher and the students, as well as students and the mathematics in an authentic, not superficial way, with a goal for students to critique society to make empowered decisions. In future uses of the rubric, teachers will evaluate tasks culturally specific to Black students, which ideally will leave little room for a vague evaluation with regard to cultural connection and cultural specificity. When teachers make these authentic connections to the students and their prior knowledge and experiences, the math in the world around them will not only be relevant but will also be specific to Black students.

CONCLUSION

At the outset, the original 11 graduate students were not asked to create culturally specific tasks. They were instructed to create culturally relevant tasks. The culture they seemed to rely on was the age of their students. They in fact did create tasks that were age appropriate and on a general level could be related to students' real lives. It was interesting, however, that very few of the teachers made specific mention to their students. In addition, very few of the teachers used race as a basis for their culturally relevant tasks. Therefore, in this chapter we were very deliberate in our intention to choose tasks that were culturally specific to Black children. In order to create these culturally specific tasks, we had to have knowledge about Black culture. As such, it is imperative that teachers who are not Black but are teaching Black students must make a special effort to address the needs of their students, not just the age of students they have. When we say "real life," we must remind the teachers that we mean their *students'* real life.

To create tasks that are culturally specific, teachers must have a place to start. The assessment rubric is but a tool providing an opportunity for this start. In addition, teachers must make the effort to go one step further and get specific information about their Black students free of stereotypes. Teachers must rely on their students to gain appropriate information on their lives and their communities to choose tasks that are relevant to them.

Because the ultimate goal is that of developing a practical way of incorporating relevance into mathematics classroom task design, a next step is to examine how teachers use this rubric and the products that result. These products will no doubt be the subject of important discussions regarding the place of culturally relevant teaching and culturally specific pedagogy in modern mathematics education reform.

REFERENCES

African-American Student Achievement Committee. (2001). Report of the African-American Student Achievement Committee and Work Groups. Presented to State Superintendent of Education Inez Tenenbaum, May 30, 2001: South Carolina Department of Education.

The American Diploma Project. (2004). *Ready or not: Creating a high school diploma that counts.* Washington, DC: Achieve, Inc.

Bartolome, L. I. (1994). Beyond the methods fetish: Toward a humanizing pedagogy. *Harvard Educational Review, 64*(2), 173–194.

The Center for Information & Research on Civic Learning & Engagement [CIRCLE] Staff. (2008). *Young voters in the 2008 Presidential election* (Fact Sheet). Medford, MA: Tufts University.

D'Ambrosio, U. (1985). *Socio-cultural bases for mathematics education.* Cumpinas, Brazil: UNICAMP Centro de Producoes.

Enyedy, N., & Mukhopadhyay, S. (2007). They don't show nothing I didn't know: Emergent tensions between culturally relevant pedagogy and mathematics pedagogy. *The Journal of the Learning Sciences,* 16(2), 139–174.

Gay. (2000). *Culturally responsive teaching: Theory, practice and research.* New York: Teachers College Press.

Guberman, S. R. (1999). *Cultural aspects of young children's mathematics knowledge.* Retrived from ERIC database (ED438892).

Gutstein, E., Lipman, P., Hernandez, P., & de los Reyes, R. (1997). Culturally relevant mathematics teaching in a Mexican American context. *Journal for Research in Mathematics Education, 28*(6), 709–737.

Herbst, P. C. (2006). Teaching geometry with problems: Negotiating instructional situations and mathematical tasks. *Journal for Research in Mathematics Education, 37*(4), 313–347.

Keeter, S., Horowitz, J., & Tyson, A. (2008, November 12). Young voters in the 2008 election. *Pew Research Center Publications.* Retrieved May 14, 2012, from http://pewresearch.org/pubs/1031/young-voters-in-the-2008-election.

Ladson-Billings, G. (1994). *The Dreamkeepers: Successful teachers of African American children.* San Francisco: Jossey-Bass, Inc.

Ladson-Billings, G. (1995a). But that's just good teaching! The case for culturally relevant pedagogy. *Theory Into Practice, 34*(3), 159–165.

Ladson-Billings, G. (1995b). Toward a theory of culturally relevant pedagogy. *American Educational Research Journal, 32*(3), 465–491.

Ladson-Billings, G. (1997). It doesn't add up: African American students' mathematics achievement. *Journal for Research in Mathematics Education, 28*(6), 697–708.

Leonard J. (2008). *Culturally Specific Pedagogy: Strategies for Teachers and Students.* New York, NY: Routledge.

Leonard, J., & Guha, S. (2002). Creating cultural relevance in teaching and learning mathematics. *Teaching Children Mathematics, 9*(2), 114–118.

Malloy, C.E. & Malloy, W.W. (1998), Issues of culture in mathematics teaching and learning. *The Urban Review,* 30(3), 245–257.

Martin, D.B. (2009). Liberating the production of knowledge about African American children and mathematics. In D.B. Martin (Ed.), *Mathematics Teaching, Learning and Liberation in the Lives of Black Children* (pp. 3–36). New York, NY: Routledge.

Matthews, L. E. (2003). Babies overboard! The complexities of incorporating culturally relevant teaching into mathematics instruction. *Educational Studies in Mathematics, 53*(1), 61–82.

Matthews, L.E. (2009). "This little light of mine!" Entering voices of cultural relevancy into the mathematics teaching conversation. In D. B. Martin (Ed.), *Mathematics Teaching, Learning, and Liberation in the Lives of Black Children* (pp. 63–87). New York, NY: Routledge.

Matthews, L. E. (2008). Lessons in "letting go": Exploring the constraints on the culturally relevant teaching of mathematics in Bermuda. *Diaspora, Indigenous, and Minority Education.*

National Assessment Governing Board (2006). *Mathematics Framework for the 2007 National Assessment of Educational Progress.* Washington, DC: U.S. Department of Education.

National Council of Teachers of Mathematics. (1989). *Curriculum and Evaluation Standards for School Mathematics.* Reston, VA: The Council.

National Council of Teachers of Mathematics. (1995). *Assessment Standards for School Mathematics.* Reston, VA: The Council.

National Council of Teachers of Mathematics. (2000). *Principles and Standards for School Mathematics.* Reston, VA: The Council.

SEF South Carolina Task Force and Advisory Committee. (2002). *Miles to go: South Carolina.* Atlanta, GA: Southern Education Foundation, Inc.

Sleeter, C. E. (1997). Mathematics, multicultural education, and professional development. *Journal for Research in Mathematics Education, 28*(6), 680–696.

Stein, M. K., Smith, M. S., Henningsen, M. A., & Silver, E. A. (2000). *Implementing standards-based mathematics instruction: A casebook for professional development.* New York, NY: Teachers College Press.

Tate, W. F. (1995). Returning to the root: A culturally relevant approach to mathematics pedagogy. *Theory Into Practice, 34*(3), 166–173.

Tate, W. F. (2004). Engineering a change in mathematics education. . In D. Nalley, K. DeMeester & T. Hamilton (Eds.), *Access and opportunities to learn are not accidents: Engineering mathematical progress in your school.* (pp. 5–7): Unpublished monograph prepared for the Southeast Regional Consortium (SERC)@ SERVE and presented at the National Alliance of Black School Educators (NABSE) and The Benjamin Banneker Association (BBA) Summit: Closing the Mathematics Achievement Gap of African-American Students, Washington, DC.

U.S. Department of Health & Human Services. (2012). *Who is at risk.* StopBullying.gov. Retrieved September 30, 2012, from http://www.stopbullying.gov/at-risk/index.html

CHAPTER 7

ADOLESCENTS LEARN ADDITION AND SUBTRACTION OF INTEGERS USING THE ALGEBRA PROJECT'S CURRICULUM PROCESS

Mario Eraso

INTRODUCTION

"Mathematics, as a subject domain, is not acultural" or "without context or purpose, including the political" (Martin, Gholson, & Leonard, 2010, p. 14). However, Black children continue to experience school mathematics as a narrow set of rules and algorithms that have little or no relevance to their lives (Leonard, 2008). The Algebra Project (AP) uses mathematics as a means to uncover Black brilliance and to promote social justice. The program targets underrepresented students, especially African American students, because its basic premise is that learning algebra is a civil right (Moses & Cobb, 2001). For AP founder and president, Robert Moses, gaining access to algebra in today's society is synonymous to Blacks gaining ac-

The Brilliance of Black Children in Mathematics:
Beyond the Numbers and Toward New Discourse, pages 151–169.
Copyright © 2013 by Information Age Publishing
151

cess to the polls to vote during the Jim Crow era. For Moses, AP is a mathematics curriculum that provides Black children with language and symbols to express their mathematical brilliance. Additionally, AP goals include increasing student confidence in mathematics and increasing enrollment in advanced mathematics courses.

I was introduced to the AP program during the 2006–2007 school year. After my university advisor retired, I was abandoned as a graduate student without a stipend or tuition waiver. Fortunately, the Center for Urban Education and Innovation (CUEI) at Florida International University (FIU), at that time directed by university professors Lisa Delpit and Joan Wynne, offered me a job as a graduate student assistant. The center's partnership with AP continued to support my education financially from 2006 until I completed my doctoral studies in mathematics education. Working at the center, I soon had the honor of co-teaching the ninth grade mathematics cohort with Robert Moses, who I will refer to in this chapter as Bob. Every week, from Monday to Thursday, I arrived at the school at 7 a.m. to finalize preparations for our 7:30 a.m. algebra class. Bob began work at 6 a.m., and, at first, I could not understand why he needed to be at school so early. I quickly learned that the work needed to accelerate learning among these students was enormous. Specifically, the 90-minute daily mathematics class to which students had committed to at a community meeting with their parents was not enough. Six-week summer institutes after high school years were also needed if we were to consider the possibility of these students being accepted to and attending college. We also knew that after-school work was required to allow students additional time to review concepts they had difficulty understanding during class time.

There are many stories I have to tell about my experiences with AP. One that I most cherish is of a Haitian-American girl who was part of the ninth grade AP cohort. By telling this story, I want readers know that my participation in AP supports my vision and work as an assistant professor of mathematics education at the university. The education principles I learned while working on the Algebra Project surpassed what I learned during my doctoral training. One morning, I noticed that one student was not focusing on the concepts discussed in class. She usually participated very well, and she was always cheerful, insightful, quick, and enjoyed working out mathematics problems and sharing with the class how she solved the problems. That morning, she did not seem the same brilliant and cheerful girl. Rather, she seemed depressed as she sat with her chin on the desk and stared out in the distance. I wanted to ask her something, but I did not really know exactly what to ask. At the end of the period, when Bob and I met as usual to debrief the class, I asked him if he had noticed this student's behavior. He had not been aware of the situation so I described my observations. My main question for Bob was, "What could I have asked her?" Bob simply and

accurately said I could have asked her, "What is going on?" as if I were "…
one of her friends." I thought his suggestion was so simple, so genuine. My
problem in not knowing what to ask was thinking too much about the behavioral techniques I learned during teacher education programs. For example, you should never turn your back to the students, you should never
treat them as friends, and you should always have clear classroom rules and
procedures. If I were to summarize, in a few words, what I learned from
AP during those days was that *all* children are our children and that Black
children are brilliant in mathematics.

THE ALGEBRA PROJECT'S CURRICULUM PROCESS

Current reform efforts to improve the quality of teaching mathematics usually includes a constructivist framework with a research base on learning
and teaching processes that promote deeper understandings of concepts
and relationships of concepts, as opposed to memorizing isolated information (von Glasserfeld, 1987). Constructivist environments are structured to
engage students in problem solving, modeling, and constructively building
conceptual understanding in student-centered classrooms (Park, O'Brien,
Eraso, & McClintock, 2002). This chapter focuses on the later action, particularly on students' use of icons and symbols to understand mathematical
concepts. This type of teaching requires an active, inquiry-based process
where students are at the center of instruction and the teacher is a chal-
lenger and facilitator of knowledge (Bigelow, 1990).

The Algebra Project has attempted to reform mathematics education
for over 25 years. Through community organizing, administration of professional development workshops for teachers, and following cohorts of
students who take 90 minutes of mathematics per day during four years of
high school, AP has established a national reputation for effective pedagogy that is supported by the National Science Foundation. The AP method is
a two-pronged approach in which constructivist theory and the application
of "ordinary discourse" (proposed by the late Harvard philosopher W. V. O.
Quine) combine to bring deeper meaning to the mathematics that students
learn. With the AP method, the teacher ensures that students transition
from *people talk* or ordinary discourse to *feature talk*, a technical language
with the precision necessary to identify characteristics of elements of the
topic under study (addition and subtraction of integers, in our case). This
method is robust, has been refined over the years, and is in close agreement
with Realistic Mathematics Education, a learning theory developed at the
Freudenthal Institute in the Netherlands and used by several mathematics educators worldwide. These methods are powerful because they involve
linking process and object conceptions through semiotic activity using
models that first record processes in situations outside of mathematics and,
subsequently, mediate activities with conventional symbols of mathematics.

In this way, meaning is given to the manipulation of objects by situating objects within contexts that are familiar to students (Cobb, Gravemeijer, Yackel, McClain, & Whitenack, 1997).

Despite the wide dissemination of several constructivist frameworks, some teachers continue to view the mathematics teaching practice as mere knowledge transmission from teacher to students. However, a constructivist model for experiential learning is frequently used because of its simplicity. The model includes four benchmarks. The first benchmark consists of a shared student experience (e.g., riding a bus during a fieldtrip from their school to a university was a shared experience for the AP students in Miami). The purpose of this benchmark is to give relevance to a selected mathematics topic by ensuring that everyone can communicate the concept by drawing, talking, and writing about the experience because everyone has lived the experience. The second benchmark involves student participation in a multi-faceted reflection. During class, students share their perspectives of the shared experience and individually draw elements of their interest that were present in the shared experience. The third benchmark includes evidence that students have developed the ability to represent an abstract concept using general qualities rather than by communicating the concrete features or elements under study. Finally, the teacher guides students through as many applications as necessary to demonstrate evidence of conceptual and procedural student understanding. This cycle repeats with another experience that is designed to introduce the next mathematics topic.

In order to change the perception of learning as knowledge transmission from teacher to students, which continues to resist educational reform, AP has developed a curricular process based on the four-benchmark model described above. This curricular process is currently used in several AP school sites across the nation. The AP curriculum process uses five steps and aims to develop students' abilities to recognize patterns, use iconic and symbolic representation, and identify relations within systems. The process enhances the four-benchmark model of experiential learning by using a consistent pedagogy that was developed by Moses (a detailed explanation is in his book *Radical Equations: Civil Rights from Mississippi to the Algebra Project*; Moses & Cobb, 2001). Of the five crucial steps in the AP curriculum process, the last four are nested between benchmarks two and three of the experiential learning model (i.e., between reflection and abstraction). As students reflect on the physical event by drawing, writing, and talking about it, the teacher works to transition the students from *people talk* to *feature talk*. It is during this transition that AP focuses on using icons and symbols; this transitioning allows students to abstract, or remove, important information present in their intuitive language. The five steps of AP include:

1. Physical event. A central experience designed by the teacher to illustrate the mathematical concept to be learned and shared by all students.
2. Pictorial representation. Students describe the experience through pictures.
3. Intuitive language (*people talk*). Students discuss and write about the physical experience in their own language.
4. Structured language (*feature talk*). Students learn the language that encodes features of interest of the objects in the physical experience.
5. Symbolic representation. Students construct symbols to represent their ideas and gradually adopt universal mathematics symbols.

By following these steps, the curriculum process advances while the teacher builds on students' pictorial representations and intuitive talk to introduce the language of the discipline and its concepts. Many students have difficulty understanding abstractions that are represented by a discipline's symbols until they experiment with creating icons and symbols from their own ideas. Only when students become familiar with the meaning of the symbols should they work on applying the concepts.

THE ALGEBRA PROJECT IN MIAMI

When I joined the organization in 2006, AP had already tried to organize a cohort of ninth grade students in Miami around demanding a quality education. Unfortunately, this effort was unsuccessful. I was fortunate to be a part of their second attempt, and, by that time, leaders of the program had gained experience in working with the local public school system. What started as a process that was full of obstacles is now a powerful story with clear success indicators, albeit a current lack of funding to sustain the initiative in Miami. One clear example of AP's successful sustainability efforts in Miami is having developed a Young People's Project (YPP) site. YPP is a spin-off nonprofit organization created by former AP students and has received funding from the National Science Foundation (for more information on YPP, see Brewley in this volume). As of summer 2011, YPP consisted of 28 math literacy workers who, after graduating from high school, continued to organize youth to demand quality education for all and change the public's dull perspective of mathematics.

Demographics

The AP cohort in Miami was comprised of a 65% majority Haitian-American student body. In initial conversations with the local school district, the CUEI/AP partnership requested that the cohort be formed with students who were performing at the bottom quartile. This request translated to performance levels 1 or 2 on the Florida Comprehensive Assessment Test (FCAT), which consists of five levels to assess student performance. A second request was that the cohort be formed with 50% male and 50% female adolescents enrolled in ninth grade. Following these guidelines, 24 students were included in the Miami cohort. However, the local school district removed four male students at the end of the first year. The local school district did not offer the CUEI/AP partnership an explanation for removing the students. We later learned that the students had moved out of the county. Of the remaining 20 students whose educational trajectories can still be tracked through the work of YPP in Miami up to six years later (Wynne & Giles, 2010), 13 self-identified as Haitian-American and spoke Kreyòl Ayisyen (Haitian Creole). Of these, 10 students also spoke English, and three were labeled as having limited English proficiency. Four other students were African American and spoke English. Finally, three were Latinas who spoke Spanish and were labeled by the school district as having limited English proficiency. Eighteen of the 20 students received free or reduced lunch.

Findings

Findings in terms of the success of the Miami AP program are presented in three parts. First, qualitative data in terms of pre-assessment are reported. Second, program deliverables that describe AP activities are presented. Lastly, post-assessments and measurable outcomes in terms of gains, graduation rates, and college attainment are reported.

PRE-ASSESSMENT OF STUDENTS' MATHEMATICS UNDERSTANDING

The first week Bob Moses and I had contact with the students, we realized that several of them were underperforming in mathematics, reading, and writing. One clear indication that the students might be challenged by ninth grade mathematics was that they had difficulty understanding how to subtract two negative integers, which is a benchmark usually found in grades six or seven, depending of the state a student is located. For example, if students cannot find the answer to $-3 - -7$, then problems will arise when they try to solve for x in an algebraic equation (e.g., $a = x + b$, when $a, b < 0$). Additionally, these students had difficulty making sense of compound instructions such as "find the location that is five stops to the left of the location that is three stops to the right of the bus stop." This problem arose in large part because students had difficulty identifying left and right when making use of reference locations. These issues indicated they had little exposure to relational and propositional knowledge, and they were not fully prepared for the algebra course in which the school district had enrolled them. The school they attended was a failing school, according to FCAT standards.

Although several students exhibited some gaps in their mathematical development, I suspected that most had been tracked for many years and enrolled in classes that did not allow them to develop and demonstrate the mathematical knowledge they possessed. As such, during the last 10 weeks of the 2006–2007 school year, when the AP curriculum was finally implemented (state testing was over), I decided to keep a journal of instances where each student demonstrated understanding of mathematics, independent of whether or not they were working on age-appropriate mathematics. This information provided me with a baseline reference to compare the students' progress in mathematics in the future.

From the journal, I extracted a description of students' learning processes while they engaged in revisiting addition and subtraction of integers and followed the AP curriculum process. The description of how this cohort learned addition and subtraction of integers followed the presentation of the four levels of understanding. Further, the description is based on a naturalistic paradigm that is frequently used as a methodology for this type of data analysis (Glaser & Strauss, 1967; Moschkovich & Brenner, 2000). This paradigm combines the linear structure of traditional quantitative research with an iterative qualitative research process in which several cycles of analysis result in trustworthy themes that are extracted from field notes.

Table 7.1 shows the students' levels of understanding based on my assessment of their work. Based on their mathematical understandings, I developed four categories: excellent, good, average, and weak understanding. These categories are defined as follows:

1. Excellent understanding: The student has clear knowledge about the relationships or foundational ideas of a topic.
2. Good understanding: The student exhibits knowledge of the rules and procedures used to carry out mathematical processes, can define the symbolism used to represent mathematics, but cannot clearly express the relationships or foundational ideas of a topic.
3. Average understanding: The student exhibits logical thought patterns, can explain work partially, and can justify answers partially.
4. Weak understanding: The student exhibits weak logical thought patterns, cannot explain work, and cannot justify answers.

Demographic information, such as gender and ethnicity, is also shown in Table 7.1 for this cohort. To my surprise, fewer than half of the students exhibited weak understanding, and I wondered why they were performing at Level 1 on the FCAT. I later confirmed my doubts when five of the 17 students who started at FCAT Level 1 moved up to Level 2, and one of the

TABLE 7.1. Demographics and Mathematical Understanding

Student Number	Gender	Ethnicity	Mathematical Understanding
1	F	Haitian American	Excellent
2	F	Haitian American	Excellent
3	M	Haitian American	Good
4	F	Haitian American	Average
5	M	Haitian American	Average
6	M	Haitian American	Average
7	M	Haitian American	Average
8	F	Haitian American	Weak
9	M	Haitian American	Weak
10	F	Haitian American	Weak
11	F	Haitian American	Weak
12	F	Haitian American	Weak
13	M	Haitian American	Weak
14	F	African American	Good
15	M	African American	Good
16	M	African American	Good
17	F	African American	Average
18	F	Latina	Good
19	F	Latina	Average
20	F	Latina	Weak

three students who began at Level 2 moved to Level 3 one year after the intervention.

Table 7.1 also reveals that my early assessment of the students' mathematical abilities was more accurate and balanced than the FCAT levels 1 and 2 assigned to the students in the cohort. Finally, the categories allowed me to assess how well the students understood the mathematical topics to which they were exposed and provided me with a direct and personal assessment rather than viewing the students through the lens of performance on a state examination.

AP Students Learn to Add and Subtract Integers: Excerpts from Weekly Activities

Week 1. Our responsibility began with an all-day fieldtrip to Florida International University where the CUEI was located. AP's belief is that providing students with such experiences allows them to dream of the possibility of someday going to college. The students had a specific task to complete while on the bus that transported them to the university. Specifically, they were required to fill out a form indicating what objects along the ride captured their attention. Each student completed a list of 20 items, stating a reason for having chosen each item. Once at the university campus, professors and graduate students, who had brought the AP cohort to the university, coordinated several activities to show students what college campus life was like. Students toured the campus and visited the library, dorms, student center, student recreation center, auditoriums, and finally prepared to have lunch at the College of Education where the CUEI is located. In the afternoon, after being given the autonomy to order the food of their preference for lunch, students painted a mural that expressed who they were, individually and collectively. The mural expressed that they were a group of students with a lot of energy, cheerful, and wanted to learn and change the way they were perceived by their community.

The next day in class, students had to negotiate their individual choices and create a class list of 20 items; learning to make decisions as a collective entity came early in the process. One item that the Haitian students wanted to include on the class list, which required them to convince their non-Haitian peers of the validity of such inclusion, was the cemetery located on 8th Street between 32nd and 33rd Avenues of southwest Miami. I was surprised that the students choose a cemetery and asked the students why they wanted it included on the list. One student said that, in Haitian culture, the dead are respected and that their nation observed All Hallows day where Haitians visit cemeteries and pay their respects to the dead. Later, in the summer of 2007, students attended a six-week residential summer institute at FIU. Some students would wake up at night saying they could not sleep because of sounds being made by ghosts dancing in the hallways. I found these stories very interesting, and asking the students about them allowed me to better understand them, thus making me feel more at ease as their instructor. This is an important observation. Teachers should understand the cultural beliefs of their students in order to work more effectively; this is especially the case for immigrant families in the United States.

Other concepts extracted from the handouts that were filled out on the bus and developed in class compared the artist with the scientist. Bob always insisted that, in both professions, what the practitioners do is simply observe. Therefore, we had students write observation sentences such as, "The girl sitting on the bench is wearing a green blouse." Students learned from this exercise that they could only abstract information that was observed. For example, they were not allowed to write, "Mary is sitting on the bench wearing an expensive green blouse." Observation is an important skill in terms of developing critical thinking.

Week 2. Bob traveled around town taking pictures and showing his dedication and actualization of the teacher/designer or teacher/researcher paradigm. The storage room inside of our classroom was always full with materials Bob had purchased. Additionally, each student was provided with a binder to keep a portfolio of his or her most valuable work. Bob used plastic covers to protect students' work and to show how much he valued the work they produced. Among the artifacts in the portfolio was a dictionary created by the students. Each entry included a mathematical term that was represented in three ways. At the top of each sheet, the student wrote the definition of the mathematical term in which an example was provided. Next, space was provided where students drew a diagram that represented the mathematical term. Finally, a small space at the bottom of the sheet was provided where students created a symbol that represented the term.

One student created an icon in which three stick men appeared next to the world globe and included a heart and peace symbol over their heads as if the stick men were thinking about love and peace as citizens of the world. I asked the student why and how he identified himself with that icon. He responded: "We humans are basically good and love the earth." Figure 7.1 shows the symbol in which the head of the person is the planet symbol used for Earth.

In another activity, students were asked to acknowledge what they already knew about the elementary aspects of algebra embedded in language. Students were exposed to the idea that pronouns in sentences act as variables in equations. For example, students might change "Mary catches the ball" to "She catches the ball" or "She catches it." One can generalize the idea

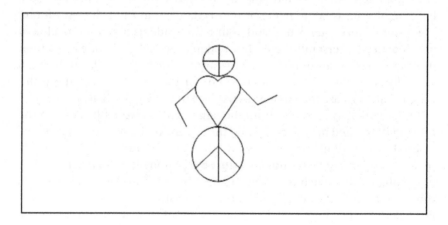

FIGURE 1. Example of Student Symbols

and write "X catches Y." Going further one can create a symbol for "catches" and write "X # Y."

Week 3. Students were exposed to objects, actions, and relationships found in mathematics. For example, we studied that Stop A is to the left of Stop B. We established that the symbols for right and left would be > and <, respectively. A student noted that the symbol < looks like a tilted L with legs at an angle less than 90 degrees and used the observation to distinguish it from the symbol >. "Less begins with L, so I can now remember that < is less than." In the Trip Line, sentences of this type are called Type A sentences, and they state a relationship between two objects. "The location of the car is to the left of the location of the pond" would be coded as © < . Using location numbers, we say that 3 is less than 8, or 3 < 8. Additionally, Type B sentences carry more meaning because they specify the number of stops to the left or right that a particular location is away from the reference location. For example, the location of the car is five stops to the left of the location of the pond.

It is important to note that, during Week 3, the teacher of record asked: "Shouldn't we mix some of the school's math with the math in your curriculum?" Bob responded, "We have been doing that type of math for three quarters of the year, and the students have not learned much. Don't you think we should try something different?" Bob understood that teachers' expectations of the program were high. However, it was not sufficient for mathematics teachers to merely teach mathematics content. Teachers need to be competent in pedagogical knowledge and cultural knowledge as well. AP provided not only content but a method or process for teaching the content.

Week 4. The process of mathematizing observation sentences consists of seven steps, with the final step identifying the conventional mathematical symbols. Again, the teacher of record made a negative comment: "The symbols were invented hundreds of years ago. Why do we have to spend time doing this?" Regardless of this opinion, students engaged in the following seven steps:

1. Identify the observation sentence.
2. Identify the names.
3. Identify the relation or action.
4. Construct an icon for the names and relations or actions.
5. Identify feature category shift; construct feature talk.
6. Construct an iconic representation of feature talk.
7. Identify the conventional mathematical symbols; construct a conventional mathematical representation. This is the abstract symbolic representation.

Students also learned how mathematical terms were related to one another. The following is an example for the case of using a relation between two names:

- The car is located to the left of the pond.
- Car and pond are the names.
- The relation is "to the left of."
- Icon for car: ©. Icon for pond: ℗. Icon for "to the left of:" ←.
- Feature category shift: from "is located" to "the location of." The location of the car is to the left of the location of the pond.
- Iconic representation: © ← ℗.
- Abstract symbolic representation: © < ℗.

Therefore, when a student is presented with the symbols © < ℗, he or she will read: The location of the car is located to the left of the location of the pond; location numbers are then assigned names. For example, if © = ⁻3 and ℗ = ⁻1, then the student reads: "On the number line, the location of negative three is to the left of the location of negative one." This later becomes "Negative three is less than negative one." This cumulative process is important when students begin adding and subtracting integers. Of note, the example ⁻4 – ⁻3 was difficult for our students to conceptualize. Therefore, we knew it was important for students to recognize, based on a convention, that in subtraction, the location of ⁻4 was compared to the location of ⁻3, with ⁻3 being the reference location. We wanted to know where ⁻4 is located with respect to ⁻3 (i.e., to the left of). Additionally, we know that, starting at ⁻3, according to the convention, we must move one step to the left to arrive at ⁻4. Therefore, the answer to ⁻4 – ⁻3 is a movement of one step to the left, which, as an abstract symbolic representation, is -1. Following this seven step process, students were able to say all of that repeatedly!

Week 5. My own difficulty learning this new way of doing mathematics was something that I needed to reflect on. When confronted with the problem ⁻7 – ⁻6, I was taught to connect the two negative signs and make them a positive, without really knowing why. Therefore, I would transform the problem to ⁻7 + 6, at which point I had the option to solve the problem directly or to rewrite it as 6 – 7. However, now that I understand the concept of comparing two vectors, what I do is based on the following convention: look at ⁻7 – ⁻6, focus on ⁻6, and ask myself how far and in what direction is ⁻7 from ⁻6. The answer is obvious: ⁻7 is to the left of ⁻6; it is actually one stop to the left. Therefore, the answer is -1. This process was difficult for me at first; because it was different from the way I was taught, my brain wanted to do it the old way. For students who see the process for the first time, it should not be as difficult because the process is common sense and students simply need to reason out the coding and conventions that are embedded in the mathematical sentences. During Week 5, the math teacher of record saw me struggling with these concepts one day and said, "You see, even you do not know the math. There is no need to confuse the students." That day, I sought comfort from Bob concerning the teacher's negativity. After talking to Bob, I saw the value of struggling to understand the new approach. In my opinion, many teachers teach how they were taught, but they do not go through the necessary pedagogical transformation needed to become expert teachers of mathematics. It is this transformation that truly impacts student learning.

Week 6. Our grading was based on a scale of C, B-, B, B+, A-, and A. The teacher of record asked us for grades, as they were required to be entered bi-weekly into a state monitored system. That same week, students were asked to go to the front of the room and demonstrate what they had learned. Some peers became frustrated because some fellow students were still having difficulty making sense of instructions such as "Make three moves to the left of the bus station." When students got rowdy, Bob would calm them down by saying softly, "Young people, calm down." He also raised his arms in the air with his hands open and, as he talked to them, he slowly lowered his arms. I believe his confidence in their brilliance and success was the impetus for his classroom management style.

Weeks 7 & 8. One of the textbooks used by AP is called the *Trip Line* and was written by Bob Moses and Ed Dubinsky, a research mathematician who values the work of mathematics educators. During Week 7, Professor Dubinsky visited the class and guided students through the dialogues in the AP textbook. All the students wanted to be the mathematician in the dialogues and took turns role-playing the mathematician: "Mathematicians do not write the mermaid statue is located to the left of the bus station. That is *people talk*. Instead, they use *feature talk* when referring to a particular feature; in this case, the particular feature is the location of the objects. We write, "The location of the mermaid statue is to the left of the location of the bus station." The expression *is located* found in *people talk* is taken out of the relationship category (is located to the left of). The idea of location is written as part of the name category (the location of the mermaid statue and the location of the bus station). Location is the feature that we are interested in and that made a category shift from the relationship category to the name category. We then constructed an iconic mathematical representation of our observation sentence. For example, © < ℗ means that the location of the car is to the left of the location of the pond.

Weeks 9 & 10. Two of the 20 items selected collectively by the students to produce their *class trip line* from the school to the university were a 35 mph road sign and a public bus station. The 35 mph sign (y) is more than seven stops to the left of the bus station (x). When I move seven stops to the left of the bus station, I then continue to move to the left ($y < x + ^{-}7$). This is an excellent example because it shows students that they cannot directly translate the sentence into symbols. Otherwise, the mathematical sentence would be: $y > x + {}^-7$, which can be written in substitution of "The 35 mph sign is more than seven stops to the left of the bus station." In other words, one would incorrectly reverse the inequality sign. The correct representation is "The location of the 35 mph sign is to the left of $x + {}^-7$." In symbols, an equivalent expression is: $© < ℗ + {}^-7$.

Example of Students' Use of Symbols

During the first three months of the intervention, students revisited adding and subtracting integers using the AP curriculum process. In this section, I show the symbols AP students created to make sense of the mathematics they were learning. To begin, we use the following symbols to represent integers 3, 5, and 8: Let $© = 3$, $n^{R/L} = 5$ and $℗ = 8$ (three and eight are location numbers; five is a vector). For the case of addition, students are expected to create symbols that represent the instruction: "Start at the location of C, move a movement of n stops to the right or left, and arrive at the location of P." Similarly, for subtraction, we have "The location of P compared to the location of C is the movement vector from C to P." A group of students created the following symbols to represent both operations:

$$LC\ M \pm S\ F\ LP \qquad\qquad LP\ Y\ LC\ D \pm A$$

The letter L indicates that C and P are locations. The letter M indicates addition and the letter Y indicates subtraction. Table 7.2 illustrates symbols created by students and those used by the mathematics community worldwide. Several students demonstrated difficulty understanding the idea that location was the feature we were interested in studying. To develop deeper understanding, I borrowed some of the symbols the AP textbook uses. The symbols I prefer to use when teaching students are:

$$© ↑ n^{R/L} ↘ ℗ \qquad\qquad ℗° / ° © ↘ n^{R/L}$$

TABLE 7.2. Sequence of Steps for Decoding Addition and Subtraction Symbols

Addition	Subtraction
LC M ± N A LP	LP Y LC I ±N
$© ↑ n^{R/L} ↘ ℗$	$℗° / ° © ↘ n^{R/L}$
$X_1 + \Delta X = X_2$	$X_2 - X_1 = \Delta X$
$3 + 5^R = 8$	$8 - 3 = 5^R$
$3 ++5 = 8$	$8 - 3 = +5$
$3 + 5 = 8$	$8 - 3 = 5$

TABLE 7.3.   Three Types of Features: Objects, Actions, and Relations

Object/Action	Student Symbol	Class Symbol	Mathematics Symbol
Move	M	↑	+
Location of Car	LC	©	X_1
Location of Pond	LP	℗	X2
Arrive at or Is	A or D	↘	=
Compare	Y	° / °	–
Movement to R/L	±S or ±A	n R/L	ΔX

The circle denotes location, so C and P are just the names of two locations. Mathematicians have agreed to use the following symbols:

$$X_1 + \Delta X = X_2 \qquad\qquad X_2 - X_1 = \Delta X$$

Table 7.2 shows a sequence of steps a student and I developed when helping another student understand the process of creating symbols.

Table 7.3 shows the symbols created by students and the symbols used in the mathematics community worldwide. Again, several students experienced difficulty understanding the idea that location was the feature we were interested in studying. To overcome this obstacle, the AP curriculum emphasizes distinguishing between mathematical objects, actions, and relations.

An example of a diagram representing the subtraction 8 – 3 on the number line and drawn on a whiteboard by a student who was preparing for a class presentation is shown in Figure 7.2.

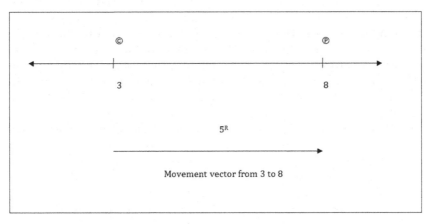

FIGURE 7.2.   Example representing 8 – 3 on the number line.

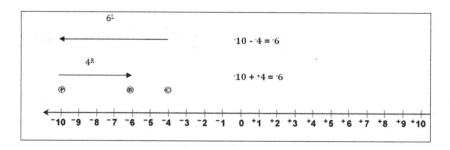

FIGURE 7.3.   Movement vectors used in addition and subtraction of integers.

Figure 7.3 shows how to draw vectors on the number line to represent n $^{R/L}$. Notice that this is different from the traditional method in which moving to the left represents subtraction. Integers ⁻10 and ⁻6 are location numbers in the addition problem, whereas ⁻10 and ⁻4 are location numbers in the subtraction problem. The movement vectors, n $^{R/L}$, are ⁺4 and ⁻6 for the addition and subtraction problems, respectively. The symbols ℗, ®, and © are to identify three different locations.

POST-ASSESSMENT OF STUDENTS' UNDERSTANDING

One measurable outcome of the intervention is the success of the 20 students who enrolled in the Miami cohort. Five of the 17 who started at FCAT Level 1 moved up to Level 2 after one year. Additionally, one of the three students who began at Level 2 moved to Level 3 one year after the intervention. This student could have requested to go back to taking the regular 45-minute mathematics class each day but decided to stay in the AP program and continue with the 90-minute daily mathematics classes, which shows incredible commitment to the AP program. This in itself is a positive outcome because the student revealed to the administrators of the program that she had already "learned more math in three months with AP than in [her] whole life," revealing her untapped brilliance in mathematics. To make as close of an experimental comparison as possible, CUEI obtained records of 20 students in a comparison group with similar characteristics such as race, gender, eligibility for free or reduced lunch, and limited English proficiency. Only one student in the comparison group had moved up an FCAT level after the same period compared to the six students in the AP cohort. Most notably, 18 of the 20 (90%) students in the Miami cohort passed the state examination and graduated from high school.

As part of the AP curriculum, students were frequently exposed to experiences that focused on developing their interest in attending college in the future. In September 2007, they visited an undergraduate class of pre-service teachers at FIU. The professor, who taught the course, Louis Rodriguez, was involved in researching how AP students developed as an effect of being in the AP cohort. Specifically, he was interested in documenting how students' perspectives toward mathematics changed and how their goals formed as they gained more exposure to college campus life. The course was part of FIU College of Education's teacher preparation program. The AP students were asked to discuss, with the pre-service teachers, their expectations of school teachers. Such candid conversations help beginning teachers understand the importance of developing positive relationships with underserved students in order to facilitate optimal learning environments.

After introducing themselves and describing the AP program, one girl explained how she had come to understand the meaning of $X = X_2 - X_1$. She described the whole process of transitioning from *people talk* to *feature talk*. She emphasized that the proper way to describe the distance between two objects on the number line was to acknowledge their locations by choosing one location as the reference location. She also explained to the pre-service teachers how icons are created by the students and how the icons then morph into symbols. She spoke with confidence and articulated the algebra concepts with mathematical precision, demonstrating complete understanding of the topics learned during the summer institute and revealing her mathematical brilliance.

As a result of the AP's success with the Miami cohort, the directors of CUEI asked two students to help them present the program at a national conference by describing success stories of the AP cohort in Miami. This was the students' first time on an airplane and they returned to tell their classmates how much fun they had on the airplane. Such experiences provide the foundation for college attainment and lifelong learning. Eight of the 20 students enrolled in Miami cohort went to college after high school graduation. Unfortunately, as of May 2012, only four were still enrolled in college. The remaining four students still desired higher education but were impeded by their immigration or economic status. It is important to note that, in addition to following the AP curriculum process, the overall success of these students can be attributed to the involvement of their parents, counseling sessions during the summer, and their participation in YPP after school—all of which are part of AP's quality education structure.

A radical and more difficult outcome to measure, qualitatively, was that students had been empowered and learned, within the first year, to demand their civil right to a high-quality education. This was clearly demonstrated when, for no apparent reason, the AP was in danger of being shut down because of a change in leadership at the school and district levels. At the beginning of the third year of the intervention, we learned that the principal and the school superintendent had been replaced. Students in the AP program went to a school board meeting to advocate for continuation of the program as well as to declare its success.

CONCLUSION

I can infer from my analysis that the AP Miami students' conceptualizations of integer addition and subtraction were deeply connected to their physical experiences, thereby minimizing memorization. Furthermore the students' collective attitude toward the curriculum processes of AP was positive. For example, the students were able to tell a story of a shared experience and represent it with icons. Their representations looked like hieroglyphs, which peers learned to decode and read fluently with understanding. Additionally, my findings suggest that the students' ample practice with iconic representations of their experiences promotes the formation of abstractions. Many students came to understand the meaning of $\Delta X = X_2 - X_1$ by the end of the first three months of the AP intervention and were able to manipulate the variables to translate a subtraction problem into an addition problem $(X_1 + \Delta X = X_2)$.

However, my analysis of student responses on how they interpreted right and left given in interviews and surveys showed that verbalizing the relationships between the locations of different objects was still difficult after the first three months. One of the main cognitive difficulties experienced by the students was learning a new method to replace a series of rules they had previously memorized without understanding. Nevertheless, most students enjoyed role-playing the mathematician and creating their own icons and symbols.

Finally, several students believed the AP was instrumental in their decision to go to college. The fact that eight students in the AP cohort described in this chapter enrolled in college after graduating from high school is another impressive outcome. The Algebra Project allowed the mathematical brilliance of students enrolled in the Miami cohort to emerge. Likewise, YPP is critical to empowering marginalized students to use mathematics for social justice. Furthermore, students' self-confidence in their ability to do abstract mathematics grew as a result of AP. Innovative curricula like AP are needed to increase the participation of Black students in mathematics, expand their opportunities to learn, and expose their mathematical brilliance. Furthermore, another major success of AP Miami was the organization of young people in Miami. The formation of a YPP site in Miami is a clear example of a sustainable outcome. Thus, I conclude that programs like the Algebra Project and the Young People's Project are needed to provide unique learning opportunities for Black students to learn and engage in advanced mathematics.

REFERENCES

Bigelow, W. (1990). Inside the classroom: Social vision and critical pedagogy. *Teachers College Press, 91*(3), 437–448.

Cobb, P., Gravemeijer, K., Yackel, E., McClain, K., & Whitenack, J. (1997). Mathematizing and symbolizing: The emergence of chains of signification in one first grade classroom. In D. Kirshner & J. A. Whitson (Eds.), *Situated cognition: Social semiotic, and psychological perspectives* (pp. 151–233). Hillsdale, NJ: Erlbaum.

Glaser, B. G. & Strauss, A. L. (1967). *The discovery of grounded theory: Strategies for qualitative research.* Mill Valley, CA: Sociology Press.

Leonard, J. (2008). *Culturally specific pedagogy in the mathematics classroom: Strategies for teachers and students.* New York, NY: Routledge.

Martin, D. B., Gholson, M. L., & Leonard, J. (2010). Mathematics as gatekeeper: Power and privilege in the production of knowledge. *Journal of Urban Mathematics Education, 3*(2), 12–24.

Moschkovich, J. & Brenner, M. E. (2000). Using a naturalistic lens on mathematics and science cognition and learning. In A. E. Kelly & R. Lesh (Eds.), *Research design in mathematics and science education* (pp. 457–486). Mahwah, NJ: Erlbaum.

Moses, R. P. & Cobb, C. E. (2001). *Radical equations: Civil rights from Mississippi to the Algebra Project.* Boston, MA: Beacon Press.

Park, D., O'Brien, G., Eraso, M., & McClintock, E. (2002). A scooter inquiry: An integrated science, mathematics, and technology activity. *Science Activities, 39*(3), 27–32.

von Glasserfeld, E. (1987). Learning as constructive activity. In C. Janvier (Ed.), *Problems of representation in the teaching and learning of mathematics* (pp. 3–17). Hillsdale, NJ: Lawrence Erlbaum.

Wynne, J. T. & Giles, J. (2010). Stories of collaboration and research within an Algebra Project context. In T. Perry, R. P. Moses, J. T. Wynne, E. Cortes Jr., & L. Delpit (Eds.), *Quality education as a constitutional right: Creating a grassroots movement to transform public schools* (pp. 146–166). Boston, MA: Beacon Press.

CHAPTER 8

THE CULTURE OF LEARNING ENVIRONMENTS

Black Student Engagement and Cognition in Math

Jamie M. Bracey

INTRODUCTION

There is tremendous urgency in the United States to increase the nation's production of American-born citizens with talent and expertise in the fields of science, technology, engineering, and mathematics (STEM). Fewer U.S. students are selecting math-intensive disciplines (National Science Foundation, 2004), and changes to pedagogy and curriculum have not yielded an increase in the number or diversity of students entering the quantitative disciplines (Jolly, Campbell, & Perlman, 2004). Although President Obama and his cabinet have made STEM an education priority since 2008, an increase in the opportunities for Black children and other minorities to enter the STEM profession is hampered by subpar retention rates of minority students in high school and college courses related to STEM. The nation's

The Brilliance of Black Children in Mathematics:
Beyond the Numbers and Toward New Discourse, pages 171–194.
Copyright © 2013 by Information Age Publishing
All rights of reproduction in any form reserved.

traditional adherence to behaviorist–reductionist teaching and learning models has not increased U.S. students' achievement in STEM domains. Jacob (1997) succinctly noted that traditional education research falls into two camps, each of which has failed to fix the continued decline in academic achievement of students in the United States. This chapter demonstrates that a culturally relevant pedagogical approach positively affects engagement, motivation, and mathematics performance for poor, underperforming middle school Black children.

RELEVANCE OF LEARNING ENVIRONMENT TO BLACK STUDENT ENGAGEMENT

Studies suggest that the level of engagement and motivation depends on the socio-emotional orientation of students during the teaching and learning process (Jarvela, Lehtinen, & Salonen, 2000). Adults help mediate that orientation through their behaviors, attitudes, and communication with students. For example, analysis of learning preferences for both African American and Latino students suggests a heightened desire for strong relationships with culturally competent teachers in the learning environment (Ladson-Billings, 1995a; Martin, 2006; Portes & Rumbaut, 2001).

There is little argument that culture is a complex, multifaceted construct that demands a richer, multidisciplinary examination of the social tools, behaviors and symbolic boundaries used by different ethnic groups in response to their macro environment (Lewin, 1946). Equally important is an understanding of the culture of institutions and environments that Black students experience for hours every day, and the impact on academic engagement. Balfanz and Legters (2004) outlined the differences in engagement, motivation, and achievement in a comparison of high school dropout rates across the United States. They found that only 11% of White students attend high schools where graduation is not the norm, compared with a staggering 46% of African American and 39% of Latino students who attend schools defined by a culture of poor performance (Balfanz & Legters, 2004).

Academic disengagement and poor achievement is disproportionately high in urban classrooms across the nation (Conchas, 2001; Ladson-Billings, 1997). As a result, cultural and linguistic minorities are disproportionately represented in learning environments that focus on low-level skills, discipline, and remediation (Ladson-Billings, 2006). Although minority students have improved on national mathematics standards, those improvements have been primarily in basic math skills (Martin, 2000). As the topics become more complex, disparity becomes more evident (e.g. fewer students of color in advanced mathematics classes; Strutchens & Silver, 2000).

The power of what happens in the classroom cannot be understated. The deep historical context of the African American community's demand for quality education (Bush, 2004; Hilliard, 1992) and more recent attention on the needs of Latino youth (Suarez-Orozco, Pimentel, & Martin, 2009), undergird the possibility that persistent underachievement may be attributed to instructional leadership that focuses on safety, discipline, and remediation to address national mandates, at the expense of teacher confidence in using instructional innovation to scaffold conceptual understanding in more challenging domains (Rittle-Johnson, Siegler, & Alibali, 2001).

A culture of prescriptive academic remediation is the hallmark of high-poverty, low-resourced schools, which do not offer the type of creative, inquiry-based learning foundational to STEM achievement. Thus, it can be argued that school cultures that serve large populations of Black children direct very few resources to advanced science, technology, engineering and mathematics learning required for the knowledge economy. The ironic rationale may be that minority student achievement does not appear to warrant doing so, completing a vicious cycle of denied opportunity. The net result is a lack of opportunity to engage Black children beyond prescriptive remediation to pass annual yearly performance (AYP) mandates.

The study described in this chapter was designed to aggressively counter policies and practices that continue to devalue Black children's participation in the *hard* subjects. To demonstrate the remarkable capacity of children to exhibit brilliance when given the opportunity, Black children from very low-performing schools were provided an opportunity to learn robotics in one of two different learning environments on a college campus. The first environment introduced robotics as a team-based project with engineering and traditional didactic mathematics instruction. The second learning environment was enriched with elements from the social culture of Black and Latino students, including high verve communal learning (Boykins et al., 2006), supportive adult relationships through expert mentoring (Green, Rhodes, Hirsch, Suarez-Orozco, Camic, 2008), and the use of cognitive apprenticeship instructional strategies (Collins, Brown, & Newman, 1989) designed to increase identity formation as early engineers with a reason to use mathematics.

ELEMENTS OF CULTURALLY RESPONSIVE PEDAGOGY

A unique feature of this study was the identification of cultural elements that may be unique to Black students' repertoires of practice (Gutierrez & Rogoff, 2003). The expectation was that there would be a difference in before and after perception of minority student engagement and motivation based on perception of a more culturally relevant learning environment.

Culturally relevant and culturally responsive are synonymous terms for instructional techniques that encourage minority student engagement using culture as a process, a resource, and a tool in the classroom (Ladson-Billings, 1995b; Leonard, Davis & Sidler, 2005; Leonard & Evans, 2008).

Proponents of culturally relevant instruction suggest that it is highly effective because it taps into the natal culture students experience in their natural (home) environment.

Ladson-Billings (1997) provided three pedagogical orientations that are congruent between home and school and contributed to success in teaching cultural and linguistic minorities: *concept of self and others, importance of social relations is very high,* and *methods to acquire knowledge is balanced.* Addressing cultural congruence in the learning environment, Leonard (2008) presented specific strategies in *Culturally Specific Pedagogy in the Mathematics Classroom* that teachers could use to infuse multicultural literature, group work, and historical references of cultural and linguistic minorities into lesson plans for K–6 students. This study bridged both culturally relevant instructional paths by focusing on the importance of social relations for African descended children. Variables for the treatment group included the involvement of a cohort of African American mentors, the use of high energy "vervistic" activities, and communal learning practices (Tyler, Boykin, Miller, & Hurley, 2006).

CONCEPTUAL FRAMEWORK

The study described in this chapter was grounded in three commonly recognized theoretical models: socio-cultural learning theory (Crawford, 1996; Hay & Barab, 2001; Wertsch, 1985), culturally relevant pedagogy (CRP) (Ladson-Billing, 1995b; Leonard & Evans, 2008), and the construct view of engagement, (Chapman, 2003). These theoretical perspectives are complementary in that each considers the social and cultural context of the learning environment as critical for both teaching and learning.

Elements of the theoretical models were integrated in a mixed-methods quasi-experimental design to examine whether minority middle school students enrolled in a STEM learning environment responded favorably to socio-cultural elements by showing higher levels of engagement and achievement in mathematics than those in a STEM environment without those elements.

Specific research questions this study sought to answer were as follows:

1. How would differences in program instructional formats affect minority middle school students' level of engagement after participating in a STEM-enriched learning environment?
2. Would there be a significant difference on motivation to persist in STEM based on the use of culturally relevant pedagogy and cognitive apprenticeship instructional methods?
3. How would the different treatments affect minority student math proficiency?

For purposes of this study, the dependent variables were changes in measures of engagement, motivation, and mathematics performance. The prediction was that the group of minority students who experienced CRP in a STEM environment would show the greatest change in before and after levels of engagement, motivation and math proficiency compared with the STEM program group without the CRP exposure.

AFRICAN AMERICAN SOCIO-CULTURAL COGNITIVE ORIENTATION

It should be noted that there is a difference of opinion about using the category *African American* to describe Black students. However, for this study it was important to establish a "universal" motif to frame the participation of Black students under one cultural umbrella. Cokley (2003) appropriately argued that using a race-homogenous approach to studying groups masks the geographic and cultural diversity within groups. For example African (Ghanian, Nigerian) and West Indian (Trinidadian and Jamaican) immigrants academically outperform indigenous, native-born Americans of African descent. Studies rarely parse out this information regarding within group variation in achievement, instead lumping all Black students together on the basis of skin color. Fortunately, because of the research of Senegalese genius Cheik Anta Diop, evidence exists to support the existence of shared cultural motifs.

Diop (1974) successfully argued at the 1974 United Nations Educational, Scientific and Cultural Organization's (UNESCO) conference that significant cultural elements exist that inform universal Black child and adolescent development in spite of the social upheaval of colonization, war, and geographic dispersion during the past 5,000 years. Those elements include a shared cosmology (spirituality), extended family social organization, matriarchy, and rhythmic linguistic patterns.

Contemporary African and African American psychologists have also contributed to increased knowledge about the socio-cultural lens through which children of African descent are socialized and prepared for identity and cognitive development (Ellison, Boykin, Tyler, & Dillihunt, 2005; Serpell, Boykin, Madhere, & Nasim, 2006; Tyler et al., 2006). Specific socio-cultural baselines transmitted from one generation to the next help influence how students cognitively receive, attend and develop knowledge schema.

Explicating the African cultural context yielded learning-related constructs such as the following:

- High-energy, sustained interaction with others
- Majority of learning in communal, group settings
- Gender role balance encouraged
- Extended family and group unity (communal relatedness)

- Hands-on manipulation of materials to solve new problems
- Goal orientation toward team achievement (competition)

Within the learning environment designed for this study, socio-cultural elements that cognitively advantage African American children included high-energy interaction (robotics competition), use of analogies by expert mentors to communicate key concepts, gender equality (equal assignment of engineering roles), and group unity (team orientation) toward preparing for the robotics competition.

CULTURALLY RELEVANT PEDAGOGY AND MATHEMATICS

A significant body of research is growing that connects culturally derived assets (communal and interpersonal relationships) to student achievement in STEM, particularly mathematics. Serpell, Boykin, Madhere, and Nasim (2006) conducted a race comparison study that explored the effect of communalism on transfer, the golden fleece of educational attainment. Participants were fourth-grade students (N = 162; 90 African American, 72 Caucasian) drawn from urban public schools with more than 80% free lunch usage and 20% or more non-Hispanic White students. All of the students were enrolled in general education courses.

Students were assigned to one of 40 three-person groups (20 African American, 20 Caucasian), balanced by gender, and a multiethnic control group of 42 students. The groups were randomly assigned to a communalism group that used a physical apparatus to solve a problem or a communal group that used computer simulations to solve the same problem. The hypotheses were that ethnicity and cultural learning context would have significant interaction on solving the problem and that the use of a tool would facilitate stronger group participation (physics apparatus) or deeper individualism (one-on-one interaction with the computer). No significant effects were found for learning condition versus ethnicity, but the dependent variables—initial learning and transfer—were significantly affected by learning condition (Serpell et al., 2006).

Students who worked communally with a physical device to help them solve problems scored better on transfer scores than either the communal computer users or the control group. African American students scored significantly higher than Caucasian students in either communal setting (mean [M] = 2.12, standard deviation [SD] = 2.15 vs. M = 1.10, SD = 1.38). The findings in this study suggest that African American children have the requisite cultural skills to negotiate a learning context that may have seemed alien to the Caucasian children.

Again, this finding is not to suggest that children can learn in only one context. That conclusion would dramatically oversimplify the case. However, from the historical and anthropological background presented earlier, it is clear that African American children share a cognitive orientation developed in large extended social networks that depend on strong interpersonal relationship and interactive communication motifs.

According to Kitayama, Duffy, Kawamura and Larsen (2003), culturally divergent thinking has been examined by using a number of measurements with the generally accepted conjecture that culturally influenced strategies for cognitive processing are differentially advantageous for tasks attended to in that person's normal cultural context. That position is supported by case studies that examined the impact of using the culture of indigenous Yup'ik Indian craftsmanship to teach mathematics to Yup'ik children. Lipka, Sharp, Brenner, Yanez, and Sharp (2005) found that a culturally competent teacher steeped in the traditions of the Yup'ik was not only able to serve as the expert mentor to her class, but used her knowledge of both culture and mathematics to engage students as novices as cognitive apprentices in the creation of indigenous crafts. The authors found that the students were more attentive and outperformed both a second treatment group and a control group in recognizing symmetry, patterns and shape congruence—elements of mathematics associated with the cultural task.

If minority students attend to information shared in contexts that promote strong interpersonal relationships, highly interactive information exchange (like the call and response practice particular to people of African descent) and scaffolding that builds on or introduces new information, then it made sense to hypothesize that the minority students in this study would benefit from a STEM enriched learning environment that incorporated their preferred cultural and cognitive orientations.

CONSTRUCT OVERVIEW OF ENGAGEMENT

An *engaged student* is generally described as a student with a positive attitude who actively participates in school-based educational and enrichment activities (Chapman, 2003). Although there is no consensus among researchers on the definition of *engagement*, research suggests that the construct represents many interrelated factors that should be considered in designing research to study engagement and mathematics achievement (Singh, Granville, & Dika, 2002).

Fredricks, Blumenfeld, and Paris (2004) provided a meta-analysis of engagement research and recommended that research on engagement view the variable as a meta-construct consisting of:

- *Affective engagement*—the emotional reaction to the learning context, including the people and academic experiences within that environment;

- *Behavioral engagement*—the level of participation in the social and extracurricular activities offered by the school or learning institution; and
- *Cognitive engagement*—students' willingness to invest the time and attention needed to master complex concepts.

When it comes to learning it is believed that: 1) engagement overlaps motivation but is not the same, 2) that both are positively correlated to achievement, and 3) student knowledge and value of the role mathematics plays in future career opportunities could increase motivation to persist at the middle grade level (Singh et al., 2002). Although a significant body of work has explored students' intrinsic motivation in the context of performance and mastery achievement goals (Dweck & Leggett, 1988), emerging research suggests that high levels of intrinsic motivation do not necessarily correlate to minority student academic self-concept or performance outcomes.

Cokley (2003) effectively outlined empirical evidence of a strong relationship between positive racial identity and achievement motivation (Fordham & Ogbu, 1986; Graham, 1997). Cokley also argued that external forces positively mediate African American student motivation; that is, African American students (particularly male students) are more likely to be extrinsically motivated by their available social networks. This finding is important given that feelings of relatedness and belonging are dimensions that affect how ethnic minority adolescents navigate and negotiate social constraints in an effort to form socially valued identities (French, Seidman, Allen, & Aber, 2006). The paradox of high self-concept, self-esteem, and poor academic achievement is not so perplexing if research considers motivation as the intersection of student engagement, the perception of belongingness and the quality of the relationships in the classroom. Researchers who have explored the culture of cognition suggest that this combination of psychological traits influences students' desire to persist (Boykin, 2002; Cokley, 2002; Green et al., 2008).

RESEARCH DESIGN

Although it is common in educational research to use previous research as a basis for protocol design, the tremendous lack of empirical evidence available examining African American and Latino/a middle school engagement in STEM required the creation of a STEM-based learning environment from which to test this study's hypotheses. To do so, the author applied for and received funds to create a STEM program to serve sixth and seventh grade middle school students from 29 of the lowest achieving middle schools in a large school district in the northeast United States. Two separate groups were formed from the district's recruiting efforts. More than 4,000 students were eligible to participate on a first-come, first-serve basis.

The first group of 50 students who completed the application was assigned to participate in a Robotics Saturday Academy with STEM expert mentors for 12 consecutive weeks. The second group of 50 students was assigned to participate in a four-week Robotics camp of 20 consecutive weekdays without expert mentors. Each group of students was from a different subset of the 29 schools; that is, none of the middle school students in Group 1 was from a school attended by students in Group 2 and vice versa.

Both the weekend and weekday formats provided mathematics support. The Robotics Saturday Academy provided mathematics tutoring 30–45 minutes per week in arithmetic and pre-algebra. It also relied on experts to demonstrate how to use arithmetic (measuring) and algebra (calculating displacement) as students engineered their underwater robots. The summer camp provided students with state standards based mathematics instruction each day for 30–45 minutes. Both groups used the same robotics' engineering curriculum, had the same dosage of STEM activities, and instructional support for the same mathematics concepts. Each program also used a district-approved mathematics gaming program, *First in Math*, to support remediation.

Participant Sample

The nature of the grant required that the study offer open recruitment to a combined cohort of 4,774 sixth and seventh grade students from working-class backgrounds across 29 middle schools. The target schools had failed to achieve adequate yearly progress (AYP) goals under the *No Child Left Behind* (2001) academic achievement guidelines and were in their second year of Corrective Action Level II as of the 2008–2009 school year. Each middle school was also designated a Title 1 school under the federal poverty guidelines, with more than 85% of the student population eligible to receive free or reduced school lunch. Table 8.1 provides information on the demographic characteristics of the program participants for each cohort of the STEM robotics program.

Group 1: The Math and Robotics Saturday Academy

For the first phase of the program (STEM with culturally relevant pedagogy), 147 students completed applications for 60 available slots. Half of the applications (73) were randomly selected; of that number, only 59 students and parents attended the mandatory orientation session.

Learning Environment. The STEM program was situated on a university campus to remove students from learning environments that might be associated with failure or frustration. Technology-rich classrooms and com-

TABLE 8.1. Demographic Characteristics across Program Groups

	Group 1 (STEM w/CRP) N = 59 n (%)	Group 2 (STEM without CRP) N = 54 n (%)
Ethnicity		
African American	41 (69.5%)	38 (69.1%)
Latino/Hispanic American	17 (28.8%)	8 (14.5%)
Asian	0 (0%)	4 (7%)
Biracial	0 (0%)	2 (4%)
White	0 (0%)	1 (2%)
Missing (no ethnicity noted)	1 (2%)	3 (5%)
Gender		
Boys	30 (50.8%)	34 (63%)
Girls	29 (49.2%)	20 (37%)
Grade		
Sixth	24 (40.7%)	33 (61.1%)
Seventh	35 (59.3%)	20 (35.2%)
Missing	0	1 (3.7%)
English spoken at home		
Yes	47 (79.7%)	47 (85.5%)
No	10 (16.9%)	3 (5.5%)
Missing	2 (3.4%)	4 (7.4%)

Note: STEM = science, technology, engineering, math; CRP = culturally relevant pedagogy.

puter laboratories were secured on campus in a location easily accessible by public transportation. Robotics kits, materials, and curricula outlines were provided by the school district's Boosting Engineering Science and Technology (BEST) program. Certified robotics teachers were hired to provide instruction and hands-on coaching to help the participants build robots that would function as underwater ROVs.

Program Structure and Staff. The first treatment consisted of ten three-hour mathematics and robotics sessions every Saturday, equal to 30 total hours of STEM exposure over two and a half months. The program culminated in the option to participate in an international ROV competition on the 12th consecutive Saturday. Three certified robotics instructors from the district's robotics office provided weekly instruction. Ten older adults (age ≥ 50), each with a minimum of 15 years of experience in a STEM discipline (engineering, technology, computer programming, physics, finance, or mathematics) participated every week for two hours as expert mentors.

The majority of the STEM mentors (80%) were African American and 20% were Caucasian; two were bilingual. The daily site manager for both treatment groups was a bilingual Latina, bringing to five the total number of staff and volunteers able to provide immediate language support. Eleven mathematics and science undergraduates from the university's pre-service teacher development program served as the primary mathematics tutors and robotics team near peer mentors. For the first treatment, all program staff ($N= 24$) participated in a group orientation about the goals of the program. They also received training in the robotics platform to be used and were presented with the pedagogical research cited on this study (Leonard, 2008).

Program staff members were asked to promote a learning community that valued social relationships through respect, open communication, and exchange of ideas across the teams. Instructors and mentors introduced new vocabulary in dynamic exchanges with students while engaging students in the activity related to that vocabulary. The practice was consistent with Mayfield and Reynolds' (1997) neural cognitive study that indicated Black children have a high word memory and sequence recall, suggesting the *call-and-response* method of teaching works best for children who attend to highly interactive modes of communication.

The expert mentors transmitted the heuristics (tricks of the trade) of working in a design-and-build shop, including demonstrating soldering, design sketching, prototype development, and short cuts in online research as examples of project-based activities the expert mentors would perform in completing the tasks. Bilingual accommodation was provided in several ways. The site coordinator for both sessions was a woman of Puerto Rican heritage fluent in English, Spanish, and several patois languages. Three of the pre-service teachers were also fluent in English and Spanish; two of the 11 pre-service undergraduates were bilingual and of Latino descent. The communal nature of the learning environment was reflected in the students' unanimous decision to travel to the international competition as one team, rather than five separate teams.

Group 2: The Math and Robotics Summer Camp

A comparison group of students also participated in a STEM-based program that used the same robotics platform as Group 1. A total of 64 completed applications were returned to the researcher by the cutoff date, which affected randomization. The summer camp program staff contacted each student, and 58 students agreed to participate in the summer program. However, only 54 actually participated.

Learning Environment Context. The program location, campus resources and access to materials were identical to the context provided for Group 1.

Program Structure and Staff. The STEM program format for Group 2 also consisted of 30 hours of STEM exposure using the same underwater robotics instruction provided to Group 1. Although the total number of hours for each program was identical, Group 2 participated for three hours per day for 10 days. The mentor experts did not participate; thus there was no support of the culturally responsive or cognitive apprenticeship model. The staffing pattern for Group 2 consisted of 18 people: a program manager, two certified robotics instructors, 10 undergraduate classroom assistants, and three graduate-level instructors. The robotics instructors were ethnic minorities, as were four of the undergraduate students. Two-thirds of the staff was White. The staff for Group 2 also participated in a program orientation; however, they were not presented with the research regarding culturally relevant strategies to improve engagement and motivation.

After completion of the robotics phase, Group 2 instructors used traditional block schedules to organize mathematics, reading, and writing classrooms for the second half (10 days) of the summer camp. Academic information was delivered in traditional lecture format using multimedia support. Instead of preparing for a competition, participants completed poster presentations for display at a culminating event.

INSTRUMENTATION

Engagement and Motivation Scales

To measure engagement, the study used the Research Assessment Package for Schools (RAPS) (Institute for Research and Reform in Education, 1998). The RAPS version for middle school (RAPS-SM) is a self-report instrument that taps engagement (ongoing including reaction to challenge), beliefs about self (perceived competence in ability and strategy use), and experiences of interpersonal support (parents, teachers). The scale has 84 items ranging from *I can't do well in school* to *I can't get my teacher to like me.* Responses were recorded using a four-point Likert scale of *very true, sort of true, not very true,* and *not at all true.* Table 8.2 provides an outline of the domains and subdomains of the RAPS.

Student engagement and motivation to persist were also measured using the Assessment of Academic Self Concept and Motivation Scales (AASCM) (Gordon Rouse & Cashin, 2002). The scale includes 80 items based on a seven-point Likert scale, repeated under four conditions that cover students' responses to questions regarding social, cognitive, personal, and extracurricular domains, including student perception of belonging, self-efficacy, and value attributed to those domains. The AASCM has been validated across ethnic groups, gender, and academic achievement levels with published Cronbach alphas for the subscales ranging from .80 to .94 (Gordon Rouse & Cashin, 2002). In the current study, six items related to social

TABLE 8.2. Domains and Subdomains for the RAPS Scale

Domain	Subdomain	Definition
Engagement	Ongoing engagement	Student adjustment to school culture
	Reaction to challenges	Coping strategies used when faced with stressful or new or novel events during learning
Beliefs about self	Perceived competence	Student knowledge about ability to use strategies to achieve academic goals
	Perceived autonomy	Student belief he or she has personal authority and resources to achieve
	Perceived relatedness	Emotional security with self and others in the learning community
Interpersonal support	Parental support	Perception of parent/caregiver involvement including structure and value for education
	Teacher support	Perception that teacher cares, provides structure and is fair

(affective) and cognitive engagement in the math domain were included in an online post-assessment survey.

The Likert scale was reduced from seven points to a four-point scale of 4 = *very true*, 3 = *sort of true*, 2 = *not very true*, and 1 = *not at all true* to increase students' likelihood of completing the survey. Reliability testing in

TABLE 8.3. Before and After Test Items from the AASCM Scale

AASCM question	Subdomains	Item No.
How important is it to you to understand the math in your class this year?	Affective engagement, cognitive engagement	4
How important is it to you to finish your math homework?	Affective engagement, cognitive engagement	5
How important is it to you to make good grades in math?	Affective engagement, cognitive engagement	6
How do you feel about your ability to understand the math taught in your classes?	Affective engagement, self-concept (perception of competence)	7
How do you feel abut your ability to learn math and science?	Affective engagement, self-concept (perception of competence)	8
Do you feel participating in the (treatment) group will increase your willingness to perform better in math?	Affective engagement, self-concept (perception of competence)	9

this study yielded a slightly diminished but very acceptable alpha of .78 for measuring engagement in the math domain.

In the post-test phase Questions 8 and 9 were modified to past tense; for example, *How do you feel about your ability to learn math after participating in the (treatment) group?* and *Do you feel being in the (treatment) group increased your willingness to perform better in math?* Only students who completed the post-test were included in the pre–post analysis of engagement and motivation using the AASCM.

The earliest available standardized math benchmark scores were collected from the district for each participant who completed the AASCM. Group 1 scores were collected in late June, six weeks after the conclusion of the program. Group 2 scores were collected in late October, eight weeks after the conclusion of the program.

DATA COLLECTION

Each group (STEM with Culturally Relevant Pedagogy, $n = 59$; STEM without Culturally Relevant Pedagogy, $n = 54$) completed the 84-item RAPS engagement survey during the pre-program orientation session. Because the RAPS scale is a more general measure of engagement, all student records for the RAPS have been included in the threshold analysis.

Analysis Strategy

Klem and Connell (2004) used threshold analysis to identify levels of engagement most likely to predict positive and negative school outcomes. Threshold analysis is an analytical strategy that moves away from comparing group means and instead focuses on understanding where people score in relation to an established standard. Using the RAPS as a threshold indicator is supported by research that identifies engaged youth as those having positive "resources and experiences" (Gambone, Klem, & Connell, 2002). The same research defines struggling youth as those with "liabilities and experiences" most likely to result in negative school outcomes. Gambone et al. (2002) established a mean score of 3.75 or higher on the RAPS as the threshold for optimal engagement for middle school students.

RESULTS

RAPS Survey

The total group mean for all items on the RAPS was 3.36, which means the students initially presented with a low-moderate initial level of engagement at the beginning of the study. Table 8.4 shows the items with the highest means for the entire sample, representing those items on which the participants are optimally engaged. Table 8.5 shows the items with the lowest

TABLE 8.4. Descriptive Data Establishing Optimal Means for School-Related Engagement

Variable	Item	Subdomain	Mean (n = 121)
24	I work really hard in school	Competence	3.92
84	How important is it to you to do the best you can in school	Ongoing engagement	3.89
70	I don't know what my parents expect of me in school	Interpersonal	3.87
20	My parents don't explain why school is important	Interpersonal expectations	3.84
33	My parents know just how well I can do in school	Interpersonal expectations	3.81
2	When I'm with my classmates I feel ignored	Relatedness emotional security	3.76
49	Trying hard is the best way to do well in school	Competence	3.75

means across the sample, representing those items on which participants were at risk for negative school outcomes.

It is interesting to note that the two groups, each from a different subset of underperforming schools, *selected 70% of the same items on the RAPS instrument as their top 10 and lowest 10 rated items.* Using those common items, means were computed and rank ordered from high to low for each group.

TABLE 8.5. Descriptive Data Establishing At-Risk Means for School-Related Engagement

Variable	Item	Subdomain	Mean (n = 121)
81	My teacher likes to be with me	Interpersonal	2.95
69	When I'm with my teacher I feel happy	Interpersonal	2.94
78	My teacher tries to control everything I do	Interpersonal	2.92
5	I do my homework because I like to do it	Ongoing engagement	2.63
51	My teacher isn't fair with me	Interpersonal expectations	2.61
72	I work on my class work because I'll feel guilty if I don't do it	Ongoing engagement	2.34
46	I do my homework because it's fun	Competence	2.32
16	I do my homework because I'll feel bad about myself if I don't do it	Ongoing engagement	2.28

TABLE 8.6. Means and Standard Deviations for the Two Groups on Math

	Pretest Mean (SD)	Posttest Mean (SD)
Culturally relevant	80.33 (13.82)	81.85 (13.77)
Not culturally relevant	84.40 (9.43)	67.18 (22.16)

Scores ranged from 1 = *not true at all* to 4 = *very true*. An independent samples *t*-test indicated Group 2 (STEM without CRP) had a higher overall total mean for engagement after the camp (*M* = 3.44) than Group1 (STEM with CRP; *M* = 3.30) after the Saturday academy. This was marginally significant at *t* = -2.021, *p* =.047.

Mathematics Achievement

The marginal statistical difference in engagement between the two groups after the camp could be attributed to a number of factors, including the school contexts the students entered in the fall. What is of particular note is the significant difference in the math post-test data. Data in Table 8.6 represent the average percentages achieved on standardized math benchmark tests assessed and reported for each participant by the school district.

A separate samples *t*-test indicated that the two groups did not differ at the pretest (*t* = 1.40, *p* = .165). Consequently, a repeated measures ANOVA was performed. The results of this analysis are contained in Table 8.7.

As shown in Table 8.7, there is a significant main effect for pre–post, and a significant interaction. The plot of this interaction is shown below in Figure 8.1.

Figure 8.1 shows that the STEM group with the culturally relevant pedagogy not only maintained performance eight weeks after the program ended, but increased slightly between the pretest and the posttest. The STEM group without this pedagogy significantly declined at the post-assessment stage, six weeks after the program ended. These are very encouraging results that warrant additional study.

TABLE 8.7. Results of Repeated Measures Analysis of Variance on Math

	df	MS	F	Significance	Partial η^2
Group	1	1017.165	2.847	.096	.039
Pre–Post	1	2231.336	15.285	.000	.177
Interaction	1	3175.172	21.751	.000	.235

Note: p < .05.

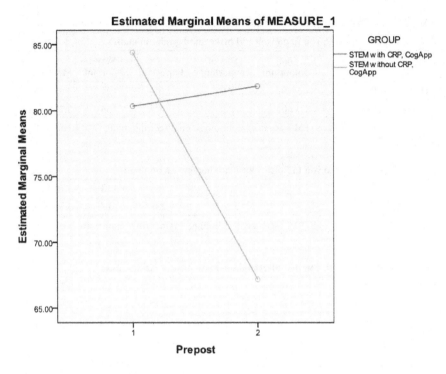

FIGURE 8.1.   Plot of the pre–post interaction of math by group and pedagogy

Analysis of Student Perceptions

Post-assessment qualitative data were collected the last full day of the program for each group. The purpose of these additional assessments was as follows: 1) to provide additional information on post-treatment engagement and motivation to persist, 2) to assess each group's perception of mathematics based on instructional format, and 3) to identify program areas with the strongest perceived value by group. The distribution of the responses for each group, and the means for each group for each of the questions in the post-assessment survey are shown in Table 8.8. An independent samples *t*-test was conducted for each question in Table 8.8, and no statistically significant differences were found when the posttest was compared for the two groups. However, Table 8.8 provides an important comparative analysis of the frequency distributions and means based on surveys of participants' attitudes toward mathematics after experiencing each instructional format of the STEM program.

Engagement. In Table 8.8A, behavioral engagement is measured as students' post assessment perception of the importance of doing math home-

TABLE 8.8.   Post-assessments of Math Engagement Based on Instructional Format

A: How important is it to you to do well and make good grades in math?

	Not important	Sort of important	Important	Very important	Mean (SD)
STEM with CRP	0	0	5	27	3.85 (.359)
STEM without CRP	0	2	8	35	3.77 (.515)

Note. t = 0.801, p = .092; STEM = science, technology, engineering, math; CRP = culturally relevant pedagogy.

B: How important is it to you to finish your math homework on time?

STEM with CRP	0	4	16	13	3.29 (.667)
STEM without CRP	0	4	23	17	3.36 (.705)

Note. t = –0.494, p = .076; STEM = science, technology, engineering, math; CRP = culturally relevant pedagogy.

C: How important is it to you to understand the math in your classes now?

STEM with CRP	0	1	10	23	3.71 (.458)
STEM without CRP	0	3	20	22	3.51 (.585)

Note. t = 1.705, p = .119; STEM = science, technology, engineering, math; CRP = culturally relevant pedagogy.

work. Both program groups indicated that they identify math homework as a priority, rating it 87.9% compared with 87.2% "important" to "very important."

In Tables 8.8B and 8.8C, students' attitudes toward understanding math are treated as a strong indicator of affective and cognitive engagement. In the post assessment, 69.7% of the minority middle school students in the STEM with CRP instructional format rated the need to understand math as "very important" compared with 55.3% who rated it "very important" in the STEM without culturally relevant pedagogy (CRP) group. In the post-assessment phase, each group rated their most recent math course as "important but not challenging" (STEM with CRP = 48.5% versus STEM without CRP = 54.2%).

Motivation to Persist. In the post-assessment, independent samples, *t*-test analysis reveals no statistically significant difference between the groups on motivation to persist. While there is no difference in their motivation, it is interesting to note that across the groups, both means are near or above the RAPS threshold of 3.75, suggesting students were positively engaged.

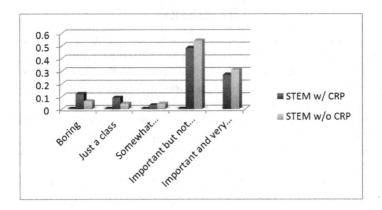

FIGURE 8.2. Do you feel the math you are learning in middle school is _____?

TABLE 8.9. Post-assessment Attitudes toward Math Academic Competence

How do you feel about your ability to understand math after participating in the program?				
	Not Sure	Anxious, Negative	Somewhat Positive	Very Positive	Mean (SD)
STEM with CRP	0	6	19	9	3.79 (.687)
STEM without CRP	0	9	26	13	3.73 (.821)

Note. (t = −1.608, p = .173). STEM = science, technology, engineering, math; CRP = culturally relevant pedagogy.

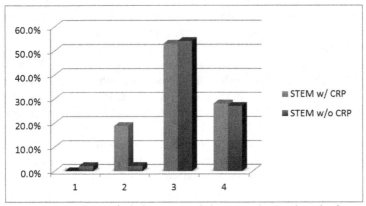

FIGURE 8.3. How do you feel about your ability to understand math after participating in the program?
Note. 1 = Not sure; 2 = anxious, nervous, 3 = somewhat more positive than before; 4 = more positive.

DISCUSSION

This study yielded important new information about the circumstances under which culturally relevant elements in a learning environment cognitively advantaged a largely Black population of middle school students. The overall increase in mathematics scores for students participating in the socially-culturally relevant weekends suggests that the treatment was directly responsible for the performance change. While encouraging, the findings on the mathematics scores should be interpreted with caution since it is not clear whether it was the strength of the social network that caused the change, or some other external factor.

However, the student responses in Table 8.5 indicate that the absence of strong interpersonal relationships decreases student engagement; thus it very possible that students in the Saturday academy maintained their performance levels because doing well was the socio-cultural expectation for the group. Those expectations were set by the eco-system of students, peer mentors, and adult experts, and the context of learning mathematics was socially and culturally relevant.

Conversely, the STEM group without culturally relevant pedagogy declined significantly on the next standardized measure of mathematics achievement, which took place eight weeks after the end of the summer program. It is difficult to determine if exposure to culturally relevant pedagogy (CRP) would have prevented the academic slide for the second group. The reality is that a growing body of research confirms Black children learn better under certain conditions, and their contemporary social culture includes elements that may still hold true as Diop (1974) reported nearly forty years ago. The mathematics performance reflected in poor children doing math as engineers suggests that creating an early sense of identity within a learning environment may motivate Black children to demonstrate brilliance in mathematics. These are very exciting avenues at the intersection of education and psychology.

One major finding is that these young students, most of them Black, understand and subscribe to the importance of mathematics achievement. A second finding is the importance of cultural pedagogy and mentoring. Clearly, these components were advantageous for the treatment group. Perhaps the most important finding, however, is the brilliance Black students demonstrated in both groups in regard to the products they produced in the robotics camps. Over-emphasis on mathematics achievement as evidenced by student performance on standardized tests often detracts from the purpose of STEM education—creation and innovation—the hallmarks of brilliance.

The implications of this study reveal that robotics can be used as a hook to attract and retain underrepresented students in STEM activities. Marginal significance on the RAPS for the STEM without CRP group in regard to engagement supports the idea that technology can be used to increase Black students' engagement and motivation to persist. Cultural pedagogy and mentoring, however, provided anchors that allowed participants in the STEM with CRP group to remember the mathematics concepts long after the robotics camps had ended. Thus, expanding opportunities to learn by drawing on social and cultural capital to build Black students' mathematics identity also allowed Black brilliance to shine. In conclusion, while the results should not be generalized beyond the confines of this study, Black brilliance manifested itself in both treatment and comparison groups - albeit in different ways. This study should be replicated using STEM with CRP and varying the learning environment to determine if differences in engagement and achievement are context specific.

REFERENCES

Balfanz, R. & Legters, N. (2004). *Locating the drop out crisis: Which high schools produce the nation's drop outs? Where are they located? Who attends them?* Technical report from the Center for Research on the Education of Students Placed at Risk (CRESPAR). Baltimore, MD: Johns Hopkins University Press.

Boykin, A. W. (2002). The effects of movement expressiveness in story content and learning context on the analogical reasoning performance of African-American children. *Journal of Negro Education, 70*(1–2), 72–81.

Boykin, A. W., Tyler, K. M., Watkins-Lewis, K. M., & Kizzie, K. (2006). Culture in the sanctioned classroom practices of elementary school teachers serving low-income African American students. *Journal of Education of Students Placed At-Risk, 11*, 161–173.

Bush, L. (2004). Access, school choice and independent black institutions: A historical perspective. *Journal of Black Studies, 34*(3), 386–401.

Chapman, E. (2003). Alternative approaches to assessing student engagement rates. *Practical Assessment, Research & Evaluation, 8*(13). Retrieved October 9, 2010, from http://PAREonline.net/getvn.asp?v=8&n=13

Cokley, K. O. (2002). Ethnicity, gender and academic self-concept: A preliminary examination of academic dis-identification and implications for psychologists. *Cultural Diversity and Ethnic Minority Psychology, 8*(4), 378–388.

Cokley, K. O. (2003). What do we know about the motivation of African American students? Challenging the "anti-intellectual" myth. *Harvard Educational Review, 73*(4), 524–558.

Collins, A., Brown, J. S., & Newman, S. E. (1989). Cognitive apprenticeship: Teaching the craft of reading, writing, and mathematics. In L. B. Resnick (Ed.), *Knowing, learning and instruction: Essays in honor of Robert Glaser* (pp. 453–494). Hillsdale, NJ: Lawrence Erlbaum Associates.

Conchas, G. (2001). Structuring failure and success: Understanding the variability in Latino school engagement. *Harvard Educational Review, 71*(3), 475–504.

Crawford, K. (1996). Vygotskian approaches to human development in the information era. *Educational Studies in Mathematics, 31,* 43–62.

Diop, C. A. (1974). *Nations negres et culture.* Chicago, IL: Lawrence Hill & Co.

Dweck, C. S. & Leggett, E. L. (1988). A social-cognitive approach to motivation and personality. *Psychological Review, 95,* 256–273.

Ellison, C. M., Boykin, A. W., Tyler, K. M., & Dillihunt, M. L. (2005). Examining classroom learning preferences among elementary school students. *Social Behavior and Personality, 33*(7), 699–708.

Fordham, S. & Ogbu, J. (1986). Black students' success: Coping with the burden of "acting White." *The Urban Review, 18,* 176–206.

Fredricks, J. A., Blumenfeld, P. C., & Paris, A. H. (2004). School engagement: Potential of the concept, state of the evidence. *Review of Educational Research, 74*(1), 59–109.

French, S. E., Seidman, E., Allen, L., & Aber, J. L. (2006). The development of ethnic identity during adolescence. *Developmental Psychology, 42*(1), 1–10.

Gambone, M. A., Klem, A. M., & Connell, J. P. (2002). *Finding out what matters for youth: Testing key links in a community action framework for youth development.* Philadelphia, PA: Youth Development Strategies, Inc. and Institute for Research and Reform in Education.

Gordon Rouse, K. & Cashin, S. (2002). The assessment for academic self-concept and motivation: Results from three ethnic groups. *Measurement and Evaluation in Counseling and Development, 33,* 91–102.

Graham, S. (1997). Using attribution theory to understand social and academic motivation in African American youth. *Educational Psychologist, 32,* 21–34.

Green, G., Rhodes, J., Hirsch, A., Suarez-Orozco, C., & Camic, P. (2008). Supportive adult relationships and the academic engagement of Latin American immigrant youth. *Journal of School Psychology, 46,* 393–412.

Gutierrez, K. D. & Rogoff, B. (2003). Cultural ways of learning: Individual traits or repertoires of practice. *Educational Researcher, 32*(5), 19–25.

Hay, K. E. & Barab, S. A. (2001). Constructivism in practice: A comparison and contrast of apprenticeship and constructionist learning environments. *Journal of the Learning Sciences, 10*(3), 281–322.

Hilliard, A. G. (1992). Behavioral style, culture, and teaching and learning. *Journal of Negro Education, 61*(3), 370–377.

Institute for Research and Reform in Education. (1998). *Research assessment package for schools (RAPS) manual for elementary and middle school assessments.*

Jacob, E. (1997). Context and cognition: Implications for educational innovators and anthropologists. *Anthropology & Education Quarterly, 28*(1), 3–21.

Jarvela, S., Lehtinen, E., & Salonen, P. (2000). Socio-emotional orientation as a mediating variable in the teaching-learning interaction: Implications for instructional design. *Scandinavian Journal of Educational Research, 44*(3), 293.

Jolly, E. J., Campbell, P. B., & Perlman, L. K. (2004). *Engagement, capacity and continuity: A trilogy for student success.* Report commissioned by the GE Foundation. Retrieved July 3, 2010 from http://www.campbell-kibler.com/trilogy.pdf

Kitayama, S., Duffy, S., Kawamura, T., & Larsen, J. T. (2003). Perceiving an object and its context in different cultures: A cultural look at new look. *Psychological Science, 14*(3), 201–206.

Klem, A. & J. Connell (2004). Relationships matter: Linking teacher support to student engagement and achievement. *Journal of School Health, 7*(7), 262–273.

Ladson-Billings, G. (1995a). But that's just good teaching: The case for culturally relevant pedagogy. *Theory into Practice, 34*(3), 159–165.

Ladson-Billings, G. (1995b). Toward a theory of culturally relevant pedagogy. *American Educational Research Journal, 32*(8), 465–491.

Ladson-Billings, G. (1997). It doesn't add up: African American students' mathematics achievement. *Journal for Research in Mathematics Education, 28*(6), 697–708.

Ladson-Billings, G. (2006). From the achievement gap to the education debt: Understanding achievement in US schools. *Educational Researcher, 35*(7), 3–12.

Leonard, J. (2008). *Culturally specific pedagogy in the mathematics classroom: Strategies for teachers and students.* New York, NY: Routledge.

Leonard, J., Davis, J. E., & Sidler, J. L. (2005). Cultural relevance and computer-assisted instruction. *Journal on Technology and Education, 16,* 23–34.

Leonard, J. & Evans, B. R. (2008). Math links: Building learning communities in urban settings. *Journal of Urban Mathematics Education, 1*(1), 60–83.

Lewin, K. (1946). Action research and minority problems. *Journal of Social Issues, 2*(4), 34–46.

Lipka, J., Sharp, N., Brenner, B., Yanez, E., & Sharp, F. (2005). The relevance of culturally based curriculum and instruction: The case of Nancy Sharp. *Journal of American Indian Education, 44*(3), 31–54.

Martin, D. B. (2000). *Mathematics success and failure among African American youth: The roles of socio-historical context, community forces, school influence, and individual agency.* Mahwah, NJ: Lawrence Erlbaum.

Martin, D. B. (2006). Mathematics learning and participation as racialized forms of experience: African American parents speak on the struggle of mathematics literacy. *Mathematical Thinking and Learning, 8*(3), 197–229.

Mayfield, J. W. & Reynolds, C. (1997). Black-White differences in memory test performance among children and adolescents. *Archives of Clinical Neuropsychology, 12*(2), 111–122.

National Science Foundation. (2004). *An emerging and critical problem of the science and engineering labor force: A companion to Science and Engineering Indicators 2004.* NSB 04-07. Arlington, VA: The National Center for Science and Engineering Statistics.

No child left behind: Reauthorization of the Elementary and Secondary Education Act of 2001. (2001). A report to the nation and the Secretary of Education, U.S. Department of Education, President Bush Initiative.

Portes, A. & Rumbaut. G. (2001). *Legacies: The story of the immigrant second generation.* Berkeley, CA: University of California Press.

Rittle-Johnson, B., Siegler, R., & Alibali, M. (2001). Developing conceptual understanding and procedural skill in mathematics: An iterative process. *Journal of Educational Psychology, 93*(2), 346–362.

Serpell, Z. N., Boykin, A. W., Madhere, S., & Nasim, A. (2006). The significance of contextual factors in African American students' transfer of learning. *Journal of Black Psychology, 32*(4), 418–441.

Singh, K., Granville, M., & Dika, S. (2002). Mathematics and science achievement: Effects of motivation, interest and academic engagement. *The Journal of Educational Research, 95*(6), 323–332.

Strutchens, M. E. & Silver, E. A. (2000). NAEP findings regarding race/ethnicity: Students' performance, school experiences, and attitudes and beliefs. In E. A. Silver & P. A. Kenny (Eds.), *Results from the seventh mathematics assessment of the National Assessment of Educational Progress* (pp. 45–72). Reston, VA: NCTM.

Suarez-Orozco, C., Pimentel, A., & Martin, M. (2009). The significance of relationships: Academic engagement and achievement among newcomer immigrant youth. *Teachers College Record, 111*(3), 712–749.

Tyler, K. M., Boykin, A. W., Miller, O., & Hurley, E. (2006). Cultural values in the home and school experiences of low-income African American students. *Social Psychology of Education, 9*, 363–380.

Wertsch, J. V. (1985). *Culture, communication, and cognition: Vygotskian perspectives.* Cambridge, MA: Cambridge University Press.

CHAPTER 9

ETHNOMATHEMATICS IN THE CLASSROOM

Unearthing the Mathematical Practices of African Cultures

Iman Chahine

INTRODUCTION

Little is known about Thomas Fuller. Proclaimed as the Virginia Calculator, Fuller (1710–1790) was a brilliant calculating prodigy with amazing arithmetic skills and extraordinary computational power. In a historical factual account of prominent mathematical prodigies in mathematics, Mitchell (1907) summarizes Thomas Fuller's biography as follows:

> Tom Fuller came from Africa as a slave when about 14 years old. We first heard of him as a calculator at the age of 70 or thereabouts, when, among other problems, he reduced a year and a half to seconds in about two minutes, and 70 years, 17 days, 12 hours to seconds in about a minute and a half, correcting the result of his examiner, who had failed to take account of the leap-years. He also found the sum of a simple geometrical progression and

The Brilliance of Black Children in Mathematics:
Beyond the Numbers and Toward New Discourse, pages 195–218.

multiplied mentally two numbers of 9 figures each. He was entirely illiterate. (p. 62)

Thomas Fuller, Benjamin Banneker, Margaret Lawrence, Walter Massey, David Blackwell, Euphemia Lofton Haynes, Arlie Petters, Richard Baker, Nathaniel Whittaker, and Trachette Jackson to mention a few, are among a large group of exceptionally brilliant Black mathematicians who broke the bonds of tradition by creating constellations of new domains for mathematicians to ponder. This fact leads me to a set of fundamental questions: Why not teach their mathematical processes in our classrooms today? Why not afford students a foresight into their life stories, the challenges they persevered, and their persistence to seize every opportunity in their journeys toward success and self-actualization? Why not capitalize on their potentials as real role models to inspire young Black students to carry on the torch of scientific discovery and mathematical breakthroughs?

It seems quite puzzling that in American society, a society obsessed with great attention to details when it comes to innovations with regards to scientific and mathematical breakthroughs (Fouche, 2003), the contributions of Black[1] mathematicians have been largely minimized. Nonetheless, I argue that this reality is the birth child of a deeply entrenched historical reduction on the part of the Western world that robbed numerous societies of their scientific and mathematical knowledge. Throughout the history of mathematics, there has been a deliberate attempt to marginalize the contributions of many societies in the production of mainstream mathematical knowledge. As an enterprise reflecting the values and practices epitomized by society at large, education mirrors those social views and behavior patterns that are inherently entrenched and valorized by the culture. This is particularly prevalent in the American educational system, where it is often claimed that integration of schools has opened up avenues of access to quality education to *all.* Yet, the realities of racial and socioeconomic isolation and segregation continue to incessantly proliferate, in many disguised forms, educational policies that valorize Western culture and dominant ways of knowing (Manning, 1999).

In such schooling regimes, the current push for accountability and the movement towards a *culturally-coded,*[2] standards–based curriculum seems to cripple every earnest attempt at rejuvenation and redemption of democratic access to educational opportunities and resources not only for youth of color but for poor Whites as well. The abysmal inequalities that we witness in schools and societies worldwide have helped to create a rather apocalyptic vision for the future of education, one that threatens education functioning as a means for critical engagement and social empowerment (OECD, 2011; Rousseau-Anderson & Tate, 2008). In the face of such threats, which transcend national and geographic boundaries, we must seek refuge and inspiration in anti-hegemonic curriculum and the perspectives of critical thinkers, such as Freire, who remind us of the ultimate goals of education—humanization and liberation.

In this chapter, I tell one epistemological and curricular tale. This tale is about the role of ethnomathematics as pedagogy of humanization that recognizes the contributions of Black communities in the production and diffusion of mathematical knowledge. Ethnomathematics is principled by an ideology that calls for reclaiming cultural identities and redeeming human dignity (D'Ambrosio, 2001). I further maintain that authentic engagement in ethnomathematics praxis permits a deeper understanding of the African conceptualization of humanity. With heightened emphasis on memorializing the nobility of mathematical practices that emerge in the daily life of Black communities, ethnomathematics shifts the gaze from focusing on Black intellect as a context-bound social construct (Stubblefield, 2009) to considering a broader emphasis on Black acuity as a steering force that monitors the constant reconstruction of cultural identities throughout history. I also argue that immersion in the ethnomathematical practices of Black cultures provides an insight into critical factors that can shape Black students' success in mathematics.

Drawing upon the vast literature on the ingenuity of indigenous African cultures, I will present a set of ethnomathematical ideas that permeates numerous African indigenous knowledge systems. These systems include folk games and puzzles, kinship relations, and divination systems.

WHAT IS ETHNOMATHEMATICS?

At its inception, ethnomathematics as an epistemology emerged in response to a longstanding history disclosing a deliberate devaluation of the mathematics developed and expanded by non-European civilizations (Powell & Frankenstein, 1997, 2009). Early accounts on the history of mathematics are fraught with chronological and epistemological trajectories depicting mathematics as a creation of Western civilizations that conquered and dominated the entire world. Bishop (1990) described Western mathematics as "one of the most powerful weapons in the imposition of Western culture" (p. 52). In a similar vein, D'Ambrosio asserted:

> [When talking about Western mathematics] especially in relation to Aboriginal's or Afro-American's or other non-European people's, to oppressed workers and marginalized classes, this brings the memory of the conqueror, the slave-owner, in other words, the dominator; it also refers to a form of knowledge that was built by him, the dominator, and that he used and still uses to exercise his dominance. (as cited in Vithal & Valero, 2003, p. 547)

However, new perspectives on the history of mathematics that challenge the "classical" Eurocentric views have recently emerged acknowledging and emphasizing the contributions of non-Western indigenous cultures to the development of science and mathematics (Ernest, 2009; Joseph, 2011). Such meta-narratives and discourses legitimize the epistemological vision advanced by ethnomathematics and have inspired many researchers and educators around the globe to recognize the need to demand respect and human dignity for communities on which Western knowledge and values have been imposed.

While a number of contested definitions and associated perspectives are increasingly emerging in the research on ethnomathematics, two overarching conceptualizations determine the orientation and underlying epistemology of the field. D'Ambrosio (1985) defined ethnomathematics as "the mathematics practiced by cultural groups, such as urban and rural communities, groups of workers, professional classes, children in a given age group, indigenous societies, and so many other groups that are identified by the objectives and traditions common to these groups" (p. 46). Additionally, D'Ambrosio (1999) described ethnomathematics as: "a program in history and epistemology with an intrinsic pedagogical action...taking into account the cultural differences that have determined the cultural evolution of human mankind and political dimensions of mathematics." (p. 150). In this respect, ethnomathematics draws on the potential of cultures in providing tools for the reconceptualization of power, thereby reconfiguring the political, epistemological, and practical dimensions of mathematics.

D'Ambrosio (2004) contended that the role of ethnomathematics in education is to: a) foster creativity by helping people capitalize on their potentials and invest in their strengths, and b) advance citizenship by promoting equitable rights and opportunities in society. Barton (2004) further prioritized the need for reconstructing a new historiography of mathematics by humanizing the field and reclaiming its ethical and moral legacy toward preserving humanity.

In the cause of humanization and liberation, ethnomathematics calls for radically revolutionizing space, content, and approaches to mathematics education. As an organic model of communal learning and mathematics-identity building, ethnomathematics emphasizes the role of the classroom as free, flexible and democratic space for reconfiguring history and empowering students as well as teachers toward the pursuit of truth. In such a space, the privileged voice of authority is deconstructed by collective critical practice where students are seen as human beings with complex lives and diverse experiences, sharing their daring voices in continuous dialogue among each other and with the teacher. Ascher (1991), on the other hand, provides a more "ideological and polemic" definition by emphasizing the mathematical ideas that were created and perfected by indigenous societies and which are inherently embedded in their social practices and daily rituals. Ascher (2002) further capitalized on the riches and complexity of numerous knowledge systems that reflect the practices of more than 6,000 eminent cultures such as the Inuit, Iroquois, and Navajo of North America; the Incas of South America; the Bushoong, Kpelle and Tshokwe of Africa; and many others.

Mukhopadhyay, Powell, and Frankenstein (2009) cited six broad categories delineating the range of emerging ethnomathematics research in conjunction with the aforementioned definitions. These include research on ethnomathematical contributions from different parts of the world, particularly in Africa (Chahine & Kinuthia, in press; Gerdes, 1994, 1995, 2007, 2009; Zaslavsky, 1973, 1999); informal and out-of-school studies (Carraher, Carraher, & Schliemann, 1985; Hoyles, Noss, & Pozzi, 2001; Jurdak & Shahin, 1999, 2001; Lave, Murtaugh, & Rocha, 1984); and the political and historical context of mathematics (Gerdes, 2009; Joseph, 2011; Powell, 1986; Struik, 1997).

AFRICAN "HUMANICITY" AND THE SUCCESS OF BLACK STUDENTS

When defining ethnomathematics, D'Ambrosio (2001) cautioned against unintentional attempts to romanticize and exoticise the field. He further advanced ethnomathematics as a system of knowledge that perpetuates an education for peace, citizenry, and reclamation of cultural dignity. As such, ethnomathematics promotes a conceptualization of culture in its widest sense to include the *authentic* humanity of the people sharing the collective beliefs, traditions, practices, and so on. To describe what I mean by authentic humanity, I propose a neologism, *humanicity*. Hence I talk about *African humanicity* to mean the authentic African life experiences and encounters that reflect genuine African cultural identity, which transcend erroneous socio-political stereotyping, artificial classifications, and color supremacy. The notion of *humanicity* occurred to me during a recent visit to South Africa in the summer of 2011.

In July of that year, I accompanied a group of students on a study abroad immersion program to KwaZulu Natal in South Africa as part of a graduate course on indigenous mathematical knowledge systems that I was facilitating that summer. The purpose of the trip was to afford graduate students firsthand experiences to interact with the local residents and to explore the riches of indigenous knowledge systems that thrive in the Zulu culture. In our everyday encounters with the Zulu speaking people and amidst the flood of quintessentially artistic African treasures that we witnessed in the Zululand province, we learned a very significant notion that closely resonates with the epistemology and philosophy of ethnomathematics. The Zulu word is *ubuntu* and it depicts a consummate perspective by the Zulu culture of what it means to be human. In his book *No Future without Forgiveness,* Tutu (1999) explained:

Ubuntu is very difficult to render into a Western language. It speaks to the very essence of being human. When you want to give high praise to someone we say, *"Yu, u nobuntu,"* He or she has *ubuntu.* This means that they are generous, hospitable, friendly, caring and compassionate. They share what they have. It also means that my humanity is caught up, is inextricably bound up, in theirs. We belong in a bundle of life. We say, "person is a person through other people" (in Zulu *Umuntungumuntungabanye*). I am human because I belong, I participate, and I share. A person with *ubuntu* is open and available to others, affirming of others, does not feel threatened that others are able and good; for he or she has a proper self-assurance that comes with knowing that he or she belongs in a greater whole and is diminished when others are humiliated or diminished, when others are tortured or oppressed, or treated as if they were less than who they are. (as cited in Lewis, 2010, p. 70)

I suggest the term *African humanicity* as a synonym to *ubuntu.* Both terms connote a conceptualization of what it means to be human in an African perspective. To this end, *ubuntu* is absolutely relevant to people of African descent, as well as to any group of people, as it empowers them toward celebrating their identities by challenging the deficit discourses that encourage demarcation and discrimination. Lewis (2010) asserted that reclaiming Black heritage and affirming cultural identity can be realized through the pursuit of *ubuntu,* which in turn empowers Black youth to maintain a positive belief in their ability toward academic success and exceptional performance. More importantly perhaps is the insight that *ubuntu* offers in understanding the *physics* of Blackness and the role that identity plays in conceptualizating Black students' academic success.

In exploring the brilliance of Black students, I argue that it is essential to understand the nature of *African humanicity,* to delve deeper into the essence of Blackness and what it really means to be Black. Grounded in the belief that the concept of Blackness extends beyond the physicality of skin color and phenotypical characteristics, *African humanicity* concurs with Lartey's suggestion that "by Black we refer in a general sense to people of African, Caribbean and Asian descent as well as people who identify with 'the Black experience' in terms of heritage, oppression and domination" (as cited in Lewis, 2010, p. 73).

BLACK STUDENTS' SUCCESS: THE ROLE OF IDENTITY

A wide range of studies has reported significant findings substantiating Black students' success in mathematics and science (Berry, 2005; Cooper, Cooper, Azmitia, Chavira, & Gullatt, 2002; Eatmon, 2009; Ellington, & Frederick, 2010; Jett, 2010; Moody, 2004; Noble, 2011; Sheppard, 2006; Stinson, 2006, 2008; Thompson & Lewis, 2005). These studies cite factors that play a significant role in perfecting Black students' success in school. Some of the factors reported in the literature include belief in self (Berry, 2005; Noble,

2011); social supportive systems such as family members, peer relations, and academic communities (Cooper at al., 2002; Ellington & Frederick, 2010; McGlamery, 2004; Walker, 2006); rate of taking advanced mathematics courses (Gutierrez, 2000; Thompson & Lewis, 2005); spirituality (Jett, 2010); and students' personal characteristics such as motivation and resilience (Berry, 2005; Sheppard, 2006). One of the critical factors most cited in the literature that is believed to contribute to Black students' accomplishments when learning mathematics is their sense of self-respect and positive mathematics self-efficacy and identity (Cokley, 2000; Fisher, 2000; Martin, 2000; Pajares, 1996).

Nasir (2005) maintains that identity "is most often defined as an amalgamation of self-concept, self-understanding, and evaluating oneself in relation to others" (p. 217). Murrell (2007) capitalized on the development of students' academic identities so that they can successfully "symbolize themselves in the cultural scenes that matter in the school" (p. 64). To better understand the mechanisms through which students' academic identities developed in relationship to success in schools, Murrell (2007) proposed Situated-Mediated Identity Theory, which is guided by the following assertions:

1. Identities are socially constructed.
2. Identities are dynamic.
3. Individuals assume roles within a community.
4. Roles taken in academic contexts influence school success.
5. When situated identities are shared, a local culture is established that influences identities and performance.
6. School success can be achieved if/when the local culture seeks to build positive academic identities.

The claim that the learner's personal identity and emotional connections with the content learned constitute steering forces leveraging students' success is undergirded by growing evidence in recent cognitive neuroscience literature (Immordino-Yang & Faeth, 2010). The view that how students think about mathematics is often best explained by reference to their emotions and dispositions toward the topic rather than by reference to their mental processes is becoming a widely accepted idea. Conchas (2006) contends that a significant factor in instilling a strong self-identity in Black students is pride in their ancestral heritage. However, in many school contexts Black children are robbed of a sense of pride in their identity and heritage as they constantly encounter disparities in opportunities afforded by educational and social enterprises. In exploring the disturbance that Black children experience when reminded of the recurring theme of slavery continuously perpetuated in schools, King (2006) explained:

Educators and parents are painfully aware that many Black students are trau-
matized and humiliated when reading about slavery. They often report that
Black students do not want to discuss slavery or be identified with Africa and
many admit that they lack the conceptual tools to intervene in this dangerous
dynamic. (p. 328)

Predictably, a history of *ethnostress* may eventually lead to a loss of sense of
self and identity confusion that can be passed on from past generations. Antone and Hill (1992) defined *ethnostress* as an internalized oppression and
a loss or confusion of identity that has been entrenched in the memory of
a community generations after the oppressive conditions have been eliminated. They further delineate four conditions that invoke ethnostress in a
society, including disruption of cultural beliefs, forcing suppressive measures on people within their own environments, negative stereotyping as a
result of *internalized racism*, and reinforced powerlessness and hopelessness
through consistent exposure to negative experiences laden with emotions
of resentment and discontent. Because of *policies of absence*[3] and *invisibiliza-
tion* tactics that thrive in the mathematics curriculum, Black students find
themselves negotiating continuously their socially constructed racial identities and vindicating their Blackness. Therefore, recognizing and acknowledging the contributions of African communities and other cultures is a
necessary step that facilitates "*Critical Studyin*," which can lead to the eman-
cipation of Black children's brilliance in mathematics (King, 2006).

A healing process can be realistically viable if we allow Black students opportunities to critically probe and question the nature of the mathematical
knowledge they encounter in schools. By transforming classroom environments into spaces where mathematical investigations provide insight into
the ingenious experiences of Black mathematicians, we create opportunities to learn, which lead to prospects for participation and inclusiveness.
The ultimate goal is to encourage Black learners to holistically experience
themselves in a cultural reality that affirms their right to reclaim success
that has long been masked by social demarcation and racial segregation
(Michelson, 1997).

In this regard, affording meaningful contexts and authentic experiences
can help students from diverse backgrounds become more *emotionally invested* in learning mathematics and develop a sense of ownership and pride
as they celebrate the prolific contributions of their ancestral heritages. Martin and McGee (2009) asserted that a major component of mathematics literacy for liberation is being knowledgeable about one's culture. Participating in learning that is situated in cultural contexts reduces the emotional
load that has been hitherto induced by limited societal affordances, thus
encouraging reaffirmation of self-identity by recognizing and valorizing the
cultural heritage.

THE ROLE OF ETHNOMATHEMATICS IN THE CLASSROOM

The myth that mathematics can be taught effectively and meaningfully without relating it to students' culture has been widely destabilized and dispelled. This claim is supported by a handful of research that has shown the importance of integrating cultural practices that resonate with students' ethnic background and everyday experiences. Murrell (2007) contended that student achievement is an activity occurring within a community, and that "school success is achievable for all students when learning is understood as the acquisition of a set of preferred cultural practices" and as we come to see teaching and learning as the "socialization of these cultural practices in educational settings" (p. 34).

Notwithstanding, there have been many success stories in the literature that report enhanced performance when cultural practices from ethnomathematics are integrated into the mathematics instruction. In a study investigating the effect of using African culture on African students' achievement in geometry, Moses-Snipes (2005) implemented ethnomathematics practices with 107 fifth grade students in a public elementary school. Classes were randomly assigned to either the Mathematics with Culture (MWC) group or the Mathematics without Culture (MWOC) group. The mathematics lessons lasted approximately 120 minutes for both groups. Both groups completed an entire geometry unit created by the researcher. The MWC group completed the geometry unit with some facet of African culture integrated into each activity. The MWOC group completed the same unit without the African culture component. Results of the study showed that students' achievement scores increased as they learned about African culture. Information from this study contributes to the growing role that culture can play in the mathematics classroom and in African American students' achievement.

In a similar study, Ensign (2003) described the efforts of integrating personal, everyday practices of students into classroom mathematics in an urban school setting. The author described the culturally related technique of teaching as *culturally connected,* which was aimed at finding creative and meaningful problems that helped to mediate the difficulty that students face in understanding mathematical concepts. The study was carried out with second, third, and fifth graders in two urban schools in New Haven, Connecticut. Results reported evidence of a raised interest on the part of students in solving their own personal mathematics problems rather than those cited in their textbooks. In addition to social and emotional effects, results of students' exams indicated a trend toward higher scores on textbook publishers' unit tests when personal experiences were included in lessons versus only when text problems were used.

Implications of the foregoing arguments for teaching Black children are highly significant. For example, Nisbett (2003) views ethnic diversity as enriching, particularly in problem solving where different cognitive orientations complement and build on the strength of each other. He further explained that the key is providing the proper setting and the appropriate materials that capitalize on and support students' abilities and interests. In such settings, good teaching also capitalizes on the students' identities as Black learners of mathematics and the cultural and social capital they bring to the classroom.

There is ample evidence in the literature supporting the claim that culture can be implemented in the mathematics classroom in various ways and with a wide range of varying resources. Kim (2000) proposed two ways to integrate ethnomathematics in the process of teaching and learning in the mathematics classroom: first, through the use of inventive ideas inspired from one's own culture; second, by the exploration of new ideas in other cultures. Kim (2000) also emphasized the role that ethnomathematics materials plays in the *enculturation* and *acculturation* processes within and across diverse cultures. Capitalizing on the riches of the African cultures and building on the ingenuity of Black people empowers African American students toward success in mathematics by providing meaningful and relevant contexts conducive to learning to further the achievement of students, particularly those from underserved populations.

TRADITIONAL AFRICAN AND AFRICAN AMERICAN GAMES

Research on the ethnomathematics of African cultures has provided a wealth of creative and thought provoking African mathematics such as number systems (Zaslavsky, 1973), indigenous games and puzzles (Laridon, Mosimege & Mogari, 2005), kinship relations (Ascher, 1991), divination systems (Eglash, 1995), symmetric strip decorations (Gerdes, 1995, 1999, 2007, 2009) and many others that can be actively used by students in the classroom to enhance their learning. Ascher (2002) argues that what makes culturally derived mathematical ideas so powerful is their inherent entrenchment in contexts in which they arise as part of the complex of ideas around them. Such contexts include indigenous games, divination, calendrics, kinship relations, decoration, and many others.

One of the most prolific cultural artifacts that unfold the creativity and rigor of the African heritage are games. It is interesting to note that while African games were initially discouraged by the colonial education authorities in favor of European originated games such as ludo and snakes and ladders (Zaslavsky, 1999), such games fervently persist in the daily life of indigenous African communities. Powell and Frankenstein (1997) asserted, "In classrooms, ethnomathematics can be implemented by investigating the mathematics of cultural products and practices, such as games, with people from that culture or by exploring the mathematics of a different culture to help students enrich their construction of mathematical ideas" (p. 249).

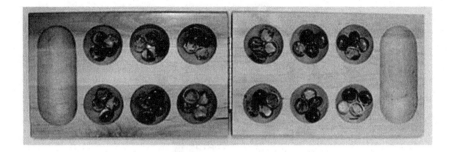

FIGURE 9.1. Oware game board played in Ghana.

Kim (2000) cited more than 200 traditional games that are inspired from African cultures.

The power of games in teaching rests in their ability to motivate learning and provide tools that bring into play many cognitively challenging strategies that hone students' mathematical skills. These games comprise a collection of mathematical investigations such as riddles, networks, arrangements, probability and chance, and so on. Powell and Temple (2001) claimed that when playing games, children develop "… intellectual frameworks that enable them further to construct and comprehend complex mathematical ideas, strategies, and theories" (p. 369). For example, *Mancala* is an African board game that is played worldwide with more than 20 different versions. The two-row versions include *oware* (Ghana), *awélé* (Ivory Coast), *ayo* and *okwe* (Nigeria); and the four-row versions, *omweso* (Uganda), *tshisolo* (Zaire) (Crane, 1982).

Powell and Temple (2001) capitalized on the potential of the *oware* board game in building and extending arithmetical ideas, thus, scaffolding strategic thinking. The w*ari* or *oware* is played by the Asante people of Ghana on a board with two rows of six holes and an additional hole or pot at each end (See Figure 9.1).

The two players face each other and begin by placing four seeds in each of six holes. The object of the game is to capture at least 25 seeds (a majority of the total of 48) into one's pot. When the opponent's hole has been cleared or has only one seed, the game has ended. Zaslavsky (1999) explained that another version of the game that is played in Uganda is called *omweso* or *mweso,* and the game involves a four-row board and requires 64 beans on a board having four rows of eight holes (see Figure 9.2). This game encourages children to count, learn simple sums and the concept of one-to-one correspondence as they drop each of their counters into con-

FIGURE 9.2.   Omweso board game played in Uganda.

secutive holes. *Omweso* also permits reverse moves, introducing the concept of negative numbers.

Other forms of board games that have been played in Africa and that relate to the teaching and learning of mathematics have been extensively documented in the literature. Zijilma (2000) cited several board games that have been played in Africa for thousands of years. For example, *Senet*, one of the oldest known board games in the world, originated in Egypt (see Figures 9.3a and 9.3b).

According to Kendall's (1978) rules, the Senet game board is composed of 30 squares that are called houses: three rows of 10 houses each with five special houses with specific names and positions are used to play. For further illustration see Table 9.1.

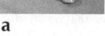

FIGURE 9.3. (a) Senet game played in Egypt; (b) Senet: oldest board game in the world.

TABLE 9.1. Symbols for Senet Game

Symbol	House Name	House Position
	House of Rebirth; the starting square	15
	House of Happiness; All the counters have to land on Square 26 before they can go any further. The exact number must be thrown to reach this square.	26
	House of water; a counter on Square 27 cannot move any other counters until a 4 has been thrown to remove the counter on Square 27 from the board.	27
	House of 3; if attacked it moves back to Square 27.	28
	House of Re-Atoum; if attacked it moves back to Square 27	29

Each of two players receives seven pawns, and the 14 pawns alternate along the 14 houses. One player puts seven counters on the odd-numbered squares, 1–13, and the other player puts seven counters on the even-numbered squares, 2–14. The pawns move according to the throw of four sticks. At the beginning of each turn, a player drops the four sticks. The player can then move any counter as many squares as the throw has scored, but the counter can only be moved to an empty square. The first player to remove all his or her counters from the board is the winner. The scores are determined by the number of decorated and plain surfaces of the thrown sticks as shown in Table 9.2.

From a game theory perspective, the Senet game stimulates thinking, teaches decision-making, and creates an opening for meaningful discussions about choices and relationships. Engaging students in learning with board games exposes the diverse strategies people use when making decisions. Senet is primarily a game of skill; chance is secondary when played by experienced players.

TABLE 9.2. Schema for Scoring in Senet Game

Number of decorated surfaces up	Number of plain surfaces up	Score
4	0	5 (gives a player an extra turn)
0	4	4
1	3	3
2	2	2
3	1	1 (starts the game; uses even-numbered squares)

Many of the African traditional board games can be played using materials found in nature, such as seeds and stones. Boards can be scratched into dirt, dug out of the ground or drawn on paper (see Figure 9.4). Furthermore, African American communities throughout the United States have appropriated and developed a constellation of ancillary games as part of their Afrocultural resources and practices. Notwithstanding the paucity of research on African American games and their syncretic impact on the mind and body, several playing activities have been cited in the literature.

FIGURE 9.4. Ancient Mancala board engraved in the sand.

FIGURE 9.5.   Dominoes' game popular among African Americans.

In analyzing the association of mathematical goals, identities, and learning in practice, Nasir (2005) explored how playing dominoes for African American children, youth and adults affords a socially-situated practice wherein the players' identities are negotiated and reconstructed in a dynamic context (see Figure 9.5). In the game of dominoes, the goal is not simply to win but to make appropriate decisions to maximize the number of points a person can score (Nasir, 2005). Thus, students make predictions about the number of points they can obtain and develop strategies simultaneously.

Moreover, emphasizing the role of engagement in games epitomizes learning, particularly in mathematics, as players forge new goals in conjunction with emerging identities. In a similar vein, Schademan (2011) capitalized on the African American game of spades and its underlying concepts as a rich resource whereby "mathematics learning becomes intertwined with youth resources and cultural practices" (p. 365) (See Figure 9.6).

Schademan contends that the cultural practice of card playing for African American communities, which dates back to the Civil War, provides a powerful artifact for examining the cognitive experiences of Black children's everyday play, which has the potential to undergird learning in the mathematics classroom.

FIGURE 9.6.   African American Spades' game.

AFRICAN KINSHIP RELATIONS

The concept of kinship relationships represents a formalized body of knowledge that can be intimately tied to mathematical structures. Anthropological research is laden with advanced mathematical networks and models that employ graph theory to solve and explain a range of culturally specifc relations. Using mathematical models to understand the wide range of kinship terminologies that exist in indigenous communities around the world yields a variety of new insights into the social structure and organization of various cultures. The Gada system of the Borana, a pastoral community, who live in the Sidamo province of Ethiopia and Southern Kenya is one interesting example of African kinship structure that involves formal linear and cyclic configurations. Gada is a generation-grading system that is characterized by the rule that all sons follow their fathers in a sequence of grades at a fixed and specific interval irrespective of their actual age. Ascher (2002) explains: "Basic to the Gada system are consecutive grades, through which, theoretically, all males pass, and named classes, which refer to those who are in the grades" (pp. 142–143).

The Borana people invented the basic principles of the system, which is a spiral of 10 grades each of eight years duration and seven class names, structured in a way that boys in the first grade are the sons of men in the sixth grade; those in the second grade are sons of men in the seventh grade, and so on. Using Ascher's (2002) configuration, the relationship between father and son can be represented through modular algebra notation by the following equation:

$$S^n(c_i) = c_{i-5n(\mathrm{mod}\,7)}$$

for n= 1,2,......; i= 1,2,3,4,5,6,7, where S(c_i) is the father of c_i .

Ascher (2002) further explained that the father-to-son descent cycle can be modeled by a permutation transforming the name cycle into the father-son cycle by interchanging c_i with $c_{2i+1(\text{mod}7)}$ (p. 149). Legesse (1973) argues that the Borana uses the terms, *gogessa or* the five-name patri-class, and *makabasa* for a cycle of seven names for the purpose of ordering the Borana chronology and to delineate the historical map of their society.

Mathematical ideas inspired by the logic underlying African kinship relations can enrich and invigorate the learning of mathematics by affording Black students the means to participate in African *humanicity*. It is important to note that in examining kinship relations across many cultures, the emphasis is not on the mathematical principles involved per se but rather on the use of mathematics as classificatory and descriptive typology to understand the genealogical relationships and internal organization of sophisticated cultural and social systems. As Hage and Harary (1983) note in their analysis of Arapesh sexual symbolism, "Mathematical models are used not because of any wish to mathematize culture but because there are ethnographic advantages for doing so. The general advantage of group models is that they preserve and exploit the richness of the data rather than obliterating it through generalizations" (p. 68). Engaging students in exploring the logical structures of diverse kinship systems and family relationships within indigenous communities sensitizes students to understand the role that mathematics plays in capturing the many ways through which culture as a complex system self-organizes and preserves its practices.

AFRICAN DIVINATION SYSTEMS

The role of African divination systems is pivotal not only in understanding the epistemology of African Peoples and their cultures but also to unfold the extensive body of indigenous knowledge embedded in those practices (Chahine & Srivastava, in press). Eglash (1997) investigated the mathematically significant aspect of doubling in African religion that occurs in the divination techniques of Bamana. He described the technique of Ifa as purely *stochastic* (i.e. operates by pure chance) and that of Bamana as systematic, highly compact oracles which follow laws of recursion.

Ascher (2002) also introduced a twin of the Bamana technique 5,000 miles to the east in Malagasy Sikidy. The rich available literature presents ample evidence that these major African divination systems were all transformations of the Arabic system of "sand science" ('ilm al-raml) or "sand calligraphy" (khatt al-raml), which spread from Abbasid, Iraq, all over the Islamic world, the Indian Ocean region, and Africa from the late first millennia Common Era (CE)[4] onwards (Van Binsbergen, 1999). Partly rooted in simple chance procedures (like hitting the earth, throwing tablets, beans, shells, etc.), the *Bamana system* is a binary system of 16 figures where each figure is four rows high and each row consists of either one dot or two dots. The figures are determined through various methods both ancient and modern. The procedure is performed through a forceful hitting of the sand with a stick, in order to produce a chance number of dot traces or marks, which can then be scored as either odd or even.

Divination systems are dynamic systems that reflect patterns of social interactions, the community's identity and cognitive processes indicative of the people's ways of thinking. Engaging students in exploring these complex routines and the mathematical structures inherent in their design will provide an insight into mathematics as a human endeavor.

TOWARD A NEW PARADIGM

There is no doubt that affording Black students the opportunities to explore the mathematical ideas of African and other cultures in the classroom is crucial for understanding the hegemonic misrepresentations and deliberate devaluation of those cultures (Alatas, 1978; Joseph, 2011; Said, 1978). I have argued that allowing purposeful engagement in existential experiences that unfold culturally embedded mathematical competences can inspire students to question enforced imminent boundaries and rediscover the joy and excitement in making transformation real and possible.

A deeper and more insightful examination of numerous valuable and unique contributions of the African cultures in the production and dissemination of mathematical knowledge has great potential in dismantling the arguments against the brilliance of African Americans in mathematics and, thus, silencing the mythical stereotyping regarding the intellectual representation of African American people. Sensitizing Black students to the inventions and contributions of Black mathematicians, such as Thomas Fuller, Benjamin Banneker, and Euphemia Lofton Haynes, highlights the richness of Black cultural experiences and affords opportunities to challenge and thereby dispute the Eurocentric view of what counts as mathematical knowledge and how to teach it to children. Needless to say, such understanding is necessary to question the veracity of claims about Black achievement in mathematics. Enriching classroom instruction with ethnomathematical investigations inspired from African cultures and Black heritage can inform, enlighten, and reconfigure the overly skewed representation of African Americans' ability in mathematics. Indeed, keeping the focus on mathematics as a human endeavor and celebrating cultural diversity are key dimensions for enacting pedagogies of transformation, liberation and transgression.

I end my discussion of ethnomathematics with this inspiring quote from an interview with Freire six months before he passed away in 1997:

I have no doubt that our presence on earth undeniably implied the invention of the world. I have been thinking a lot that the decisive step that made us capable of being human, as women and men, was precisely the step by which the support in which we found ourselves became "the world" and the life we lived began to become existence. It is during this passage from support to the world when history begins to unfold and when culture is born...language, thought that is enriched by the possibility of communication...of conveyance. I believe at this point we have also become mathematicians. (Freire, D'Ambrosio, & Mendonça, 1997, p. 7)

NOTES

1. Throughout the chapter, "Black" refers to African and African diaspora people. The African Union defined the African diaspora as "[consisting] of people of African origin living outside the continent, irrespective of their citizenship and nationality and who are willing to contribute to the development of the continent and the building of the African Union." Its constitutive act declares that it shall "invite and encourage the full participation of the African Diaspora as an important part of our continent, in the building of the African Union."
2. Where cultural assumptions are embedded in curriculum content and instruments
3. Marginalization and dismissal
4. Common Era also known as *Current Era* or *Christian Era*, abbreviated as CE and is traditionally identified with *Anno Domini* (abbreviated AD).

REFERENCES

Alatas, S. H. (1978). *Myth of the lazy native.* London, England: Frank Cass and Company Limited.

Antone, R. A. & Hill, D. (1992). *Ethnostress: The disruption of the aboriginal spirit.* Ontario, Canada: Tribal Sovereignty Associates.

Ascher, M. (1991). *Ethnomathematics: A multicultural view of mathematical ideas.* Pacific Grove, CA: Brooks/Cole.

Ascher, M. (2002). *Mathematics elsewhere: An exploration of ideas across cultures.* London, UK: Princeton University Press.

Barton, B. (2004). Moving forward. *Proceedings of the 10th International Congress of Mathematics Education Copenhagen.*

Berry, R. (2005). Voices of success: Descriptive portraits of two successful African American male middle school mathematics students. *Journal of African American Studies, 8,* 46–62.

Bishop, A. (1990). Western mathematics: The secret weapon of cultural imperialism. *Race & Class, 32*(2), 51–56.

Carraher, T. N., Carraher, D. W., & Schliemann, A. D. (1985).Mathematics in the streets and in schools. *British Journal of Developmental Psychology, 3,* 21–29.

Chahine, I. C. & Kinuthia, W. (in press). Juxtaposing form, function, and social symbolism: An ethnomathematical analysis of indigenous technologies in the Zulu culture. *Journal of Mathematics and Culture.*

Chahine, I. C. & Srivastava, N. (in press). The mathematics of Cleromancy: The role of subjective probabilities and optimism bias in outcome prediction across divination systems. *American Anthropologist.*

Cokley, K. (2000). An investigation of academic self-concept and its relationship to academic achievement in African American college students. *Journal of Black Psychology, 26,* 148–164.

Conchas, G. (2006). *The color of success: Race and high-achieving urban youth.* New York, NY: Teacher College Press.

Cooper, C. R., Cooper, R. G., Azmitia, M., Chavira, G., & Gullatt, Y. (2002). Bridging multiple worlds: How African American and Latino youth in academic outreach programs navigate math pathways to college. *Applied Developmental Science, 6*(2), 73–87.

Crane, L. (1982). African games of strategy, a teaching manual. *African Outreach Series, 2,* 1–63.

D'Ambrosio, U. (1985). Ethnomathematics and its place in the history and pedagogy of mathematics. *For the Learning of Mathematics, 5,* 44–48.

D'Ambrosio, U. (1999). Literacy, matheracy, and technocracy: A trivium for today. *Mathematical Thinking and Learning, 1*(2), 131–153.

D'Ambrosio, U. (2001). *Etnomatemática: Elo entre as tradições e a modernidade [Ethnomathematics: Link between tradition and modernity].* Belo Horizonte, MG: Autêntica.

D'Ambrosio, U. (2004). Preface. *Proceedings of the 10th International Congress of Mathematics Education Copenhagen.*

Eatmon, D. (2009). A case of resilience based on mathematics self-efficacy and social identity. In S. L. Swars, D. W. Stinson, & S. Lemons-Smith (Eds.), *Proceedings of the 31st annual meeting of the North American Chapter of the International Group for the Psychology of Mathematics Education* (pp. 129–134). Atlanta, GA: Georgia State University.

Eglash, R. (1997). Bamana sand divination: Recursion in ethnomathematics. *American Anthropologist, 99*(1), 112–122.

Ellington, R. & Frederick, R. (2010).Black high achieving undergraduate mathematics majors discuss success and persistence in mathematics. *Negro Educational Review, 61*(1), 61–84.

Ensign, J. (2003). Helping teachers use students' home cultures in mathematics lessons: Developmental stages of becoming effective teachers of diverse students. In A. J. Rodriguez & R. S. Kitchen (Eds.), *Preparing mathematics and science teachers for diverse classrooms: Promising strategies for transformative pedagogy* (pp. 225–242). Mahwah, NJ: Lawrence Erlbaum Associates, Publishers.

Ernest, P. (2009). The philosophy of mathematics, values, and Keralese mathematics. In P. Ernest, B. Greer, & B. Sriraman (Eds.), *Critical issues in mathematics education* (pp. 189–204). Charlotte, NC: Information Age Publishing Inc.

Fisher, T. (2000).Predictors of academic achievement among African-American adolescents. In S. Gregory (Ed.), *The academic achievement of minority students: Perspectives, practices and prescriptions* (pp. 307–334). New York, NY: University Press of America.

Fouche, R. (2003). *Black inventors in the age of segregation.* Baltimore, MD: The Johns Hopkins University Press.

Freire, P., D'Ambrosio, U., & Mendonça, M. (1997). A Conversation with Paulo Freire. *For the Learning of Mathematics, 17*(3), 7–10.

Gerdes, P. (1994). *Sona geometry: Reflections on the sand drawing tradition of peoples of Africa south of the equator.* Maputo: Universidade Pedagogica.

Gerdes, P. (1995). *Women and geometry in Southern Africa.* Universidade Pedagogica, Ethnomathematics Project.

Gerdes, P. (1999). *Geometry from Africa: Mathematical and educational explorations.* Washington, DC: The Mathematical Association of America.

Gerdes, P. (2007). *Drawing from Angola: Living mathematics.* Maputo, Mozambique: CEMEC.

Gerdes, P. (2009). *Introducing Paulus Gerdes' ethnomathematics books.* Maputo, Mozambique: CEMEC.

Gutierrez, R. (2000). Advancing African-American urban youth in mathematics: Unpacking the success of one math department. *American Journal of Education, 109*, 63–111.

Hoyles, C., Noss, R., & Pozzi, S (2001). Proportional reasoning in nursing practice. *Journal for Research in Mathematics Education, 32*(1), 4–27.

Hage, P. & Harary, F. (1983). *Structural models in anthropology.* Cambridge, England: Cambridge University Press.

Immordino-Yang, M. & Faeth, M. (2010). The role of emotion and skilled intuition in learning. In Sousa, D. (Ed.), *Mind, brain, and education* (pp. 46–69). Bloomington, IN: Solution Tree Press.

Jett, C. (2010). "Many are called but few are chosen": The role of spirituality and religion in the educational outcomes of "chosen" African American male mathematics majors. *The Journal of Negro Education, 79*(3), 324–334.

Joseph, G. G. (2011). *The crest of the peacock: Non-European roots of mathematics* (3rd ed.). Princeton, NJ: Princeton University Press.

Jurdak, M. & Shahin, I.C. (1999). An ethnographic study of the computational strategies of a group of young street vendors in Beirut. *Educational Studies in Mathematics, 40*(2), 155–172.

Jurdak, M. & Shahin, I.C. (2001). Problem solving activity in the workplace and the school: The case of constructing solids. *Educational Studies in Mathematics, 47*, 297–315.

Kendall,T. (1978). *Passing through the Netherworld: The meaning and play of Senet, an ancient Egyptian funerary game.* Belmont,MI: The Kirk Game Company.

Kim, S. H. (2000). Development of materials for ethnomathematics in Korea. In H. Selin (Ed.), *Mathematics across cultures* (pp. 455–465). Dordrecht, The Netherlands: Kluwer Academic Press.

King, J. E. (2006). Perceiving reality in a new way: Rethinking the Black/White duality of our times. In A. Bogues (Ed.), *Caribbean reasonings: After man toward the human: Critical essays on Sylvia Wynter* (pp. 25–56). Kingston, Jamaica: Ian Randle Publishers.

Laridon, P., Mosimege, M., & Mogari D. (2005). Ethnomathematics research in South Africa. In R. Vithal, J. Adler, & C. Keitel (Eds.), *Researching mathematics education in South Africa: Perspectives, practices and possibilities* (pp. 133–160) Pretoria: HSRC Press.

Lave, J., Murtaugh, M., & Rocha, O. (1984). The dialectic of arithmetic in grocery shopping. In B. Rogoff & J. Lave (Eds), *Everyday cognition:Its development in social context* (pp. 67–94). Cambridge, MA: Harvard University Press.

Legesse, A. (1973). *Gada.* New York, NY: Free Press.

Lewis, B. (2010). Forging an understanding of black humanity through relationship: An ubunto perspective. *Black Theology, 8*(1), 69–86.

Manning, K. (1999). *Can history predict the future? The African American Presence in Physics.* Atlanta, GA: Ronald E. Mickens.

Martin, D. (2000). *Mathematics success and failure among African-American youth: The roles of sociohistorical context, community forces, school influence, and individual agency.* Mahwah, NJ: Lawrence Erlbaum Associates.

Martin, D. & McGee, E. (2009). Mathematics literacy and liberation: Reframing mathematics education for African American children. In B. Greer, S. Mukhopadhyay, S. Nelson-Barber, & A. Powell (Eds.), *Culturally responsive mathematics education* (pp. 207–238). New, York, NY: Taylor and Francis.

McGlamery, S. (2004). *What factors contribute to the success of African American women in science and mathematics: Do teaching techniques matter?* Paper presented at the Ninth Annual Conference POCPWI. Retrieved from http://digitalcommons.unl.edu/pocpwi9/37

Michelson, E. (1997). Multicultural approaches to portfolio development. *New Directions for Adult and Continuing Education, 75,* 41–53.

Mitchell, F. D. (1907). Mathematical prodigies. *American Journal of Psychology, 18,* 61–143.

Moody, V. R. (2004). Sociocultural orientations and the mathematics success of African American students. *The Journal of Educational Research, 97,* 135–146.

Moses-Snipes, P. (2005).The effect of African culture on African American students' achievement on selected geometry topics in the elementary mathematics classroom, *Negro Educational Review, 56* (2 & 3), 147–166.

Mukhopadhyay, S., Powell, A., & Frankenstein, M. (2009). An ethnomathematical perspective on culturally responsive mathematics education. In B. Greer, S. Mukhopadhyay, S. Nelson-Barber, & A. Powell (Eds.), *Culturally responsive mathematics education* (pp. 65–84). New, York, NY: Taylor and Francis.

Murrell, P. (2007). *Race, culture, and schooling: Identities of achievement in multicultural urban schools.* New York, NY: Routledge.

Nasir, N. (2005). Individual cognitive structuring and the sociocultural context: Strategy shifts in the game of dominoes. *Journal of the Learning Sciences, 14*(1), 5–34.

Nisbett, R. E. (2003). *The geography of thought.* New York, NY: Free Press.

Noble, R. (2011).Mathematics self-efficacy and African American male students: An examination of models of success. *Journal of African American Males in Education, 2*(2), 188–213.

Organisation for Economic Cooperation and Development (OECD). (2011). *Divided we stand: Why inequality keeps rising*. Paris, France: Author. Retrieved from http://dx.doi.org/10.1787/9789264119536-en

Pajares, F. (1996). Self-efficacy beliefs and mathematical problem solving of gifted students. *Contemporary Educational Psychology, 21*, 325–344.

Powell, A. (1986). Economizing learning: The teaching of numeration in Chinese. *For the Learning of Mathematics, 6*(3), 20–23.

Powell, A. & Frankenstein, M. (1997). *Ethnomathematics: Challenging Eurocentrism in mathematics education*. New York, NY: State University of New York Press.

Powell, A. & Temple, O. (2001). Seeding ethnomathematics with *Oware: Sankofa*. *Teaching Children Mathematics, 7*(6), 369–374.

Rousseau-Anderson, C. & Tate, W. (2008). Still separate, still unequal: Democratic access to mathematics in U.S. schools. In L. D. English (Ed.), *Handbook of International Research in Mathematics Education* (pp. 299–318). New York, NY: Taylor & Francis.

Said, E. (1978). *Orientalism*. New York, NY: Vintage.

Schademan, A. R. (2011). What does playing cards have to do with science? A resource-rich view of African American young men. *Cultural Studies of Science Education, 6*, 361–380.

Sheppard, P. (2006). Successful African American mathematics students in academically unacceptable high schools. *Education, 126*(4), 609–625.

Stinson, D. W. (2006). African American male adolescents, school (and mathematics): Deficiency, rejection, and achievement. *Review of Education Research, 76*, 477–506.

Stinson, D. W. (2008). Negotiating sociocultural discourses: The counter-storytelling of academically (and mathematically) successful African-American male students. *American Educational Research Journal, 45*, 975–1010.

Struik, D. J. (1997). Marx and mathematics. In A. B. Powell & M. Frankenstein (Eds.), *Ethnomathematics: Challenging Eurocentrism in mathematics education* (pp. 173–193). Albany, NY: State University of New York.

Stubblefield, A. (2009). The entanglement of race and cognitive disability. *Metaphilosophy, 40*(3–4), 531–551.

Thompson, L. R. & Lewis, B. F. (2005). Shooting for the stars: A case study of the mathematics achievement and career attainment of an African American male high school student. *The High School Journal, 88*(4), 6–18.

Tutu, D. (1999). *No future without forgiveness*. New York, NY: Random House, Inc.

Van Binsbergen, W. (1999). *Board-games and divination in global cultural history a theoretical, comparative and historical perspective on mankala and geomancy in Africa and Asia—Part II*. Retrieved from http://www.shikanda.net/ancient_models/gen3/mankala/ mankala1.htm

Vithal, R. & Valero, P. (2003). Researching mathematics education in situations of social and political conflict. In A. J. Bishop & M. A. Clements (Eds.), *Second international handbook of mathematics education* (pp. 545–591). Dordrecht, Netherlands: Kluwer academic Press.

Walker, E. N. (2006). Urban high school students' academic communities and their effects on mathematics success. *American Educational Research Journal, 43*, 43–73.

Zaslavsky, C. (1973). *Africa counts*. Boston, MA: Prindle, Weber & Schmidt.

Zaslavsky, C. (1999). *Africa counts: Number and pattern in African cultures.* Chicago, IL: Lawrence Hill.

Zijilma, A. (2000). African culture and people. Retrieved from http://goafrica. about.com/od/peopleandculture/tp/African-Games_Played-In_Africa.htm

SECTION IV

STUDENT IDENTITY AND STUDENT SUCCESS

CHAPTER 10

COUNTERSTORIES[1] FROM MATHEMATICALLY SUCCESSFUL AFRICAN AMERICAN MALE STUDENTS

Implications For Mathematics Teachers And Teacher Educators

David W. Stinson, Christopher C. Jett, and Brian A. Williams

INTRODUCTION

Over three decades ago, motivated in part by the publications of *An Agenda for Action: Recommendations for School Mathematics of the 1980s* (National Council of Teachers of Mathematics [NCTM], 1980) and *A Nation at Risk: The Imperative for Educational Reform* (National Commission on Excellence in Education, 1983), a new, often-repeated phrase emerged within the discourses[2] of mathematics education research and policy: *mathematics for all.* Coupled with this mathematics for all rhetoric (Martin, 2003) has been the proliferation of research studies that documents the mathematics "achieve-

The Brilliance of Black Children in Mathematics:
Beyond the Numbers and Toward New Discourse, pages 221–245.
Copyright © 2013 by Information Age Publishing

ment gap" between Black children[3] and their White counterparts (e.g., Strutchens, Lubienski, McGraw, & Westbrook, 2004). These studies primarily focus on the aggregated "achievement outcomes" of African American and other historically underserved children but rarely, if ever, explore how schooling experiences and race and racism contribute to these outcomes (Lubienski & Bowen, 2000; Parks & Schmeichel, 2012). Absent from many, if not most, of these studies is a critical examination of the socio-cultural and historical inequities experienced by African American children (and other marginalized groups), inside and outside the mathematics classroom, that inhibit the possibility of mathematics for all (Martin, 2003). Furthermore, many, if not most, of the subsequent policy documents derived from this "gap-gazing" research consistently position African American and other historically underserved children as being somehow mathematically deficient (Gutiérrez, 2008). In many ways, these research studies and policy documents sustain the mathematics education enterprise as a White institutional space (Martin, 2008, 2010). Intentionally or not, collectively, they reify the "White male math myth," marking the perceived mathematical "abilities" of the White, male child as *the* point of reference toward which all children should aim (Stinson, 2010).

Within the past decade or so, however, there has been a growing number of researchers who are producing a different body of knowledge that has begun to trouble the waters (Morris, 2009), making salient the racialized mathematics experiences of African American (and all) children (Martin, 2006). These researchers explicitly explore the multilayered and complex lived experiences of African American children within the mathematics domain (e.g., Berry, 2008; Jett, 2010; Martin, 2000; McGee & Martin, 2011; Moody, 2000; Nzuki, 2010; Stinson, 2010; Walker, 2006; see also the edited volumes Martin, 2009, and Strutchens, Johnson, & Tate, 2000). On the whole, their research seeks to find ways that every child, regardless of "race"[4] or ethnicity, might be provided learning opportunities to aim toward levels of excellence (Hilliard, 2003). Unique to this group of researchers is a different perspective of the African American child. Rather than beginning with the misguided perspective of the Black child in need of "fixing"—a perspective too often found in gap-gazing research—these researchers, if not explicitly, then certainly implicitly, begin with the mathematics brilliance of the Black child.

To contribute to this growing body of knowledge, in this chapter we discuss empirical findings from our three dissertations studies (see Jett, 2009; Stinson, 2004; Williams, 2003); each examined Black young men (and women, in Williams's case) who experienced success in school mathematics (K–16). These findings, which reify the brilliance of Black children, are used as a point of departure for discussing what characteristics an effective mathematics teacher of Black children might possess and for making recommendations to those who are charged with teaching teachers of Black children. We divide the discussion into three sections. First, we provide a brief description of each of our individual research projects, revealing the similarities and differences. Second, we highlight and discuss three factors of intersection across the three studies that specifically relate to mathematics teachers, arriving at a baseline description of an effective mathematics teacher of African American children. And third, we conclude with a discussion of the three factors of intersection in the context of teacher education, making explicit recommendations to mathematics teacher educators.

THREE DIFFERENT STUDIES, ONE SAME PURPOSE

Before we briefly describe our individual studies it is important to provide a context of why each of us chose to study the mathematical achievement and persistence of Black young men (and women). Our individual and collective decision to do so was not intended to somehow romanticize away the numerical evidence of a gap in aggregated standardized measures of "achievement" between Black students and their White and Asian counterparts. Decades of limiting and decontextualized research studies and policy documents continue to report and reify the existence of such a gap (for a discussion about the vital importance of contextualizing research see Martin, Gholson, & Leonard, 2010). Each of us, however, understands the reported gap not as the problem but only as a mere symptom of a systemic "education debt" problem that limits the opportunities to learn for too many Black (and other historical underserved and marginalized) children (Ladson-Billings, 2006). Choosing to document further, as others have done (e.g., Darling-Hammond, 2005; Kozol, 1992), this systemic problem in search of probable solutions presented each of us with a viable direction for our dissertation research. But there were important life experiences as or with mathematically talented and successful Black children that provided another viable, although less explored, direction.[5] These experiences of success clearly showed that the whole story was not being told. In short, the reported findings from the then existing research just didn't add up (Ladson-Billings, 1997). Therefore, rather than work within the more commonly found discourses of deficiency (e.g., Black children lack cognitive and behavioral abilities) or rejection (e.g., Black children oppose schooling and academics), each of us took a new direction, a new turn in our research, the turn toward a new discourse: the discourse of achievement (Stinson, 2006).

Jett's Turn

Jett's (2009; see also 2010, 2011a, 2011b) turn toward achievement chronicled the mathematical experiences of four African American male graduate students in mathematics or mathematics education who experienced academic success as undergraduate mathematics majors. Here *success* was defined as young African American men who earned an undergraduate degree in mathematics and continued in the mathematics pipeline to pursue a graduate degree in either mathematics or mathematics education. During data collection, the participants completed a demographic survey instrument; participated in three semi-structured interviews; read, reflected on, and responded to research literature regarding male African Americans' schooling experiences; and shared personal artifacts pertaining to their mathematical experiences. More specifically, the young men provided counterstories—stories that expose, analyze, and challenge the majoritarian stories of racial privilege (Solórzano & Yosso, 2002)—as they reflected on how they purposefully gained access to and achieved in college mathematics and how they effectively negotiated race and racism during their racialized, K–16 school mathematics experiences.

Foregrounding race and voice throughout the project, critical race theory (e.g., Bell, 1992; Tate 1997) and case study (e.g., Merriam, 1998) were employed. Within this theoretical and methodological framework, an analysis of the data revealed that the participants' achievement and persistence in mathematics could be explained, in part, by their a) internal characteristics such as strong cultural identities as African American men, persistent attitudes, and spiritual connections; b) abilities to negotiate racial injustices as African American men; c) positive mathematics identities[6] developed and nurtured as undergraduate mathematics majors at historically Black colleges and universities; and d) positive outlooks concerning the participation of male African Americans in mathematics. These four categories, present in varying degrees in each participant's counterstories, were earmarked as defining characteristics that strengthened their identities as African American men as well as assisted in establishing strong connections to mathematics, thereby aiding in their mathematical achievement at the K–16 level and beyond.

Stinson's Turn

Stinson's (2004; see also 2008, 2010) turn toward achievement aimed to shed light on the mathematics experiences of academically successful African American male high school students—specifically, the influence of sociocultural discourses on the agency of four young African American men who had demonstrated conventional achievement and persistence in K–12, school mathematics was retrospectively explored. *Agency* here was defined as the young men's ability to effectively negotiate sociocultural discourses that most often characterize African American male students as a) incapable of "measuring up" to predetermined goals and objectives of schools and/or b) lacking in behavioral and social skills and life experiences to be academically successful.

Using participative inquiry with its emphasis of testing theory, experiential knowing, and engagement with others (Reason, 1994), through a series of four semi-structured and narrative interviews, the young men were asked to read, reflect on, and respond to some of the most often-cited research regarding the schooling experiences of African American students (e.g., Fordham & Ogbu's [1986] "burden of 'acting White' theory"; see Stinson [2011]). To analyze the participants' responses, an eclectic theoretical framework that included poststructural theory (e.g. St. Pierre, 2000), critical race theory (e.g., Tate, 1997), and critical theory (e.g., Kincheloe & McLaren, 1994) was applied (see Stinson, 2009). This critical postmodern (Stinson & Bullock, 2012) analysis revealed that each participant had acquired a robust mathematics identity as a component of his overall efforts toward success. How these young men acquired such "uncharacteristic" mathematics identities was to be found in part in how they understood sociocultural discourses of U.S. society and how they negotiated the specific discourses that surround male African Americans. Although their responses were never monolithic, present throughout the counterstories of each participant was recognition of himself as a discursive formation (cf. Foucault, 1969/1972) who, as a *self*-empowered subject, could and did negotiate sociocultural discourses as a means to subversively repeat (cf. Butler, 1990/1999) his constituted "raced" self.

Williams's Turn

Williams's (2003) turn toward achievement explored the schooling and life experiences of African American men (*n*=16) and women (*n*=16) who demonstrated success in science, technology, engineering, and mathematics (STEM); *success* here was defined as enrollment and continued study in a STEM doctoral degree program. Specifically, the study was designed to identify the critical life elements related to the participants' achievement and persistence in STEM. *Critical life element* was operationally defined as an experience, idea, and/or relationship that had a significant effect on a participant's life choices and trajectories regarding STEM. Data for the study were derived from two, semi-structured interviews with each of the 32 participants.

Using interpretive grounded theory (Glaser & Strauss, 1967) as a conceptual framework, an analysis of the interview data revealed four major themes related to the participants' success in STEM. All participants a) had early and ongoing encounters that allowed a significant level of participation in STEM, b) experienced some form of positive personal intervention by other people in their lives in regards to STEM, c) held perceptions of STEM that were linked in some way to positive outcomes (both immediate and eventual), and d) believed that they possessed intrinsic qualities that qualified and prepared them for success in STEM. While all of these themes were evident in the lives of each participant, they exhibited themselves in different ways over the course of participants' lives. Therefore, the findings were organized by three overlapping, developmental periods or phases: the interest-building phase (birth–age 12), the knowledge-acquisition phase (ages 12–20), and the career-building phase (age 20 and beyond). The findings provided new insights into how to effectively prepare all children for a science and technology driven society, and for some, induction into the scientific community. Most importantly, the findings provided valuable information to those directly involved in the education of African American children concerning the critical factors at different phases that increase opportunities for achievement and persistence in STEM.

Summary of Turns

Refuting reductive, decontextualized numerical methods, in each of our studies we employed qualitative methods to contextualize and "get at" the multilayered and complex schooling and life histories and mathematics identities of young African American men (and women). In each of our studies, through a series of interviews, participants provided powerful counterstories to existing research, specifically in regards to the schooling and mathematics experiences of male African Americans. In many ways, individually and collectively, our discursive turns toward achievement demonstrate that at the core of male African Americans' experiences in school and society at large is persistence and triumph (Polite & Davis, 1999). By using academically and mathematically successful African American male voices (Secada, 1995), we took a different and sorely needed trajectory within mathematics education research (Ladson-Billings, 1997). Individually and collectively, our intentions for conducting the studies were to contribute to the expanding discussion about African Americans' successes in and contributions to mathematics, as well as to make salient factors that either narrow or broaden African American children's access to or opportunities for achievement and persistence in mathematics.

Our explicit focus on achievement in each study was purposeful and deliberate, and directly counters the vast body of quantitative research on Black children's schooling experiences in general and mathematics experiences in particular. By ignoring systemic inequities, this vast body of quantitative research is most often rooted in assumed deficiencies or predicted underachievement in mathematics by certain groups based on racial classifications. Strategically employed, these assumptions and predictions only serve to normalize socio-historical and socio-political discourses that produce and reproduce the very disparities under investigation. The pervasive hegemony of this misdirected research becomes increasingly harmful to children when policymakers and educators design interventions aimed at addressing the "achievement gap" based on findings and recommendations from such research. Unfortunately, more times than not, reform efforts in mathematics education have been (are) predicated on research that begins with assumptions and predications that African American children have not achieved nor cannot achieve in mathematics. We believe that ours and other similar studies (e.g., Berry, 2008; Martin, 2000; McGee & Martin, 2011; Moody, 2000; Nzuki, 2010; Walker, 2006) are marking the beginning of a new direction in mathematics education research in regards to historically underserved and marginalized student groups. The proliferation of such studies that take the discursive turn toward the brilliance of "other people's children" (Delpit, 1995) has the potential of ultimately replacing critically unexamined assumptions and predictions with a different discourse about other people's children and mathematics—a discourse of achievement.

MATHEMATICS TEACHERS AND AFRICAN AMERICAN CHILDREN

As we read across our three discursive turns toward achievement, there were many clear commonalities. These commonalities were found not only in the findings from each study but also in the collective participant narratives of each study. Within and across the three studies, participants often used strikingly similar language, metaphors, and examples to tell and clarify their counterstories of success. Given that all three studies were conducted in the southeastern United States, such similarities would be expected (Morris & Monroe, 2009). But these striking similarities go beyond the southeast; strikingly similar narratives or counterstories are reported in the literature from young Black men (and women) living on the west (Martin, 2000) and east (Berry, 2008) coasts and in the northeastern (Walker, 2006) and mid-western (McGee & Martin, 2011) regions of the United States. In many ways, it could be argued that the validity and reliability of the findings from each study are verified not only by the triangulation of data within each respective study but also by the triangulation of data from across the nation. Individual and collectively, in these studies the Black male is no longer "a problem" (Du Bois, 1903/1989, p. 2),[7] as fictions about mathematics and Black children are replaced with facts (Stinson, 2010). Here, we discuss and highlight three overlapping key factors (or facts) identified across our three studies that relate specifically to how mathematics teachers (K–16) were instrumental to the participants' achievement and persistence. Teachers noted as instrumental were those who a) developed caring relationships that reached beyond the mathematics classroom and set high expectations for academic success, b) accessed and built on out-of-school experiences and community funds of knowledge during mathematics instruction, and

c) disrupted school mathematics and mathematics in general as a White institutional space. As we discuss each factor in turn, we not only draw on the collective datasets of each of our respective studies but also on our collective experiences as mathematics (and science) teachers of Black children and youth as well as our experiences as teacher educators and education researchers. It is important to note that throughout the discussion, due to limitations of space, we provide few direct participant quotes. The quotes that are present, however, represent "power in reserve" (Olson, 1991, p. 249).[8]

Developing Caring Relationships and Setting High Expectations

Mathematics teachers' caring relationships with students can be a means not only of making mathematics relevant and meaningful but also of engaging students more deeply in mathematical reasoning and sense making. Generally speaking, teachers who develop *caring relationships* learn about students' ethnic and cultural backgrounds, past and present experiences, interests and disinterests, and dreams and aspirations beyond the mathematics classroom as well as assist students (and themselves) in developing into more ethical selves (Noddings, 1988). More specifically, teachers who develop *mathematical caring relationships* engage in an ever dynamic process of harmonizing with and making interpretations of students' mathematical systems and activities and base future instruction and modeling on those interpretations so that mathematical tasks might be more sensible to students (Hackenberg, 2010). Mathematics teachers who *care with awareness*—somewhat of a combination and extension of caring and mathematical caring relationships—are those who know their students well mathematically, racially, culturally, and politically; make connections with students' cultures and communities; assist students in developing positive racial, cultural, and political identities; reflect critically on their own assumptions about children and communities (e.g., colorblind—i.e., "I don't see color, I just see children"—and deficit perspectives); and labor to neutralize status differences within and beyond the classroom walls (Bartell, 2011). These teachers, in turn, use this knowledge to set expectations of and provide support to students in reaching *their* unique level of mathematics excellence, culturally connecting and scaffolding mathematical tasks based on where students are as they extend and deepen students' knowledge (Bartell, 2011). This pedagogical approach of caring, in many ways, is aligned with culturally responsive teaching where teachers who know their students well solicit ideas from them concerning their interests and experiences both inside and outside of the classroom as a means to build instruction and extend and deepen their knowledge (Gay, 2010). Not surprisingly, caring relationships extend beyond academics in general and mathematics in particular, *self*-empowering students (and teachers) to achieve success on many fronts, assisting in the development of a more humanizing experience for both (Bartolomé, 1994; Freire, 1970/2000).

Throughout the participants' narratives of success, participants often spoke about mathematics teachers who developed caring relationships and set high expectations as instrumental to their mathematics achievement and persistence. One participant spoke explicitly about the importance of a teacher–student relationship that went beyond just the classroom.

> ...[A relationship] where you could go into their...classroom after class and not necessarily talk about academics, but you can still learn from them, learn about life, and learn about just different aspects....I think that it is very important to have a relationship with a teacher...engaging in other things, things as academic, but not necessarily just the subject, where you develop a trust. (Stinson)

The nature of these caring and trusting relationships enables teachers to better understand their students' norms, values, and ethics and use this knowledge to set expectations and to tap into their students' ways of being and knowing—and learning. Consequently, some participants stated that these caring relationships were transformative, shifting their mathematical identities from one of "I can't" to one of "I can":

> Actually, I was really struggling [in math]. And I had [an African American male professor], he was a math *teacher*; he took me under his wings and stuff. I mean I was trying to do well and work hard at it....Stuff wasn't clicking. But after he got with me and showed me how to do things, the ins and outs and stuff like that...basically, he just worked with me, and it was easy. (Jett)

> I had a teacher in eighth grade, an African American male... he seemed really interested in our well-being and seeing us [succeed]. He...wouldn't accept us being mediocre....He...always wanted [us] to do our best...to see us strive to succeed....When he was at school, he was there for us, that was always the sense we got from him. That is the sense I got from the teachers who really, really seemed interested in being there; it is like, they were there for you and they let you know that, too. (Stinson)

It is interesting to note that in both excerpts above, achievement in mathematics was encouraged and piqued by an African American male educator who "worked with me" and "was there for us," bringing to the fore the mathematics brilliance of his African American students. Nonetheless, participants also spoke about non-African American mathematics professors and teachers who formed caring relationships and became instrumental to their achievement and persistence in mathematics. But regardless of the racial and cultural synchronization (Irvine, 1991), or lack thereof, for these relationships to be established and sustained, teachers must allow students to be the experts of their own lives (Gay, 2010). Teachers who develop equitable and just classroom learning environments listen to and learn from the voices of their African American students (and other marginalized groups), working against the silencing that too often occurs (Secada, 1995). By drawing on students' accounts of their own lives, teachers can develop innovative ways to infuse students' lived experiences into their pedagogical practices to meet both affective and cognitive needs of students. In general, effective mathematics teachers of Black children develop caring relationships, setting high expectations as they identify and capitalize on the various emotional and intellectual gifts that Black children bring to the mathematics classroom.

Accessing and Building on Out-of-School Experiences and Community Funds of Knowledge

Successful mathematicians often begin their careers through explorations of the ways in which mathematics was presented in their homes and communities and lived experiences in general during childhood. Often these early mathematics experiences are connected to dramatic play and mediated by friends and family and community members. Generally speaking, throughout life, children and youth, and people of all ages, engage in a variety of formal and informal out-of-school mathematics activities. Effective mathematics teaching requires that teachers access and make connections to these activities; such connections are especially important in the early years of schooling when children have spent more of their life outside of school than in school (McCulloch & Marshall, 2011). Drawing on these out-of-school mathematics experiences increases students' identities with and attitudes toward mathematics as it becomes part of and relevant to them and their lives (e.g., Gutstein, 2003; Nasir, 2007). Equally important is accessing and connecting school mathematics to students' community funds of knowledge: the historically accumulated bodies of knowledge, skills, and attitudes learned and developed in and through interactions among and between community members (González, Andrade, Civil, & Moll, 2001). Effective teachers of African American children scaffold or build bridges between the knowledge students carry with them from their homes and community and what is being taught in the classroom in an effort to facilitate and enhance the learning process (Ladson-Billings, 1994, see also 1995). And when teachers use cultural referents to impart new knowledge, skills, and attitudes, they are not only seeking to empower students academically but also socially, emotionally, and politically (Ladson-Billings, 1994).

Throughout the participants' narratives of success, they provided several explicit examples that demonstrated the importance of their out-of-school mathematics experiences to their in-school learning. One young man described the ways in which his mother developed (consciously or not) the skills he would use as an engineer later in life by insisting that he design and build his own toys (Williams). This experience, while indirectly related to mathematics, provided an informal opportunity through play to develop a degree of familiarity with both mathematics and engineering early in his cognitive development. But in addition to play, the participants also spoke about links between the ways in which they observed mathematics being used in their homes and the work they pursued later in their STEM careers. One participant explained how his mother's work as a seamstress influenced the ways he thought about his own research in computer science. He drew a direct connection of how his research on a developmental programming model called adaptive infrastructures was based on his mother's use of mathematics as a seamstress:

> My mother looks at a person, takes five measurements, then she sews them something to fit. If it doesn't fit, she puts it on them, adjusts in a few strategic areas and it just fits perfectly. My whole research area is built on this adaptive system. You write a program that will run on all these machines and it doesn't run right. But you don't care because the whole idea is it starts at point A and you want to get to point B. [You] adjust certain different pieces based on parameters to let it adjust to get to where you want. (Williams)

These two examples demonstrate the importance of teachers assisting students in bridging the mathematics they acquire in their out-of-school experiences to the mathematics they are taught during their formal schooling. As one participant stated, "It is dire important, that teachers...those who are involved in the child's life, show how [mathematics] is relevant to that life, show how it is applicable to that life" (Stinson). These early out-of-school experiences provided young men with direct access (consciously or not) to mathematics as they applied prior knowledge, honed developing skills, and cultivated new knowledge. In both the play and sewing examples, these young men not only experienced early participation in mathematics but also engaged in the direction and development of the activities that facilitated their learning. In many ways, these mathematical experiences provided opportunities for them to be active creators of mathematics rather than passive recipients. All in all, effective mathematics teachers of Black children use culturally specific pedagogical practices (see, e.g., Leonard, 2008) throughout instruction that build on out-of-school experiences and community funds of knowledge while positioning Black people as doers and creators of mathematics.

Disrupting School Mathematics and Mathematics in General as a White Institutional Space

Societal discourses surrounding mathematics continue to position it as a discipline primarily reserved for elite White men. Children and youth, and people of all ages, internalize these discourses and, in turn, imagine the mathematician as a White, middle-aged, balding or wild-haired man (Picker & Berry, 2000). Publishers of mathematics textbooks and curriculum materials more times than not echo these images by trivializing non-European or ethnomathematics (see, e.g., Powell & Frankenstein, 1997) by making only cosmetic changes (e.g., including culturally diverse names in word problems and pictures of boys and girls with different skin color) (Sleeter, 1997). These trivialized efforts sidestep how mathematics has been historically constructed cross-culturally and how it is reflected differently in different cultural contexts (Sleeter, 1997). In many ways, school mathematics and mathematics in general are masked in a somewhat invisible yet most powerful strategic rhetoric of Whiteness that influences the identities of both those within and without its domain (Nakayama & Krizek, 1995). Too often mathematics education researchers and policymakers mark the perceived mathematics abilities of the White, male child as the point of achievement toward which to aim. Implicit in this marking are messages that Black and other marginalized children are somehow deficient (i.e., they do not meet the predetermined White norms and expectations) and are in dire need of intervention. With all things considered, mathematics education as an enterprise or institution can be characterized as a White institutional space (Martin, 2008): a space that produces and reproduces White privilege and power relations by permitting an exclusively White construction of the institution's norms, values, and ideologies (Moore, 2008). This characterization of Whiteness not only attempts to cloak mathematics education research and policy—as well as mathematics teaching and learning—in apolitical neutrality but also establishes who is allowed (and not allowed) to speak on issues such as teaching, learning, curriculum, and assessment (Martin, 2008).

Mathematics teachers who are effective with African American children build on the brilliance of the non-European roots of mathematics (see, e.g., Joseph, 1990), pushing against this institutional space of Whiteness as they diligently work to erase the "Whites only" face of mathematics (see, e.g., Strutchens et al., 2000). Throughout the participants' narratives, both African American and non-African American teachers were recognized for their efforts. Effective teachers of African American children aggressively put in the minds of their students that African Americans can and *do*, historically and currently, mathematics and push them to *their* mathematical limits, not some predetermined White limit:

> My African American teachers, they did aggressively try to employ the mentality that as an African American we did fit in, they aggressively tried to, not necessarily brainwash, but try to help us realize, put into our minds that we do fit in…we can do the same things [e.g., achieve in mathematics] that [White students] do. (Stinson)

I guess they [two White mathematics teachers] saw some sort of potential in me specifically, because in a [mathematics] classroom full of bright, talented young students, I felt that I was always expected to be the best, and they both pushed me to my limits. (Stinson)

Evidently, effective mathematics teachers of Black children see the potential for mathematics brilliance and contribution in every child. Most often they work from a social constructivism perspective, encouraging students to understand that for all its wonder, mathematics, like everything else, remains a set of human practices grounded in the material world we co-inhabit and co-construct (Ernest, 1998). Throughout mathematics instruction, contributions to mathematics made by all humans are interwoven to make an integrated whole rather than setting "indigenous," "non-school," or non-European (ethno-) mathematics apart for "special projects." Further erasing the boundaries between "academic" mathematics and ethno-mathematics, effective teachers expand the definition of the latter to include socio-historical and socio-political investigations of how mathematics has become what it is today (Pais, 2011). Such teachers, in many ways, have taken the sociopolitical turn (Gutiérrez, 2010), adopting a degree of social consciousness and responsibility in seeing the wider social and political picture of mathematics teaching and learning (Gates & Vistro-Yu, 2003). Understanding and acknowledging that both mathematics and mathematics teaching and learning are profoundly political, effective mathematics teachers of Black children push against mathematics as a White domain by highlighting and building on the mathematics brilliance of Black people.

Summarizing the Effective Mathematics Teacher

So who is this effective mathematics teacher of African American children? Based on our cross readings of the datasets from three discursive turns toward achievement and our collective experiences as teachers, teacher educators, and education researchers, we assert that she (or he) cares with awareness as she develops and nurtures mathematical caring relationships with students. Through knowing students well culturally, emotionally, cognitively, and mathematically, she sets high academic expectations for all students, assisting each in reaching her or his own unique level of excellence in schooling generally and mathematics specifically. She understands the importance of accessing and connecting out-of-school experiences to in-school learning, drawing on these experiences and community funds of knowledge in general throughout mathematics instruction. By connecting and relating mathematics to their experiences and communities, she positions students as contributing producers of mathematical knowledge rather than mere consumers. And when Black children identify themselves as doers of mathematics and creators of knowledge—mathematical or otherwise—both historically and currently, the effective mathematics teacher of Black students is pushing against (consciously or not) an institutional space of Whiteness as she literally changes the face of mathematics. Most importantly, however, the effective mathematics teacher of African American students begins each school day with one simple question: How will I pull the always already to be awakened mathematics brilliance out of each of my children today?

MATHEMATICS TEACHER EDUCATION AND AFRICAN AMERICAN CHILDREN

In this concluding section, we return to our three factors of intersection and make explicit recommendations to those charged with the teaching of teachers of Black children. Because the context of every teacher education program is unique, our intentions here are not to provide prescriptive details of how these recommendations might be carried out, but rather to open up mathematics teacher education to new possibilities. We do, however, provide references to research, when available, that supports our recommendations. Moreover, although there were several recommendations for teacher education that each of us suggested based on the findings from our respective studies, here we stay focused on the three intersecting factors discussed in this chapter and address each one in turn. Recapping the three factors: effective mathematics teachers of African American children a) develop caring relationships that reach beyond the mathematics classroom and set high expectations for academic success, b) access and build on out-of-school experiences and community funds of knowledge during mathematics instruction, and c) disrupt school mathematics and mathematics in general as a White institutional space.

Recommendation #1: Focus on Developing Caring Relationships

The first factor, developing caring relationships and setting high expectations, provides a most direct recommendation: to develop an explicit focus on the three Rs of education in mathematics teacher education, both pre-service and in-service. The three Rs that we are referring to, however, are not reading, writing, and arithmetic but "relationships, relationships, and relationships" (T. J. Smith, personal communication, April 2004). In speaking about their experiences in schools, there was no other single factor identified throughout the participants' narratives that matched the influence on their achievement and persistence as did experiencing committed (and knowledgeable) mathematics teachers who developed caring relationships and set high expectations. These relationships and expectations however must go hand in hand; without first developing caring—mathematical or otherwise—relationships, the setting of any expectation is disingenuous

(Noddings, 1988). How can a teacher set expectations for her students if she does not *know* her students? We believe that developing and nurturing caring relationships is essential for all teachers but critically so for mathematics teachers, given that many students have an aversion toward the discipline. Teaching teachers how to develop these teacher-student relationships is not a substitute for teaching content knowledge, pedagogical content knowledge, curricular knowledge (Shulman, 1986), nor mathematical knowledge for teaching (Hill, Rowan, & Ball, 2005) but rather a crucially needed addition to these somewhat limiting forms of knowledge. With the exception of a few mathematics educators such as Bartell (2011) and Hackenberg (2010), discussions of caring relationships in the context of mathematics teaching and learning are nonexistent. Nevertheless, the individual and collective participant narratives support our recommendation that effective teaching and learning—even mathematics teaching and learning—begins with developing caring relationships.

Recommendation #2: Focus on Culturally Specific Pedagogy

The second factor, related to the first, accessing and building on out-of-school experiences and community funds of knowledge, has deep roots in education that reach back to the early twentieth century (see, e.g., Dewey, 1902/1990, 1938/1998). The idea of connecting out-of-school experiences and community knowledge to curriculum and instruction gained strength in the multicultural movement of the 1970s and 1980s (Banks, 2002), and began to take center stage in the 1990s and 2000s with culturally relevant pedagogy (Ladson-Billings, 1994, 2001) and culturally responsive teaching (Gay, 2010). The importance of connecting experience and community knowledge to the learning of mathematics for Black (and all) children is clearly illustrated in the literature. For instance, the Algebra Project begins with mathematizing an out-of-school physical experience (Moses & Cobb, 2001), and the Bridge Project begins with mathematizing everyday household and community practices (Gonzalez et al., 2001). We recommend that mathematics teacher educators provide pre-service and in-service teachers opportunities to learn about and understand the long history and tangible benefits of experiential and culturally specific teaching; they need opportunities to engage with and struggle through the literature of the past and present. But exposing teachers to literature—theoretical or otherwise—is not enough. Mathematics teachers need opportunities for extended engagement with children in out-of-school, community-based field experiences (see Leonard & Evans, 2008); they need opportunities to become researchers with the goal of learning about the richness of intellectual resources in children's homes and communities (see Civil, 2007). Teaching teachers how to become culturally specific mathematics teachers is an arduous task that demands long-term commitment and personal (and oftentimes uncomfortable) introspection by both teacher and teacher educator (McCulloch & Marshall, 2011). It demands persistence through trial and error. Because, undeniably, both teacher and teacher educator will make multiple missteps as they jointly seek how to best transform out-of-school experiences and community knowledge into culturally specific mathematics teaching in the classroom (see Chu & Rubel, 2010).

Recommendation #3: Focus on the Hegemony of Whiteness

The third factor, disrupting mathematics as a White institutional space, is directly related to the first two and, in many ways, is the linchpin factor. It is the linchpin because it is most doubtful that a mathematics teacher can develop meaningful caring relationships with children or use children's experiences and cultural referents to build instruction if she has not critically examined the uses and *misuses* of school mathematics. Mathematics teachers, for example, too often do not critically question why in a racially integrated school that opportunities to learn are most often split among racial lines, creating "schools within schools": White and Asian children in honors and Advanced Placement courses and Black and Hispanic children in remedial and "low-level" courses (Solórzano & Ornelas, 2004). When teachers—White, Black, or any race—accept such inequities and injustices as the natural order of things without critical examination, they are participating in dysconscious racism (King, 1991). We recommend in order to move preservice and in-service mathematics teachers away from this "uncritical habit of mind" (King, 1991, p. 135), teacher educators need to provide them with learning opportunities to explore Whiteness as a cultural, historical, and social construction, acknowledging it as a racial category with power, privilege, and ideology (Giroux, 1997). Such an exploration and acknowledgement of Whiteness will reveal that White is often the hidden norm by which all others are measured (Roman, 1993). The hegemony of this omnipresent White ideology therefore becomes the major obstacle in developing racially and culturally equitable and just schools and (mathematics) classrooms (Hilliard, 2001). So that teachers might no longer be complicit in the continued mis-education of Black children (Woodson, 1933/1990), mathematics teacher educators must problematize White race and cultural identity in such a way that questions of White advantage, White privilege, and White ways of knowing and learning that dominate U.S. schools are addressed in their courses (McIntyre, 1997). Coupled with disrupting the hegemony of Whiteness, teacher educators need to engage teachers in critical discussions about the historical and continuing struggle for education as freedom that is at the core of the African American experience (Perry, 2003). Conceivably then, the mathematics teacher will begin to see how much of the current policy in mathematics education and how much of her current practices in the mathematics classroom continue to reify the

"Whites only" face of mathematics. And then, just perhaps, she will begin to strategize with her colleagues and students in ways of dismantling the master's house with other people's tools (Lorde, 1979/2007).

Summarizing Recommendations for Mathematics Teacher Educators

So what does the teacher educator who assists her (or his) pre-service and in-service teachers in becoming effective mathematics teachers of Black children do in her courses? She places the value of teachers learning how to develop caring relationships on equal par to them learning content knowledge, pedagogical content knowledge, and curricular knowledge. She emphasizes that teachers must know children culturally, emotionally, cognitively, and mathematically before setting expectations. She exposes teachers to the body of literature, theoretical and practical, on the historical development and tangible benefits of experiential and cultural specific pedagogy. She creates opportunities for teachers to engage with children in extended, out-of-school field experiences so that they may learn about and come to understand children's ways of knowing. She encourages teachers to research how both teacher and student might see everyday home and community activities through a mathematical lens. She supports teachers in determining ways to use community funds of knowledge in general to build in-school curriculum and instruction. She blurs the distinction between academic mathematics and ethnomathematics, challenging teachers' fundamentals beliefs about mathematics as they jointly explore questions such as: What is mathematics? Who decides? Who creates mathematics? Who does mathematics? How is mathematics used? Where is mathematics used? Including the omnipresent but rarely asked question: Why mathematics? And she has explicit, more times than not uncomfortable, classroom discussions about the permanency of race and racism in U.S. schools and society in general and the harmful effects of White hegemony on children, specifically, as it relates to the racialized mathematics experiences of all children and assumptions about who is "good" (or not) at mathematics.

CLOSING REMARKS

Our explicit recommendations to mathematics teachers and teacher educators are not intended to be yet another outside entity telling teachers and teacher educators what they must do. But rather, as insiders, we make the recommendations in the spirit of solidarity with mathematics teachers and teacher educators who, in seeing the potential of mathematics brilliance in every child, believe that all children "regardless of their personal characteristics, backgrounds, or physical challenges must have opportunities to study—and support to learn—mathematics" (NCTM, 2000, p. 12). We believe that our recommendations would conceivably move the mathematics education community closer to achieving the leading NCTM principle: *equity*. Our recommendations are somewhat easily implemented; they require no realignment of curriculum or infusion of additional funds. They do, however, and most importantly, require *will*—a will to excellence in educating all children (Hilliard, 1991).

NOTES

1. Following Terry's (2011) lead, here we use the term *counterstory* rather than the more common hyphenated term *counter-story* (e.g., *counterstory-telling* rather than *counter-storytelling*). More than semantics, the displacement of the hyphen honors the teller's agency by shifting emphasis on the construction and telling of her or his own story rather than merely being a story that is in response to or counter to the often-told stories.

2. Foucault (1969/1972) argued that discourses "in the form in which they can be heard or read, are not, as one might expect, a mere intersection of things and words . . . [but are] practices that systematically form the objects of which they speak" (pp. 48–49). Gee (1999) claimed, capital-D Discourses are innumerable and "are always language plus 'other stuff'" (p. 17); they "are out in the world and in history as coordinations ('a dance') of people, places, times, actions, interactions, verbal and non-verbal expression, symbols, things, tools, and technologies that betoken certain identities and associated activities" (p. 23).

3. Throughout this chapter the terms *Black* and *African American* are used somewhat interchangeably; however, the term *Black* is more inclusive of all people from/within the African diaspora. Additionally, the terms *children* and *child* are a proxy for school-aged children and youth.

4. See the American Anthropological Association's (1998) complete "Statement on 'Race,'" as adopted by the Executive Board May 17, 1998 at http://www.aaanet.org/stmts/racepp.htm.

5. Both Jett and Williams are themselves mathematically talented African Americans, and Stinson taught in an all Black mathematics and science magnet high school for five years, where he encountered hundreds of mathematically talented Black children.

6. Martin (2000) defines *mathematics identity* as students' beliefs about their mathematics abilities, the instrumental importance of mathematics, and the opportunities and constraints that affect their participation in mathematics, as well as their motivations to obtain mathematics knowledge; it is this definition used throughout.

7. Du Bois (1903/1989) opened his collection of essays *The Souls of Black Folk*, writing: "Between me and the other world there is ever an unasked question....How does it feel to be a problem? I answer seldom a word" (pp. 1–2).

8. In an interview conducted by Gary Olson (1991), Clifford Geertz, the renowned anthropologist, suggested that when a researcher uses an anecdote or example in her or his research it should not be an exception to the data but rather representative of fifty other similar anecdotes or examples (i.e., "power in reserve"). The limited quotes provided here do, indeed, represent power in reserve; we encourage readers to explore the complete studies. All three studies (dissertations) are available online at no cost (see Jett, 2009; Stinson, 2004; Williams, 2003). Moreover, where the data were extracted is noted parenthetically by author: Jett, Stinson, or Williams.

REFERENCES

American Anthropological Association. (1998). *American Anthropological Association statement on "race."* Retrieved from http://www.aaanet.org/stmts/racepp.htm

Banks, J. A. (2002). Race, knowledge construction, and education in the USA: Lessons from history. *Race Ethnicity and Education, 5*, 7–27.

Bartell, T. (2011). Caring, race, culture, and power: A research synthesis toward supporting mathematics teachers in caring with awareness. *Journal of Urban Mathematics Education, 4*(1), 50–74. Retrieved from http://ed-osprey.gsu.edu/ojs/index.php/JUME/article/view/128/84

Bartolomé, L. I. (1994). Beyond the methods fetish: Toward a humanizing pedagogy. *Harvard Educational Review, 64*, 173–194.

Bell, D. A. (1992). *Faces at the bottom of the well: The permanence of racism.* New York, NY: Basic Books.

Berry, R., Q. III. (2008). Access to upper-level mathematics: The stories of successful African American middle school boys. *Journal for Research in Mathematics Education, 39*, 464–488.

Butler, J. (1999). *Gender trouble: Feminism and the subversion of identity.* New York, NY: Routledge. (Original work published 1990)

Chu, H. & Rubel, L. H. (2010). Learning to teach mathematics in urban high schools: Untangling the threads of interwoven narratives. *Journal of Urban Mathematics Education, 3*(2), 57–76. Retrieved from http://ed-osprey.gsu.edu/ojs/index.php/JUME/article/view/50/60

Civil, M. (2007). Building on community knowledge: An avenue to equity in mathematics education. In N. Nasir & P. Cobb (Eds.), *Improving access to mathematics: Diversity and equity in the classroom* (pp. 105–117). New York, NY: Teachers College Press

Darling-Hammond, L. (2005). New standards and old inequalities: School reform and the education of African American students. In J. E. King (Ed.), *Black education: A transformative research and action agenda for the new century* (pp. 197–223). Mahwah, NJ: AERA/Erlbaum.

Delpit, L. (1995). *Other people's children: Cultural conflict in the classroom.* New York, NY: The New Press.

Dewey, J. (1990). The child and the curriculum. In J. Dewey, *The school and society; The child and the curriculum* (pp. 179–209). Chicago, IL: University of Chicago Press. (Original work published 1902)

Dewey, J. (1998). *Experience and education: The 60th anniversary edition.* West Lafayette, IN: Kappa Delta Pi. (Original work published 1938)

Du Bois, W. E. B. (1989). *The souls of Black folk.* New York, NY: Bantam Books. (Original work published 1903)

Ernest, P. (1998). *Social constructivism as a philosophy of mathematics.* Albany, NY: State University of New York Press.

Fordham, S. & Ogbu, J. U. (1986). Black students' school success: Coping with the "burden of 'acting White.'" *The Urban Review, 18,* 176–206.

Foucault, M. (1972). *The archaeology of knowledge* (A. M. Sheridan Smith, Trans.). New York, NY: Pantheon Books. (Original work published 1969)

Freire, P. (2000). *Pedagogy of the oppressed* (M. B. Ramos, Trans.). New York, NY: Continuum. (Original work published 1970)

Gates, P. & Vistro-Yu, C. P. (2003). Is mathematics for all? In A. J. Bishop, M. A. Clements, C. Keitel, J. Kilpatrick, & F. K. S. Leung (Eds.), *Second international handbook of mathematics education* (Vol. 1, pp. 31–73). Dordrecht, The Netherlands: Kluwer.

Gay, G. (2010). *Culturally responsive teaching* (2nd ed.). New York, NY: Teachers College Press.

Gee, J. P. (1999). *An introduction to discourse analysis: Theory and method.* New York, NY: Routledge.

Giroux, H. A. (1997). Rewriting the discourse of racial identity: Toward a pedagogy and politics of Whiteness. *Harvard Educational Review, 67,* 285–320.

Glaser, B. G. & Strauss, A. L. (1967). *The discovery of grounded theory: Strategies for qualitative research.* Chicago, IL: Aldine.

Gonzalez, N., Andrade, R., Civil, M., & Moll, L. (2001). Bridging funds of distributed knowledge: Creating zones of practices in mathematics. *Journal of Education for Students Placed At Risk, 6,* 115–132.

Gutiérrez, R. (2008). A "gap-gazing" fetish in mathematics education? Problematizing research on the achievement gap. *Journal for Research in Mathematics Education, 39,* 357–364.

Gutiérrez, R. (2010). The sociopolitical turn in mathematics education. *Journal for Research in Mathematics Education, 41.* Retrieved from http://www.nctm.org/publications/article.aspx?id=31242

Gutstein, E. (2003). Teaching and learning mathematics for social justice in an urban, Latino school. *Journal for Research in Mathematics Education, 34,* 37–73.

Hackenberg, A. J. (2010). Mathematical caring relations in action. *Journal for Research in Mathematics Education, 41,* 236–273.

Hill, H. C., Rowan, B., & Ball, D. L. (2005). Effects of teachers' mathematical knowledge for teaching on student achievement. *American Educational Research Journal, 42,* 371–406.

Hilliard, A., G. III. (1991). Do we have the will to educate all children? *Educational Leadership, 49*(1), 31–36.

Hilliard, A. G., III. (2001). "Race," identity, hegemony, and education: What do we need to know now? In W. H. Watkins, J. H. Lewis, & V. Chou (Eds.), *Race and education: The roles of history and society in educating African American students* (pp. 7–33). Boston, MA: Allyn & Bacon.

Hilliard, A. G., III. (2003). No mystery: Closing the achievement gap between Africans and excellence. In T. Perry, C. Steele, & A. G. Hilliard, III, *Young, gifted, and Black: Promoting high achievement among African-American students* (pp. 131–165). Boston, MA: Beacon Press.

Irvine, J. J. (1991). *Black students and school failure: Policies, practices, and prescriptions.* Westport, CT: Praeger.

Jett, C., C. (2009). *African American men and college mathematics: Gaining access and attaining success.* Unpublished doctoral dissertation, Georgia State University. Atlanta, GA. Retrieved from http://etd.gsu.edu/theses/available/etd-07162009-103245/

Jett, C. C. (2010). "Many are called, but few are chosen": The role of spirituality and religion in the educational outcomes of "chosen" African American male mathematics majors. *The Journal of Negro Education* [Special issue], *79,* 324–334.

Jett, C. C. (2011a). HBCUs propel African American male mathematics majors. *Journal of African American Studies,* Online first. Retrieved from http://dx.doi.org/10.1007/s12111-011-9194-x

Jett, C. C. (2011b). "I once was lost, but now am found": The mathematics journey of an African American male mathematics doctoral student. *Journal of Black Studies, 42,* 1125–1147.

Joseph, G. G. (1990). *The crest of the peacock: Non-European roots of mathematics.* London, UK: Penguin Books.

Kincheloe, J. L. & McLaren, P. (1994). Rethinking critical theory and qualitative research. In N. K. Denzin & Y. S. Lincoln (Eds.), *Handbook of qualitative research* (pp. 139–157). Thousand Oaks, CA: Sage.

King, J. E. (1991). Dysconscious racism: Ideology, identity, and the miseducation of teachers. *The Journal of Negro Education, 60,* 133–146.

Kozol, J. (1992). *Savage inequalities: Children in America's schools.* New York, NY: Harper Perennial.

Ladson-Billings, G. (1994). *The dreamkeepers: Successful teachers of African American children.* San Francisco, CA: Jossey-Bass.

Ladson-Billings, G. (1995). Toward a theory of culturally relevant pedagogy. *American Educational Research Journal, 32,* 465–491.

Ladson-Billings, G. (1997). It doesn't add up: African American students' mathematics achievement. *Journal for Research in Mathematics Education, 28,* 697–708.

Ladson-Billings, G. (2001). *Crossing over to Canaan: The journey of new teachers in diverse classrooms.* San Francisco, CA: Jossey-Bass.

Ladson-Billings, G. (2006). From the achievement gap to the education debt: Understanding achievement in U.S. schools. *Educational Researcher, 35*(7), 3–12.

Leonard, J. (2008). *Culturally specific pedagogy in the mathematics classroom: Strategies for teachers and students.* New York, NY: Routledge.

Leonard, J. & Evans, B. R. (2008). Math links: Building learning communities in urban settings. *Journal of Urban Mathematics Education, 1*(1), 60–83. Retrieved from http://ed-osprey.gsu.edu/ojs/index.php/JUME/article/view/5/5

Lorde, A. (2007). The master's tools will never dismantle the master's house. *In sister outsider: Essays and speeches* (pp. 110–113). Berkley, CA: Crossing Press. (Original work 1979)

Lubienski, S. T. & Bowen, A. (2000). Who's counting? A survey of mathematics education research 1982–1998. *Journal for Research in Mathematics Education, 31*, 626–633.

Martin, D. B. (2000). *Mathematics success and failure among African-American youth: The roles of sociohistorical context, community forces, school influence, and individual agency.* Mahwah, NJ: Erlbaum.

Martin, D. B. (2003). Hidden assumptions and unaddressed questions in *mathematics for all* rhetoric. *The Mathematics Educator, 13*(2), 7–21.

Martin, D. B. (2006). Mathematics learning and participation as racialized forms of experience: African American parents speak on the struggle for mathematics literacy. *Mathematical Thinking & Learning, 8*, 197–229.

Martin, D. B. (2008). E(race)ing race from a national conversation on mathematics teaching and learning: The National Mathematics Advisory Panel as White institutional space. *The Montana Mathematics Enthusiast, 5*, 387–398.

Martin, D. B. (Ed.). (2009). *Mathematics teaching, learning, and liberation in the lives of Black children.* New York, NY: Routledge.

Martin, D. B. (2010). Not-so-strange bedfellows: Racial projects and the mathematics education enterprise. In U. Gellert, E. Jablonka, & C. Morgan (Eds.), *Proceedings of the sixth International Mathematics Education and Society conference* (Vol. 1, pp. 42–64). Berlin, Germany: Freie Universitat Berlin.

Martin, D. B., Gholson, M. L., & Leonard, J. (2010). Mathematics as gatekeeper: Power and privilege in the production of knowledge. *Journal of Urban Mathematics Education, 3*(2), 12–24. Retrieved from http://ed-osprey.gsu.edu/ojs/index.php/JUME/article/view/95/57

McCulloch, A. W. & Marshall, P. L. (2011). K–2 teachers' attempts to connect out-of-school experiences to in-school mathematics learning. *Journal of Urban Mathematics Education, 4*(2), 44–66. Retrieved from http://ed-osprey.gsu.edu/ojs/index.php/JUME/article/view/94/92

McGee, E. O. & Martin, D. B. (2011). "You would not believe what I have to go through to prove my intellectual value!": Stereotype management among academically successful Black mathematics and engineering students. *American Educational Research Journal, 48*, 1347–1389.

McIntyre, A. (1997). *Making meaning of Whiteness: Exploring racial identity with White teachers.* Albany, NY: State University of New York Press.

Merriam, S. B. (1998). *Qualitative research and case study applications in education* (2nd ed.). San Francisco, CA: Jossey-Bass.

Moody, V. (2000). African American students' success with school mathematics. In M. E. Strutchens, M. L. Johnson, & W. F. Tate (Eds.), *Changing the faces of mathematics: Perspectives on African Americans* (pp. 51–60). Reston, VA: National Council of Teachers of Mathematics.

Moore, W. L. (2008). White space. In *Reproducing racism: White space, elite law schools, and racial inequality* (pp. 9–45). Lanham, MD: Rowman & Littlefield.

Morris, J. E. (2009). *Troubling the waters: Fulfilling the promise of quality public schooling for black children.* New York, NY: Teachers College Press.

Morris, J. E. & Monroe, C. R. (2009). Why study the U.S. South? The nexus of race and place in investigating Black student achievement. *Educational Researcher, 38,* 21–36.

Moses, R. P. & Cobb, C. E. (2001). *Radical equations: Math literacy and civil rights.* Boston, MA: Beacon Press.

Nakayama, T. K. & Krizek, R. L. (1995). Whiteness: A strategic rhetoric. *Quarterly Journal of Speech, 81,* 291–309.

Nasir, N. S. (2007). Identity, goals, and learning: The case of basketball mathematics. In N. S. Nasir & P. Cobb (Eds.), *Improving access to mathematics: Diversity and equity in the classroom* (pp. 132–145). New York, NY: Teachers College Press.

National Commission on Excellence in Education. (1983). *A nation at risk: The imperative for educational reform.* Washington, DC: U.S. Department of Education.

National Council of Teachers of Mathematics. (1980). *An agenda for action: Recommendations for school mathematics of the 1980s.* Reston, VA: National Council of Teachers of Mathematics.

National Council of Teachers of Mathematics. (2000). *Principles and standards for school mathematics.* Reston, VA: National Council of Teachers of Mathematics.

Noddings, N. (1988). An ethic of caring and its implications for instructional arrangements. *American Journal of Education, 96,* 215–230.

Nzuki, F. M. (2010). Exploring the nexus of African American students' identity and mathematics achievement. *Journal of Urban Mathematics Education, 3*(2), 77–155. Retrieved from http://ed-osprey.gsu.edu/ojs/index.php/JUME/article/view/45/68

Olson, G. A. (1991). The social scientist as author: Clifford Geertz on ethnography and social construction. *Journal of Advanced Composition, 11,* 245–268.

Pais, A. (2011). Criticisms and contradictions of ethnomathematics. *Educational Studies in Mathematics, 76,* 209–230.

Parks, A. N. & Schmeichel, M. (2012). Obstacles to addressing race and ethnicity in the mathematics education literature. *Journal for Research in Mathematics Education, 43,* 238–252.

Perry, T. (2003). Up from the parched earth: Toward a theory of African-American achievement. In T. Perry, C. Steele, & A. G. Hilliard III, *Young, gifted, and Black: Promoting high achievement among African-American students* (pp. 1–86). Boston, MA: Beacon Press.

Picker, S. & Berry, J. (2000). Investigating pupils' images of mathematicians. *Educational Studies in Mathematics, 43,* 65–94.

Polite, V. C. & Davis, J. E. (1999). Introduction. In V. C. Polite & J. E. Davis (Eds.), *African American males in school and society: Practices and policies for effective education* (pp. 1–7). New York, NY: Teachers College Press.

Powell, A. B. & Frankenstein, M. (Eds.). (1997). *Ethnomathematics: Challenging Eurocentrism in mathematics education.* Albany, NY: State University of New York Press.

Reason, P. (1994). Three approaches to participative inquiry. In N. K. Denzin & Y. S. Lincoln (Eds.), *Handbook of qualitative research* (pp. 324–339). Thousand Oaks, CA: Sage.

Roman, L. G. (1993). White is a color! White defensiveness, postmodernism, and anti-racist pedagogy. In C. McCarthy & W. Crichlow (Eds.), *Race, identity, & representation in education* (pp. 71–88). New York, NY: Routledge.

Secada, W. G. (1995). Social and critical dimensions for equity in mathematics education. In W. G. Secada, E. Fennema, & L. B. Adajian (Eds.), *New directions for equity in mathematics education* (pp. 146–164). Cambridge, UK: Cambridge University Press.

Shulman, L. S. (1986). Those who understand: Knowledge growth in teaching. *Educational Researcher, 15*(2), 4–14.

Sleeter, C. E. (1997). Mathematics, multicultural education, and professional development. *Journal for Research in Mathematics Education, 28,* 680–696.

Solórzano, D. G. & Ornelas, A. (2004). A critical race analysis of Advanced Placement classes: A case of educational inequality. *Journal of Latinos and Education, 1,* 215–229.

Solórzano, D. G. & Yosso, T., J. (2002). Critical race methodology: Counter-storytelling as an analytical framework for education research. *Qualitative Inquiry, 8,* 23–44.

St. Pierre, E. A. (2000). Poststructural feminism in education: An overview. *International Journal of Qualitative Studies in Education, 13,* 467–515.

Stinson, D. W. (2004). *African American male students and achievement in school mathematics: A critical postmodern analysis of agency.* Unpublished doctoral dissertation, University of Georgia, Athens, GA. Retrieved from Dissertations & Theses: A & I. (ATT 3194548).

Stinson, D. W. (2006). African American male adolescents, schooling (and mathematics): Deficiency, rejection, and achievement. *Review of Educational Research, 76,* 477–506.

Stinson, D. W. (2008). Negotiating sociocultural discourses: The counter-storytelling of academically (and mathematically) successful African American male students. *American Educational Research Journal, 45,* 975–1010.

Stinson, D., W. (2009). The proliferation of theoretical paradigms quandary: How one novice researcher used eclecticism as a solution. *The Qualitative Report, 14,* 498–523. Retrieved from http://www.nova.edu/ssss/QR/QR14-3/stinson.pdf

Stinson, D. W. (2010). Negotiating the "White male math myth": African American male students and success in school mathematics. *Journal for Research in Mathematics Education, 41.* Retrieved from http://www.nctm.org/eresources/article_summary.asp?URI=JRME2010-06-2a&from=B

Stinson, D. W. (2011). When the "burden of acting White" is not a burden: School success and African American male students. *The Urban Review, 43,* 43–65.

Stinson, D. W. & Bullock, E. C. (2012). Critical postmodern theory in mathematics education research: A praxis of uncertainty. *Education Studies in Mathematics* [Special issue], *80,* 41–55.

Strutchens, M. E., Johnson, M. L., & Tate, W. F. (Eds.). (2000). *Changing the faces of mathematics: Perspectives on African Americans.* Reston, VA: National Council of Teachers of Mathematics.

Strutchens, M. E., Lubienski, S. T., McGraw, R., & Westbrook, S. K. (2004). NAEP findings regarding race and ethnicity: Students' performance, school experiences, and attitudes and beliefs, and family influences. In P. Kloosterman & F. K. Lester (Eds.), *Results and interpretations of the 1990 through 2000 mathematics assessment of the National Assessment of Educational Progress* (pp. 269–304). Reston, VA: National Council of Teachers of Mathematics.

Tate, W. F. (1997). Critical race theory and education: History, theory, and implications. In M. Apple (Ed.), *Review of research in education* (Vol. 22, pp. 195–247). Washington, DC: American Educational Research Association.

Terry, C. L., Sr. (2011). Mathematical counterstory and African American male students: Urban mathematics education from a critical race theory perspective. *Journal of Urban Mathematics Education, 4*(1), 23–49. Retrieved from http://ed-osprey.gsu.edu/ojs/index.php/JUME/article/view/98/87

Walker, E. N. (2006). Urban high school students' academic communities and their effects on mathematics success. *American Educational Research Journal, 43*, 43–73.

Williams, B. A. (2003). *Charting the pipeline: Identifying the critical elements in the development of successful African American scientists, engineers, and mathematicians.* Unpublished doctoral dissertation, Emory University, Atlanta, GA. Retrieved from Dissertations & Theses: A & I. (AAT 3080373).

Woodson, C. G. (1990). *The mis-education of the Negro.* Trenton, NJ: Africa World Press. (Original work published 1933)

CHAPTER 11

GROWING UP BLACK AND BRILLIANT

Narratives of Two Mathematically High-Achieving College Students

Ebony O. McGee

INTRODUCTION

In this chapter, the author exposes how mathematical brilliance and racial identity operates for two high-achieving African American students, through an analysis of two in-depth narratives. The two upper-level college students, Cory and Tiffany (all names are pseudonyms), were chosen from participants of a larger study of 23 high-achieving mathematics and engineering undergraduate students. Because of similarities and differences, they viewed the significance or operation of race as constraining and/or contributing to their academic and life chances (McGee, 2009). While they do hold different frames on who they are racially, they both struggle to respond to racial discrimination on the part of peers and educational institutions, and they both see mathematics and engineering as a way to prove

The Brilliance of Black Children in Mathematics:
Beyond the Numbers and Toward New Discourse, pages 247–272.
Copyright © 2013 by Information Age Publishing
All rights of reproduction in any form reserved.

the stereotypes wrong. Additionally, they both struggle to articulate and respond to the subtle (and sometimes not so subtle) racism within White peer groups.

Although both students self-identified as African American or Black, they differed in gender (Cory is male and Tiffany a female) and in their understandings of the challenges inherent in functioning within the mainstream society; Tiffany felt that her racial group was undervalued within mainstream culture, while Cory subscribed to sometimes conflicting levels of what he described as "color-blindness" and racial awareness. Their narratives show how our inequitable educational system, *both directly and indirectly,* upholds the marginal position of Black students and provides them with particular challenges, even for students like Cory, who has contrasting viewpoints about his own marginalization. As their quotes below demonstrate, they both shared racialized experiences on the path toward earning their degrees. However, the translations and thus actions, in part based on how they made meaning of these racialized experiences, were unique.

Cory's excerpt below speaks to the lack of Black students and teachers in the field of mathematics and his responsibility to represent positively for his race. Tiffany's illuminates a new reality as she perceives it, racism and racialized experiences in college.

> I have an obligation to work as hard as I can and take things as far as I can just because there aren't that many of us [African Americans]. It's not that I believe I'm better than anybody else; it's just that there aren't that many math professionals to begin with and there are even fewer [who are] African Americans. So I figure if that's what I'm into, if those were the gifts that I was given, I should go ahead to pursue that. *Cory, a 20-year-old Black male, high-achieving mathematics senior college student*

> I mean, I've had my share of "What did she just say to me?" or "Did that teacher just roll his eyes at me?" experiences but nothing like what I faced when I got to college. *Tiffany, 24-year-old Black female, high-achieving electrical engineering senior student*

The excerpts presented above come from two separate interviews obtained in a research study designed to explore the voices and experiences of a select group of academically successful Black college students majoring in mathematics and engineering. At the time of the initial interviews, Tiffany was a senior student in electrical engineering and Cory was a senior mathematics major, both attending urban universities in the Midwest. Although the time between the first interview and the second interview was nine months, Cory was beginning to formulate new and more critical conceptions of what it means to be a Black male in his predominantly White and Asian mathematics classrooms. During our first interview, Cory denied any forms of racial marginalization, but the second interview revealed subtle racial tensions he experienced by being the "only one," meaning the only Black male in most of his upper-level mathematics classrooms. Tiffany highlighted her experiences in the context of growing up "all Black" until she attended her first postsecondary institution, which she described as "lily White." This was in sharp contrast to her K–12 schooling, where she attended mostly Black schools, except for her tracked high school honors classes that where mostly White.

Cory and Tiffany had related yet distinct self and societal constructions of what it means to "be Black" in the contexts of mathematics and engineering participation and what it means to persist in contexts where Black students are few in number and where negative societal beliefs about their abilities and motivation endure. Such considerations around race and science, technology, engineering, and mathematics (STEM) achievement are infrequently raised in current mathematics education research on Black students. Thus this research model conceptualizes important factors regarding school experiences and the role of race and racial identity among mathematically high-achieving Black students.

THEORETICAL FRAME: BLACK RACIAL IDENTITY

There are several approaches used by researchers to better understand the roles that race plays in the education of African American students. The role of racial identity—that is, the extent to which race influences a person's self-concept and consequent behavior (Cross, 1991, 1995; Sellers, Shelton, Cooke, Chavous, Rowley, & Smith, 1998) in the lives of mathematically successful African American students is an understudied phenomenon. The research on academic achievement among African American students suggests that racial identity plays a protective role in their lives (Cross, 1991; Phinney & Alipuria, 1990). A great deal of theoretical and empirical work has demonstrated links between racial identity and positive psychosocial adaptation (Cross, 1991; Miller, 1999). The general literature on development in African American students suggests that racial identity plays a protective role in their lives; students who identify strongly with their racial group are better able to negotiate potentially negative environments and deal with discrimination and prejudice. However, there are several conceptual issues that complicate the discourse regarding the influence of racial identity on academic achievement. One issue is that racial identity is dynamic across situations—that race is not salient to all African Americans and that racial identity should be examined within the larger social and environmental context.

Although racial identity had been traditionally viewed within the context of oppression in America, there is little consideration of the influence of the Black culture in mediating some of these oppressive encounters. Newer models of African American racial identity focus less on generic conceptions of the universality of racial identity (Helms, 1990, 1995). For example, Cross's Nigrescence model, conceptualizes healthy identity development as five stages that culminate in the ability to view the positives and negatives of the Black and White races. In arguing for the interchangeability of identity, racial, ethnic, and cultural identities overlap at the level of lived experience where different layers of identity interact (Cross, Parham, & Helm, 1991).

Newer conceptualizations of racial identity incorporated the varied influence of the Black culture on African American racial identity (Cross, 1991; Sellers, Rowley, Chavous, Shelton, & Smith, 1997). Sellers, Smith, Shelton, Rowley, and Chavous (1998) assert that African Americans are very heterogeneous in the ways in which they define themselves in the context of race and developed a conceptual model that delineate four different dimensions of racial identity. Two dimensions deal with the significance of race. The salience dimension looks at the extent to which race is a defining characteristic at any given moment in time, which is very situationally specific and driven by cues. The centrality dimension is the extent to which an individual normatively defines race. According to these theorists, racial identity in African Americans is the significance and qualitative meaning that individuals attribute to their membership within the Black racial group and their self-concept (Sellers, Smith, Shelton, Rowley, & Chavous, 1998).

Most models of African American racial identity do assume that African Americans' attitudes and beliefs about the role of race in defining themselves has a significant impact on how African Americans experience schooling (Cross, 1991; Miller, 1999; Phinney, 2006; Phinney & Alipuria, 1990; Sellers, et al. 1998a; Sellers, et al., 1998b). The racial identity of Cory and Tiffany play an important role in how they identify, reflect upon, and respond to racialized encounters, and their subsequent actions and life choices. One goal in this study is to explicate how issues of mathematics achievement and racial identity interact for these two high-achieving advanced undergraduate mathematics and engineering students.

NARRATIVES

This study sought to better understand, through their *first-person* accounts, how a select group of African American mathematics and engineering college students maintain, interpret, and frame high academic achievement and success within educational arenas where African American presence is scarce. In particular:

- How do these successful students give meaning to and negotiate what it means to "be African American" in the context of doing mathematics and engineering?

- What meaning(s) is used to describe their racial identity and to what extent do they attribute racial identity to their subsequent success in mathematics and engineering?
- To what extent do academically successful African American college students characterize and respond to learning and participation in mathematics and engineering as racialized forms of experience?

Although each of these questions is quite complex and worthy of study on its own, their interconnectedness and the likelihood of the participants addressing each question led to a rich analysis and powerful results and serve as data for ongoing research.

METHODS

The main purpose of this phenomenological qualitative study is to investigate the life stories and experiences of high-achieving mathematics and engineering Black college students—to unpack their reflections and subsequent actions related to their academic success; and to show what role, if any, race plays. This study includes using a semi-structured interview protocol and a structured debriefing protocol process to identify perceptions and explore the insights and beliefs of the participants.

Participants and Context

Soho University, an institution known for its STEM programming in the Midwest, with a high international population, allowed use of a room housed within their Multicultural Affairs Office for this study's on-campus interviews. At the time of the first interview, Cory was a second-semester junior majoring in applied mathematics.

Medium University is a public institution located in a large urban city in the Midwest that serves the local population. Medium University is primarily a commuter campus and has a student population of about 25,000. About 90 percent of Medium's students are residents of the state. At the time of the first interview, Tiffany was a senior majoring in electrical engineering.

Tiffany and Cory self-identified as African American or Black based on a voluntary response to solicitation to participate in the study, which suggests an acknowledgement of some of the common experiences that exist within and across Black culture. College transcripts indicated their junior and senior academic status, mathematics or engineering program major, and maintenance of at least a 3.0 grade point average (on a 4.0 scale) in mathematics and engineering courses. The junior and senior program status of these upper-level collegiate students serves as an indicator that they have successfully negotiated many of the academic, political, and social obstacles traditional in the pursuit of earning a college degree in mathematics or engineering. The criteria noted emphasize conventional quantitative measures of academic achievement and persistence outcomes (Ewell & Wellman, 2007). However, the analysis of first-person data provided a number of other psychosocial factors that expand the notion of success and resilience for African American students in these disciplines (McGee, 2009, in progress).

My research commitments lie within exploring the voices and experiences of African American students in ways that give primacy to first-hand accounts of individual life experiences. This strategy is especially useful when there is minimal understanding of developmental and contextual processes in a particular field of study or for a particular group of respondents. Knowledge concerning the predictors of African American students' academic successes in mathematics and mathematics based fields, particularly through the K–16 pipeline, remains scant. I conducted three one-and-a-half hour interviews with both students between 2007–2008. All interviews were video and audio recorded, transcribed and coded for analysis. A narrative-based approach was used, which required speaking with students in depth about their lives. The narratives were obtained through the integration of *counter-storytelling* and *life-stories* methodologies.

Delgado and Stefancic (2001) define counter-storytelling as the telling of stories of and by people whose experiences are not often told, such as African American and Latina/o students in city schools. Counter-storytelling can serve as a tool for exposing, analyzing and challenging the stories of those in power, which are often a part of dominant discourse. Counter-storytelling is not only a tool for telling the experiences of marginalized individuals but challenges the dominant stories of those in power, whose stories are regarded as normative (Solorzano & Yosso, 2002a, 2002b; Stovall, 2005).

In addition to counter-storytelling, I incorporated a condensed form of *life-story* method to better understand the construction of African American students' life experiences. Life stories assist in reconstructing the past, present, and perceptions of the future, through each individual's internalized narrative. Although these narratives are evolving and often being renegotiated by new life experiences or new reflections, they still serve as the foundation for making sense of one's life and establishing identities. Life stories position the individual as the expert in reconstructing his or her own life experiences (McAdams, 2008a, 2008b). Life stories do not go through the fact-checking processes that scholarly biographies do; this methodology acknowledges that there is a complex relationship between factual happenings and a person's perception of those events (McAdams, Josselson, & Lieblich, 2006). In sum, these methods serve as a means of fashioning identity, and therefore assist efforts to understand the identities of Tiffany and Cory, from the subjective interpretation of the storyteller.

Coding and Analysis of Interview Data

Interviewing is typically employed as a qualitative research tool, but here I embraced the concept put forth by Schostak and Schostak (2008), where the interview is broken apart as *inter* and *view*, focused on both the divisions and commonalities associated with two people coming together face to face. Its importance is the ability to harness alternative views, for understanding identity, decision-making, and action (Schostak, 2006). Data analysis process is a way to discover "patterns, coherent themes, meaningful categories, and new ideas and in general uncovers better understanding of a phenomenon or process" (Suter, 2007). Data were reviewed after all interviews were transcribed and read twice, analyzed, and interpreted into themes and meanings to lay the foundation of codification. Creswell (2005) suggested that content analysis categorizes, synthesizes and interprets qualitative text data description.

Neuman (2003) described the process of data analysis as a means for looking for patterns to explain the goal of the studied phenomena. The analysis of data used responses from the interviews. From these sources, the emerging themes were categorized and coded. This study used an open-coding system to analyze participants' narrative responses. The unit of analysis was mostly phrase-by-phrase; however there were a smaller subset of codes that were line-by-line and word-by-word (Creswell, 2005; Suter, 2007). Analysis of the interview data incorporated an iterative coding scheme. This process of sorting and resorting, coding and recoding of data led to emergent categories of meaning. As Cory and Tiffany retrospectively detailed their K–12 experiences and college experiences, and speculated about their future lives, I initially focused on these three time periods, knowing that the stages of human development are significant to the progression of their STEM/racial identities. Tiffany's and Cory's narratives are presented with the following themes that resulted from the data analysis: childhood experiences, racialized mathematics experiences, racialized college experiences, and future aspirations.

CORY'S NARRATIVE

Cory's narrative reflected an acceptance of the ideologies and practices of the dominant society. Although he attended K–12 schools with racially diverse populations, his school's tracking policies and practices afforded him few opportunities to interact with fellow Black students. Cory thrived in the traditional educational networks of the cultural mainstream—more particularly, those of the White, heterosexual, middle-class male student; however, he expressed a drive to achieve based on his sense of responsibility to represent the Black community more equitably. Cory at times claims colorblindness, by explicitly stating, "I don't really see color," yet his social networks overwhelmingly reflect a White, middle-class male orientation. Mathematics and soccer consume Cory's academic and social life; and his mannerisms, dialect, and choice of friendships, music, and clothing style are all attributes that are deemed "highly acceptable" within the White culture. Cory, however, does resist the temptation to be critical or judgmental of other Black students who may not be as academically successful as he is. On the rare occasion that he shares a class with another Black student, he often serves as a mathematical mentor to him or her. Cory is very humble about his mathematical achievements and considers his success the result of a gift that he was given, which he has an obligation to cultivate. Although Cory is currently engaged in complex number theory, he jokes about his difficulties with simple mathematics tasks, like calculating a tip or estimating the sales tax on an item.

Cory's Childhood Experiences

Cory described his family life as very stable, where academic achievement was the most important family value, and excelling in education was promoted as the key to lifelong success. Mom, Dad, and two older sisters provided Cory with an early and solid mathematics foundation. Cory's father is an engineer, and Cory's mother is a mathematics teacher who supplemented his mathematics education with several dozen mathematics workbooks "until a math teacher told my mom she needed to stop." Cory regards his K–12 school experiences very positively. He attributes his early success in mathematics in part to his family support (emotional and physical), which cultivated Cory's intellectual abilities in mathematics and science.

Cory recollects his having "only a few" Black friends throughout his life. Although he went to a high school that at the time was 30 percent Black, his honors and AP classes kept him isolated from the Black school population. He spent a significant amount of time ("24/7" by his account) with his high school mathematics club, where he was the only Black student "consistently" in the group. Cory has fond memories of participating in mathematics competitions and referred to his drive for them as "a male thing." Upon further inquiry Cory suggested that his male-dominated mathematics competitions were the result of "guys being more competitive than girls."

Cory recalled not giving too much attention to the fact that he was one of only a few Blacks at his elementary school and high school, and could not recall any experiences of discrimination or racism in his elementary school years. However in his second interview, he declared that, there were instances of racism but, "I didn't care because I was good in mathematics, so nobody was going to get to me." It appeared that his robust mathematics identity did not allow racialized experiences to impact his achievements for motivation to persist in the field. In college contexts, Cory's racial identity offered an appreciation for White mainstream cultural ideologies and limited but supportive contact with Black students.

Racialized College Experiences

Campus engagement is closely related to feeling a part of a campus community and may also serve as a mediator between past and future academic self-concept and achievement. Tinto (1993) suggests that students of color may be more likely than their White peers to struggle with social integration. In Cory's case, he is engaged in campus activities mostly through his fraternity. Although Cory notes that students of color have relatively fewer options for campus activities—and thus socialization (Tinto, 1993)—he seemed to diminish the impact of isolation and lack of social engagement by joining a fraternity that mirrors the racial demographics on Soho's campus. Although, Cory describes himself as "colorblind," when he begins to describe the fraternity his awareness of race appears:

> I'm actually in a fraternity, and the makeup of the fraternity is 90 percent White and 10 percent other. Of course, I'm in that 10 percent and, um, I don't know, we [members of Cory's fraternity] don't even think about it. So, the people I hang out with, it's not actually racially diverse, but it's not something that bothers us [members of Cory's fraternity] because, like I said, we don't even see it.

In Cory's narrative above, he points out not only his stance on diversity, but also his fraternity brothers' perspective and rationale for not associating with a diverse group of people. Although it seems apparent that Cory's emergent identity includes rationalizing racial discrimination, he still maintains a conflicted sense of two selves: one immersed in acceptance of White male mainstream values and one where he privately regards his Blackness as a source of strength. All of his friends hail from the fraternity, which he describes as "an island." However, Cory claims that his White fraternity brothers don't care about race but joke about race in very stereotypical ways:

> They [White fraternity members] crack Black jokes sometimes, and I just kind of look at them funny. About Black people stealing, shooting, smelling bad, you know, the stereotypical stuff, and sometimes it gets a little bit annoying, and, you know, I'm just like, "Yeah, okay," but, I mean, they don't do it all the time. It's not a regular thing. But they kind of figure I'm a brother, and at the same time it is a fraternity; it's a bunch of guys; we give each other a hard time. It is still part of the camaraderie; that's basically what guys do.

The fact that Cory is recounting these racialized situations gives credence to his ability to see and make sense of his racial experiences. However, Cory had insisted several times during both interviews that race is not something that he thinks or cares about. The fact that he "hangs out" with peers who do not look like him does not bother Cory. However, Cory did not suggest that he wanted to or conspired to divorce himself from the larger Black community; instead it appears that he was given too few opportunities to connect with the Black community, and may be socialized to not seek out those relationships. Despite his limited contact with other Black youth, however, there are reflections within Cory's interview that align directly with the values of the Black community.

Cory's Racialized Mathematics Experiences

Cory acknowledged that he enjoys proving his value in his mathematics classes specifically within the context of his Black racial identity:

> Ebony: So you mentioned in your first interview that some students might be thinking, "Who's that Black kid in the back of the class?" How does that make you feel?

> Cory: It's definitely been a driving force of mine because I get a certain amount of satisfaction sometimes being the only African American in the class because inside my own head—I don't even know if there's people in the class that think like this—but, inside my own head, I figure that there are people like, "Who's the Black kid at the back of the class? What's he doing here? He doesn't belong here," and, like I said, it drives me forward knowing that there are people out there that think like that and I get to prove them all wrong. I think it's motivation for me to do better. I mean, when I walk into a class, I'm like, "Oh yeah, okay, now I get to prove something to these people." I mean, basically it's the motivation to myself. So, like I said, even if they don't say it, I'm still thinking it to myself. I'm probably a competitive person, and that is definitely motivating; it's a huge motivator to me—someone thinking that I can't do something—so I wanna go out there and prove them wrong.

Cory armed himself against suggestions or thoughts of in-classroom bias and discrimination by working to "prove them all wrong." McGee's and Martin's (2011) study revealed that many successful Blacks have been and remain very aware of the necessity of proving themselves qualified, particularly in science, technology, engineering, and mathematics (STEM) fields. Although Cory was successful in mathematics, he still felt the personal necessity to struggle to prove or to distinguish himself against negative stereotypes. This phenomena of proving oneself academically within educational environments where negative stereotypes exist, is a recurring theme in the lives of academically high-achieving Black mathematics and engineering college students (McGee, 2009). As a result, McGee and Martin (2011) introduced *stereotype management* to explain high achievement and resilience among 23 Black mathematics and engineering college students. Characterized as a tactical response to ubiquitous forms of racism and racialized experiences across school and non-school contexts, stereotype management emerged along overlapping paths of racial, gender, and mathematics identity development. As demonstrated in the dialogue above, Cory, like many others within the Black culture, have responded to racism and discrimination in ways that have promoted educational attainment and academic excellence (McGee & Martin, 2011). Additionally, Cory expressed an obligation to "work as hard as he can" in order to be recognized for his achievement within the context of being a Black in mathematics. Cory's productivity and success in mathematics appears to stem from his love of the field. However, he expresses a sense of satisfaction in adapting strategies that prove his worth in mathematics classes. Part of his mathematics identity incorporates defense for an environment in which he assumes proof of ability is necessary.

Cory's Passion for Mathematics

In college Cory first took up mechanical engineering because he enjoyed working with his hands; but in engineering the mathematics requirements went only as far as trigonometry, and he ran out of the mathematics electives he would have needed to pursue additional mathematics coursework. At the end of his sophomore year Cory dove headfirst into a new major—mathematics—once he realized that he was more interested in abstract mathematical concepts than the physical aspects of engineering. Despite the heavy emphasis on engineering at his school, Cory's love for mathematics and his strong mathematics identity helped him make the switch to his new major:

> I loved mathematics. It's what I do. Basically, of all the different classes I've taken, the only ones I've enjoyed going to are the math classes. I mean, I just enjoy learning new things and new concepts, the way numbers work with each other—it just interests me. I don't really have to—it's almost—it's still work, but if I had to choose between doing homework in any other class, I mean, I'm gonna do the math because I enjoy it.

Cory's mathematics activities increased in breadth and frequency, as did his success in building his knowledge of mathematics:

> I'm in a number theory class, and number theory is my favorite topic of math; and there's a lot of stuff I don't understand in it. It's probably one of the hardest classes I've ever been in. I took it once in high school, and now I'm taking it again; and it should be easy for me, but, no, it's still difficult, and I don't always get the concepts. But it really doesn't matter because it's just what I love doing. So, yeah, when I don't get something in number theory, it doesn't even bother me. I just keep working at it; I just keep trying. Now, in certain other subjects, like probability or statistics, when I don't get something, I just get frustrated, or I just give up because I don't like doing it.

Cory remains undaunted by the difficulty of his number theory class; and, while he loves abstract and discrete mathematics, he also takes other mathematics classes because he is determined to secure a complete foundation in mathematics. Cory does not have highly sophisticated study strategies: he crams for tests but rarely pulls all-nighters; sometimes he has to stare at a problem for two hours before he tries to tackle it, and he mostly studies alone.

Cory's Future Aspirations

When questioned about the low numbers of Blacks in mathematics, he focused on the lack of financial prosperity that a mathematics career entails. Additionally, Cory spoke of the limited career options for mathematics degree holders; he anticipated that he would end up working in education or as a government actuary.

> I feel it's because it's not really a career choice that makes a lot of money. And it's something that I had to struggle with when I made the switch from engineering, because I came to Soho University doing mechanical engineering and my thought was, "Well, four years from now I'm guaranteed to have a job, and I'm pretty much guaranteed to have a starting salary between thirty and fifty thousand dollars a year." That's, like I said, that's pretty much guaranteed as an engineer. But then when I made the switch to math, it actually took me a while because I had to keep thinking to myself, "What am I going to do when I graduate?"...[B]ut I love math so much I didn't even care. Actually, I asked an actual math teacher before I made the switch what he thought about doing an applied math major and his response to me was "I hope you like math." [I said,] "Yeah, I do," [and he said,] "No, I hope you like math a lot." [*laughs*] I understood what he meant, and I had to make that decision that I may not know where I'm gonna end up when I'm done with this, but this is what I love to do.

Cory's love of mathematics, particularly number theory, outweighed his concerns about job placement. Cory plans to continue his mathematics education by enrolling in graduate school upon completion of his undergraduate program, and after earning his master's degree he will then go on to work as a research and development analyst. Cory's self-appraisal process interacts productively with mathematics and impacts his life outcomes in his continued mathematics educational and job pursuits. Cory's dream career is a job in cryptography, which he described as very high level "puzzle-solving." Even with his impressive skill level in mathematics, Cory still worried about whether he is on par with the other mathematics scholars currently thriving in the field:

> I'm taking all my classes, fulfilling all my requirements, but at the same time I feel like the level of mathematics in the real world is so far out there. Um, I'm

kind of thinking, "Wow, how am I gonna catch up with that?"...So when you ask will I be prepared, I mean, yes, I guess I'm about as prepared as I can be, but I don't know; I'm still concerned about it.

The fact that he is more knowledgeable in mathematics than most college students does not comfort Cory, who continues to work diligently toward his expertise in the field.

Cory's mathematics identity is one of the central psychological elements that defines and maintains his individual sense of self. In the course of experiencing situational stress in his racialized experiences inside of the mathematics classroom, Cory has coped with these encounters through meaning-making processes that overwhelmingly minimized racism. Cory stated that his race was not central to his life; however, he feels positively toward the Black community and defines his racial group membership as Black. From an ecological perspective, Cory has a complex mix of both downplaying race/racism and having "quiet" pride in his Blackness; both seem to serve as protective factors in maintaining a racial identity with which he is comfortable.

Cory excuses the racist jokes and behavior frequently exhibited by his White fraternity brothers. Strong probes for Cory to reveal more details about these racialized incidents were required, as he later admitted that he might be breaking a fraternal code of silence by revealing these racial truths. For example, Cory retold his two race-related jokes of his fraternity brothers about the Obama presidency that suggested the White House would become "more ghetto" with "barbecues on the front lawn" by suggesting that these were examples of innocent banter "between the guys."

Cory is not bicultural (LaFromboise, Coleman, & Gerton, 1993) and does not seamlessly move back and forth between his predominately White world and a Black frame of reference; he appears to be more comfortable operating within the dominant mainstream culture and associating with its architects. Cory has had very little opportunity to develop relationships with other Black people outside his immediate family. Overall, Cory exclaimed that he is "fine" with being Black, while mainly associating with White people, places, and spaces.

TIFFANY'S NARRATIVE

At the time of Tiffany's last interview she was days away from graduating magna cum laude in electrical engineering from Medium University. I first interviewed Tiffany in the house she was raised in by her mother, grandmother, and grandfather in a Black community within an urban city in the Midwest.

Childhood Experiences

Tiffany's childhood is filled with memories of her mother, Dorva, providing her with all the educational opportunities she could get her hands on, and a stringent reward and punishment system that kept her "in check." Tiffany's mother exposed her to a host of academic and mathematics materials through videotapes, audiotapes, workbooks, and summer and after-school programs. Tiffany arranged her evening schedule around her favorite TV show, the *Cosby Show*. Of all the characters, she liked Rudy the most because "she was the closest one to my age, and she was the only dark-skinned kid on TV." Holt (2000) boldly suggests that the formation of Black racial identity is, in part, due to the media's portrayal of Blackness, often through stereotypically negative images and Black celebrities. Since the media regulates the portrayal of Black images, it often perpetuates a Black ideal with images of Blacks who have very light skin and Eurocentric features. At the time of the Cosby Show, there were very few images of dark-skinned people (youth or adults) portraying positive, healthy roles (Frisby, 2004).

Tiffany recalled early memories of exposure to mathematics materials in the home, with flashcards serving as a daily mathematics-learning tool. Dorva, who had attended college in Early Childhood Education and English, pushed Tiffany to excel early in mathematics, as she herself had "never really developed a good background in math." Those early skills paid off and allowed Tiffany, in the sixth grade, to take algebra and trigonometry.

Tiffany was bused to a school in the "good neighborhood," for sixth through eighth grades. She took notice of the differences between her education and the education of her neighborhood friends who attended school close to home. She described the stark disparities between her "privileged" elementary school education and the elementary school on the "Black side of town" where her mother taught. Tiffany pointed out the obstacles of Black students receiving a "decent" education in America.

> ...My mom worked at Carter City School, and I went to the school where she worked at,...and it was like night and day [to my current middle school]. My old school had two different classrooms for one teacher and they had two different grades for one teacher versus my new school. At my new school you would have a teacher and a teacher's aide. And we had a lot of extracurricular activities, like we did swimming....We had a lot of stuff. We had a lot of resources there.

Tiffany attributed her educational and social opportunities to her mother's diligence and skill in creating and maneuvering her through better educational opportunities. Dorva's background as an elementary school teacher created social capital that strategically negotiated educational, cultural, and social opportunities for Tiffany. For example, Dorva was able to help Tiffany secure a scholarship to a top-tier school high school:

My neighborhood school was Corbert and she didn't really want me to go to Corbert. She also didn't want to pay for me to go to a private school. So she knew somebody who knew the swimming coach at College Pathways High School. So they were like, "If she swims on our swim team she can go to our school," and that's basically how I got in.

For African American female adolescents, being connected to others within the community, especially mothers, may assist in developing their abilities to integrate academic and life selves. Although mother and daughter relationships are dynamic and complex, literature shows a strong influence of a mother's expectations for her daughter (Chope, 2006). Rigsby, Stull, and Morse-Kelley's (1997) study showed that a mother's expectation that her daughter would attend college was the strongest predictor of the daughter's own educational expectations.

In high school, Tiffany was enrolled in all regular classes, except for mathematics, which were all honors. For the first two years of high school, most of Tiffany's friends were in the regular classes. She admittedly was influenced by some negative school behavior they demonstrated, and she believed her grades suffered as a result. In her junior and senior year, after a serious punishment from her mother for cutting class, Tiffany began to adopt a more positive academic identity. She was placed in additional honors classes and began to have friends who also had strong academic identities.

Tiffany's first two years for high school were a challenge for her academically. Her grades started to improve by the end of her sophomore year, and her friends changed as she adopted a stronger academic identity. Her mom was able to redirect some of Tiffany's academically debilitating behaviors through punishment, which helped Tiffany to maintain and develop her academic identity during some rough spots. Tiffany's mathematics achievements remained solid through her high school education, and she credits her success in mathematics as a catalyst for excelling in her classes. Tiffany viewed herself as studious, competent, and capable, intelligent, and multi-talented. She discussed the importance of choosing friends wisely:

When I first got into College Pathways High School, the group of friends... weren't necessarily honors students; they weren't as studious. They weren't as motivated and that's who I hung out [with]. And then like by my junior year when I started taking more honors classes, 'cause I like the way it worked in College Pathways High School—if you get an A in like regular classes, you can get promoted to the honors class. So when I started taking, like my junior or senior year when I was taking honors classes, my group of friends changed. And they were more studious and they thought about college....I guess they were taught to think differently.

Tiffany demonstrates that the relationship between teen students and their peers can affect attitudes about the value of school, academic capabilities, motivation and drive, and subsequent academic attainment (Eccles, 2004; McGee & Tyler, in progress).

Tiffany's Racialized Mathematics Experiences

Tiffany spoke of blossoming academically in high school, but not until her junior year when she started interacting with the "smart kids." She began to excel in classes outside of mathematics, moving into more honors classes, and began to like "nerdy people." Although Tiffany's high school was around eighty percent Black, the honors classes were predominantly White. Tiffany's first two years of "regular" classes were full of Blacks, Latinos and other non-Asian students of color, and for Tiffany the classroom racial make-up "was no accident":

> When I got there it just seemed like they had this program that was an honors program, but the honors program was mostly White. I think that was their way of trying to retain the few White people they had at College Pathways High School. That's what I thought at the time.

Research on tracking suggests that in schools with diverse populations most historically marginalized students are tracked into remedial or less stimulating courses, thereby leaving the advanced placement and honors courses to predominantly White and Asian (College Board, 2012). For the few students like Tiffany who are placed in advanced high school mathematics, Black students often experience racial isolation (McGee & Spencer, 2012).

Tiffany refers to particular experiences on her high school swim team as "a little racist," which, upon reflection, she decided not to call racist. She felt something racially was happening, but didn't know how to categorize it. Similarly, Tiffany described many of her mathematical experiences with the same recollection. She would not call it "outright racist" because in her mind it means something blatant and undeniable. She did discuss her mathematics and engineering classes with this sentiment, "its not in your face but its [racism] kinda still there." Tiffany concluded that the lingering stereotypes being in the air did not bother her too much and she tried to focus on the "bigger picture."

Tiffany's Racialized Experiences in College

After graduating high school with a 3.7 GPA in mathematics and an overall 3.0 GPA, Tiffany enrolled in a large, predominantly White, out-of–state Midwestern university, in part because she started the application process late and in part because she "was tired of Black people." Tiffany remembers college as the first time she ever experienced racism, and in two words, she "hated it." Tiffany experienced racial segregation as the only Black person in her dorm and most of her classes.

The transition from Tiffany's majority Black community to a majority White university in a majority White surrounding rural community was, in her words, "isolating and alienating," and she experienced frequent bouts of non-acceptance and "quiet" rejection. For Tiffany, sometimes this rejection manifested as subtle disapproving glances from classmates, which she translated as, "What are you doing here?" and the invisibility of her presence:

> And I get there and I did not like it. It was White kids around. I didn't meet no Black people! I was the only Black person on my floor and my roommate ain't never been around a Black person before. It was horrible!

Another stark reality that Tiffany experienced in her second year at this predominantly White university was a Black female roommate who did not associate with Black people and who had no problem expressing her negative views about Blacks. Her roommate spoke stereotypically (e.g., "Ghetto Black people," "I'm not cool with Black people"), and only had White friends. Tiffany ended up in a physical altercation with this Black roommate:

> My second year I lived with two Black girls. One of the girls…was from Minnesota and we are still friends. And the other girl was from like a town in Ohio. And the girl that was from Ohio, she was one of those girls that had never really been around Black people either. And the thing is, when we was moving in together, I was like okay, this is way better than living with, you know, some White chick. But the thing is like that to me she was just as racist as a White person. Because she came at me like, "I don't like Black girls, see, because they petty, they like to fight." So, yeah, I got in a fight with her, too, because she said "I hate Black bitches."

Some Black individuals describe others within their same race in ways that idealize Whites and devalue Blacks (Anglin & Wade, 2007). Tiffany's roommate seemed to be very critical of the Blacks that she perceived were not like her and relied on negative conceptions of Black individuals, to make herself appear more astute and worthy. Her sentiments were sharply condescending of African Americans who are economically "poor" and "uneducated." Tiffany's racial pride was hurt by this Black student's negative descriptions of Black people and thus felt like she had to defend not only herself but her race. Tiffany also described this incident as one of the first times she ever had to "deal with" her identity as a Black woman. She expressed feeling an extra sense of pain because this student was not just Black, she was a female. Six months after this incident, Tiffany transferred out of that university and went back home to attend school at a local community college. She sums up her experiences at the predominantly White university as follows:

> ...People was looking at me like "you're gonna fail, why are you here?" That's the feeling that I got. It wasn't like a lot of outreach, it wasn't a lot of encouraging people or whatever. So I don't know. It was a big regret for me that I went out of state.

Tiffany expressed that, in order to maintain her academic and racial identity, she had to leave her predominantly White rural institution and attended a college locally within her home city.

Similarly, Leonard (2009) vividly retells her initial college experiences of attending a predominately White institution, Boston University. Leonard's time at Boston University during the mid-1970s was filled with racialized instances, including switching from a White roommate to a Black one, enduring racist jokes and other expressions, and experiencing racialized and gendered assaults inside college classrooms, even as the University made strides in their initiatives to support Black students. During the summer, a St. Louis brokerage firm did not hire her even though she passed their mathematics test because she was Black. Leonard, too, transferred out of that college environment to a more affirming one. Leonard (2009), like Tiffany, in part indicts predominately White universities for not going far enough and, as a result, leaving some Black students racially and socially stranded.

Tiffany enrolled in a community college because of the affordable tuition and her limited financial resources. Although the community college was majority African American, followed by Latinos, she described mathematics classes as "full of foreigners," which she later defined as Africans, Arabs, and Indians:

> It was weird. I wouldn't say it was a bad or a good [school], but it was weird because when I moved to Upper Community College, um, all you see is like Black people. A large percentage of Upper Community College is Black. But then when you take the calculus classes, it's full of foreigners, and I'm like what's this about? It was kind of weird.

The majority of students in Tiffany's calculus classes at the community college intended to major in electrical engineering at Medium University, a local mid-sized university, once they had completed the requirements at the community college level. Since most of the students in her calculus class were expected to major in electrical engineering, she did too, although, in her words, she "really didn't know what electrical engineering is and what it does." In her junior year at the community college, she transferred to Medium University.

Tiffany discussed the highly competitive mathematics and engineering culture that existed when she transferred to this commuter, research-driven university. Many of the professors and students in mathematics and engineering adamantly practiced the "old" competitive "drill and kill," which contributed to a lack of collaboration between students in test sharing and homework study, particularly between the races. For Tiffany this competitive environment made the establishment of natural-forming mentoring and peer relationships difficult, partly because she felt marginalized from the mainstream mathematics population who were, again, mostly "foreigners:"

> This one Indian kid, he did me dirty. We had a math class together. When I first got to math class I was cool with him. I had some old math exams, so he was like "Can I get a copy of it 'cause I don't have a copy of the second exam." So I'm like "Cool sure." Then I asked him for a copy of the second exam. I wanted to see what he was going to say and then he's like, "No, no, I don't have anything, do you have anything?" So then I walk in the library and I see him and all his little Indian buddies studying for the math exam. And I busted him out right in front of his buddies. I was like, "I thought you didn't have the second exam," and then he tried to say, "Man I just found it, this isn't mine, it's my friend's." And I was like, "You got my number, you could've called me and told me you all was going to study."

Contrastingly, along with many of the students I interviewed, Tiffany spoke of organizational mentorship (but not individual mentorship) through the National Society of Black Engineers (NSBE). The NSBE is a nonprofit organization with the purpose of increasing the number of African American engineers who excel academically, succeed professionally, and positively impact the community. Tiffany's peers in the NSBE provided emotional and intellectual support. She found solace in the NSBE as they provided her the community she desired. Although she had no formal individual mentors, she knew a number of people who she admired, such as the individuals at her church.

Tiffany stated that she "really" didn't experience racism until college, she appeared to accept some stereotypes about the lack of Black students (besides herself) in mathematics. Embedded within Tiffany's perspectives were negative associations resulting from these stereotypes perpetuated in society, such as "Black people just don't like math." Ironically, she felt a personal responsibility to prove this stereotype wrong:

> Ebony: Do you know of any stereotypes about African Americans participating mathematics?
> Tiffany: Yeah they're at the bottom.
> Ebony: How do you feel about that?
> Tiffany: I try to just prove them wrong.

Similar to Cory, Tiffany's high school experiences and her incorpora-tion of what McGee and Martin (2011) describe as stereotype management facilitated success in her engineering classes; however, she maintained an intense and perpetual state of awareness that her racial identity and Black-ness are undervalued and constantly under assault.

Tiffany's Future Aspirations

Tiffany is currently pursuing a Master's degree in Electrical Engineer-ing at an institution in the Midwest known for its prestigious engineering programs. Her concentration is in Electrical/Electronic Manufacturing, and she recently worked as a research assistant at an optics laboratory. Tif-fany now has a more realistic and critical understanding of the functions of "being a Black woman" in engineering. She sighs when she exclaims that, "After a while I just got used to it!" She further describes the "it" as sometimes subtle and sometimes not so subtle demonstrations by her peers, colleagues, teachers, researchers, and employers who let her know that her position as a high-achiever in electrical engineering is constantly under cri-tique.

Tiffany's mother had social capital in terms of navigating a host of educa-tional opportunities for Tiffany. Tiffany also benefited from living with her grandparents, who offered ample dosages of wisdom and common sense, which helped her to stay grounded in times of turmoil. Tiffany admitted to not giving too much attention to racial attitudes in her K–12 years. Tiffany did reflect on striking racial boundaries (e.g., her swim team was predomi-nately White, her neighborhood predominately Black, her AP mathemat-ics classes being mostly White, her only Black peers in the remedial high school mathematics classes). Tiffany explained, "Although it was wrong, it was kinda normal," referring to the Black-White dynamics at her high school. It did not start to be "real wrong" until she arrived at her rural historically White university. Tiffany could not escape the racism she expe-rienced there. It was in the demographics of the students, faculty, and ad-ministration and expressed in their discriminatory treatment. Even some of the Black students, like Tiffany's roommate, did not feel a sense of solidar-ity although they, too, were Black. Tiffany's racial identity was challenged in this environment, and she seemed psychologically unable to handle the racialized experiences on this campus, even though her engineering grades stayed above a 3.0 GPA. She found solace with a more diverse campus envi-ronment and expressed her racial and engineering identity through a com-bination of Black-centered and engineering-centered organizations, most notably the National Society of Black Engineers. However, she has come to the unfortunate conclusion that her future as an electrical engineer will be filled with moments where her intellect and abilities will be on trial.

DISCUSSION

As a subgroup of a larger study, Tiffany's and Cory's social background characteristics and mathematics experiences varied from one another, reflecting the diversity of high-achieving Black student experience. Additionally, these two students provided a unique opportunity to explore the factors that spur Black students' mathematics and engineering perseverance and buffer them against academic underachievement and the factors that lead other similar students to lose hope and give up.

In different ways both students were socialized early on to adopt a healthy mathematics identity through interpersonal contexts. Explicit considerations revealed in their narratives are the key roles that their families played in fostering positive mathematics socialization and identities. This socialization served dual purposes as both a formal educational resource (e.g., knowledge of mathematics, help with homework) and informal support (e.g., conversation about the math and engineering fields, mentorship). As a result, they had early on bought into the idea that mathematics and engineering were important and became cognizant of the fact that mathematics knowledge could bring new opportunities to their lives. Tiffany's new peer relations influenced her decision to go to college and her college selection. While Cory appeared to carve a more autonomous mathematics trajectory, Tiffany was influenced by her mother's guidance and support. Tiffany frequently mentioned her mother requiring and assisting in establishment of a strong mathematics foundation. Cory, too, said his mother was instrumental, but he stressed his own early passion to pursue mathematics knowledge.

The identity-related considerations offered a more robust understanding of the major factors potentially influencing Tiffany's and Cory's mathematics developments. Although parents, siblings, educators, and so on aided in establishing their mathematics and life identities, how Tiffany and Cory understand and think about events and circumstances is critically important in determining the impact of these factors.

Tiffany achieved, academically and socially, in spite of conflicting racial ideologies, while Cory embraced the mainstream ideology for his personal and academic success. Tiffany used her knowledge of the dominant cultural capital in co-construction of her own Black cultural capital (Carter, 2003). Cory uses his dominant cultural capital without the co-construction of Black cultural capital because he believes the dominant capital is in line with his identity as a "colorless man." Tiffany and Cory also used their awareness of racial discrimination as an extra motivation to succeed (McGee & Martin, 2011). However, constant acts of discrimination and racial isolation at Tiffany's first college campus, even the subtle acts, caused her to experience diminishment of her confidence and led her to feel a sense of isolation and alienation, dissonance, and eventually a discontinuance of her education at that particular university. The two narratives show that racism and racialized experiences have strong implications for how these students navigate the educational terrain.

These two narratives reveal several issues about the mathematics education of high-achieving African American students. Through Cory's story we learn that some mathematically talented Black students, at K–12 schools with little diversity, are tracked out of significant opportunities to interact with other Black students. In college both students experience increasingly less diverse environments as they progress in mathematics and engineering. This structural aspect of their academic experiences has serious implications for their lives in school, and for opportunities to have Black peers. It also has implications for their opportunities to develop and express Black racial identity. Tiffany relies on organizations such as the National Society of Black Engineers as a place to openly exercise her racial identity. Although Cory does not appear to possess anti-Black sentiments, his interpretations of racial situations reveal much more than they hide; as he continues to be affected by racism, he at times seems to diminish its existence, and he very well may use this as a strategy to protect his academic and racial identity. Inequities within the educational system have somewhat inhibited Cory's and Tiffany's involvement and exposure to building healthy relationships with other Black students. One cannot help but ponder the role of structural racism that impacts the development of some high-achieving Black students in providing few opportunities to interact with people that look like them.

Although these two students grew up knowing and capitalizing on their brilliance and are by most standards a success, unemployment rates remain higher for historically underrepresented scientists and engineers than for White scientists and engineers, and there are even greater disparities by gender (National Science Foundation, 2011). As they enter the STEM workforce they will face additional forms of racialization and discrimination, first by the sheer numbers of STEM professionals, who are not representative of the U.S. population nationally (National Science Foundation, 2011). African Americans only represent about seven percent of all computer and mathematics workers (although they are roughly 12.5 percent of the population); Hispanics comprise about five percent of computer and mathematics workers (although they are roughly 30 percent of the population). Nevertheless, the STEM field has remained fairly homogenous over the past ten years (National Science Foundation, 2011). So even if you grow up Black and brilliant and find academic and personal success in mathematics and engineering, racial and gender biases will continue to plague you along your life and career trajectories. This fact makes Cory's and Tiffany's narratives sobering analysis of Black brilliance in mathematics and engineering brilliance.

CONCLUSION

The stories of Tiffany and Cory represent their ongoing sacrifices and a variety of public and private forms of resistance, resilience, accommodation, and struggle as both students have negotiated their lifelong achievements and success in mathematics and engineering. These students remind us, as researchers, that their lives and the world they live in are not one-dimensional. There is no one meta-story that defines Black students in mathematics and engineering. Examining a broader array of individual, contextual, and social factors is crucial to unpacking and discerning students' meaning-making processes from their narratives.

As high-achieving African American college mathematics and engineering students, Tiffany and Cory have impressive academic credentials and internalized beliefs about their abilities in mathematics and engineering. Some of their narratives are tied to larger stories of oppression, resistance, and liberation, provide a voice for those who are historically marginalized in the literature, and demonstrate how students like Tiffany left a predominately White university but not her educational pursuits, to protect her psyche against overt forms of racism. Taken in total, these stories seek to contribute to the growing body of literature that portrays achievement from the eyes of the talented Black students. Their narratives can also contribute to the advancement of theories addressing how conceptualizations of race with their structural and cultural manifestations impact and shape high-achieving Black students' lives in STEM fields.

NOTES

1. I owe a debt of thanks to the general editors, Jacqueline Leonard and Danny Martin, as well as *Na'ilah Nasir*, for very helpful and supportive comments.

REFERENCES

Anglin, D. M., & Wade, J. C. (2007). The effects of racial socialization and racial identity on black students' adjustment to college. *Cultural Diversity and Ethnic Minority Psychology, 13*, 207–215.

Carter, P. L. (2003). Black cultural capital, status positioning, and the conflict of schooling for low-income African American youth. *Social Problems, 50*(1), 136–155.

Chope, R. C. (2006). *Family matters: The influence o f the family in career decision making.* Austin, TX: Pro-Ed.

College Board. (2012). *The 8th annual AP Report to the nation.* Reston, VA: College Board

Creswell, J. W. (2005). *Educational research: Planning, conducting, and evaluating qualitative and quantitative research* (2nd ed.). NJ: Pearson Education.

Cross, W. E., Jr. (1991). *Shades of Black: Diversity in African-American identity*. Philadelphia: Temple University Press.

Cross, W. E., Jr. (1995). The psychology of nigrescence: Revising the Cross model. In J. G. Ponterotto, J. M. Casas, L. A. Suzuki, & C. M. Alexander (Eds.), *Handbook of multicultural counseling* (pp. 93–122). Thousand Oaks, CA: Sage.

Cross, W., Parham, T., & Helms, J. (1991). The stages of Black identity development: Nigrescence models. In R. Jones (Ed.), *Black psychology* (3rd ed., pp. 319–338). Berkley, CA: Cobb & Henry

Delgado, R., & Stefancic, J. (2001) *Critical race theory: An introduction*. New York: New York University Press.

Eccles J. S. (2004). Schools, academic motivation, and stage-environment fit. In R. M. Lerner & L. Steinberg (Eds.), *Handbook of adolescent psychology*, (pp.125–53). Hoboken, NJ: Wiley.

Ewell, P., & Wellman, J. (2007). *Enhancing student success in education: Summary report of the NPEC Initiative and National Symposium on Postsecondary Student Success. National Postsecondary Education Cooperative (NPEC)*. Available at: http://nces.ed.gov/npec/pdf/Ewell_Report.pdf.

Frisby, C. M. (2004). Does Race Matter? Effects of idealized images on African American women's perceptions of body esteem. *Journal of Black Studies, 34*(3), 323–347.

Holt, T. C. (2000). *The problem of race in the twenty-first century*. Cambridge, MA: Harvard University Press.

Helms, J. E. (Ed.). (1990). *Black and White racial identity: Theory, research, and practice*. Westport, CT: Greenwood Press.

Helms, J. E. (1995). An update of Helms' White and people of color racial identity development models. In J. G. Ponterotto, J. M. Cases, L. A. Suzuki, & C. M. Alexander (Eds.), *Handbook of multicultural counseling* (pp. 188–198). Thousand Oaks, CA: Sage.

LaFromboise, T., Coleman, H. L., & Gerton, J. (1993). Psychological impact of biculturalism: Evidence and theory. *Psychological Bulletin, 114*, 395–412.

Leonard, J. (2009). "Still not saved": The power of mathematics to liberate the oppressed. In D. B. Martin (Ed.), *Mathematics teaching, learning, and liberation in the lives of Black children* (pp. 304–330). New York: Routledge.

McAdams, D. P. (2008a). Personal narratives and the life story. In O. John, R. Robins, & L. Pervin (Eds.), *Handbook of personality: Theory and research* (pp. 241–261). New York: Guilford Press.

McAdams, D. P. (2008b). American identity: The redemptive self. *The General Psychologist, 43*, 20–27.

McAdams, D. P., Josselson, R., & Lieblich, A. (Eds.). (2006). *Identity and story: Creating self in narrative*. Washington, DC: APA Books.

McGee, E. O. (2009). *Race, identity, and resilience: Black college students negotiating success in mathematics and engineering*. Unpublished doctoral dissertation, University of Illinois, Chicago.

McGee, E. O. (in progress). Fragile & robust mathematics identities: An emerging framework for mathematically high-achieving Black college students.

McGee, E. O., & Martin, D. B. (2011). "You would not believe what I have to go through to prove my intellectual value!": Stereotype management among academically successful Black mathematics and engineering students. *American Education Research Journal, 48*(6), 1347-1389.

McGee, E. O., & Spencer, M. B. (2012). Theoretical analysis of resilience and identity: An African American engineer's life story. In E. Dixon-Román & E. W. Gordon (Eds.), *Thinking comprehensively about education: Spaces of educative possibility and their implications for public policy* (pp. 161–178). New York, NY: Routledge.

McGee, E. O., & Tyler, A. T. (in progress). The early lives of mathematically high-achieving Black high school students in urban schools. (For submission to *Urban Education*).

Miller, D. (1999). Racial socialization and racial identity: Can they promote resiliency for African American adolescents? *Adolescence, 34,* 493–501.

National Science Foundation, Division of Science Resources Statistics. (2011). *Women, minorities, and persons with disabilities in science and engineering: 2011. Special Report, NSF 11-309.* Arlington, VA. Available at http://www.nsf.gov/statistics/wmpd/.

Neuman, W. L. (2003). *Social research methods: Qualitative and quantitative approaches* (5th ed.) Boston: Allyn and Bacon.

Phinney, J. S. (2006). Ethnic identity in emerging adulthood. In J. J. Arnett & J. L. Tanner (Eds.), *Emerging adults in America: Coming of age in the 21st century* (pp. 117–134). Washington, DC: American Psychological Association Books.

Phinney, J. S., & Alipuria, L. (1990). Ethnic identity in older adolescents from four ethnic groups. *Journal of Adolescence, 13,* 171–183.

Rigsby, L. C., Stull, J. C., & Morse-Kelley, N. (1997). Determinants of student educational expectations and achievement: Race/ethnicity and gender differences. In Taylor, R. D., & Wang, M. C. (Eds.), *Social and emotional adjustment and family relations in ethnic minority families* (pp. 201–224). Erlbaum, Mahwah, NJ.

Schostak, J. F. (2006) *Interviewing and representation in qualitative research.* Open University Press: London, UK.

Schostak, J. F., & Schostak J. R. (2008). *Radical research designing, developing and writing research to make a difference.* Routledge: London, UK.

Sellers, R. M., Smith, M. A., Shelton, J. N., Rowley, S. A. J., & Chavous, T. M. (1998a). Multidimensional model of racial identity: A reconceptualization of African American racial identity. *Personality and Social Psychology Review, 2*(1), 18–39.

Sellers, R. M., Shelton, J. N., Cooke, D. Y., Chavous, T. M., Rowley, S. A. J., & Smith, M. A. (1998b). A multidimensional model of racial identity: Assumptions, findings, and future directions. In R. L. Jones (Ed.), *African American identity development: Theory, research, and intervention* (pp. 275–302), Hampton, VA: Cobb & Henry.

Sellers, R. M., Rowley, S. A. J., Chavous, T. M., Shelton, J. N., & Smith, M. (1997). Multidimensional inventory of Black identity: Preliminary investigation of reliability and construct validity. *Journal of Personality and Social Psychology, 73*(4), 805-815.

Solórzano, D. G., & Yosso, T. J. (2002a). A critical race counterstory of race, racism, and affirmative action. *Equity and Excellence in Education, 35*(2), 155–168.

Solórzano, D. G., & Yosso, T. J. (2002b). Critical race methodology: Counter storytelling as an analytical framework for educational research. *Qualitative Inquiry, 8*(1), 23–44.

Suter, G. W., II. (2007). *Ecological risk assessment* (2nd ed.). Boca Raton, FL: CRC Press.

Stovall, D. (2005). A challenge to traditional theory: Critical race theory, *African-American community organizers and education. Discourse, 26*(1), 95–108.

Tinto, V. (1993). *Leaving college: Rethinking the causes and cures of student attrition.* Chicago: The University of Chicago Press. 2nd ed.

Tinto, V. (1993). *Leaving college: Rethinking the causes and cures of student attrition* (2nd ed.). Chicago, IL: The University of Chicago Press.

CHAPTER 12

MATHEMATICS LITERACY FOR LIBERATION AND LIBERATION IN MATHEMATICS LITERACY

The Chicago Young People's Project as a Community of Practice

Denise Natasha Brewley

The mission of the Young People's Project is to use mathematics literacy as a tool to develop young leaders and organizers who radically change the quality of education and quality of life in their communities so that all children have the opportunity to reach their full human potential.

—YPP Members, 2006

INTRODUCTION

In view of research that has examined the underachievement and limited persistence of Black students in mathematics (College Board, 1999; Jencks & Phillips, 2001), the accumulation of these findings has created master-

The Brilliance of Black Children in Mathematics:
Beyond the Numbers and Toward New Discourse, pages 273–294.
Copyright © 2013 by Information Age Publishing
273

narratives that normalize failure (Giroux, Lankshear, McLaren, & Peters, 1996). As a result of these master-narratives, Black students are marginalized and portrayed with less than adequate success. They are portrayed as passive when it comes to their education (Perry, 2003) and mathematics achievement, and absent of agency and voice (Martin, 2007). Some argue that this is a "gap-gazing" fetish (Gutiérrez, 2008) on the part of researchers, which does not attend to the advancements, excellence, or gains made by students from marginalized communities (Gutiérrez & Dixon-Román, 2011). These master-narratives fail to highlight acts of empowerment by Black students (Walker, 2006), particularly those in urban settings like Chicago, who have worked to further develop mathematics literacy, not just for themselves, but also for others in their community through out-of-school student-led initiatives, such as the Young People's Project (YPP). Further, these master-narratives fail to show that Black students do, in fact, understand that the work of acquiring mathematics literacy is an act of liberation in itself, for personal fulfillment and for citizenship.

This chapter draws from a broader study and extends the context of mathematics learning environments. Through a focus on excellence, I highlight the mathematical brilliance of Black children.[1] Frustrated with the master-narratives described above, I wanted to investigate the role that young people have played in the mathematics teaching of their peers. To begin this inquiry, I considered the transformative work of the Young People's Project as a social context for engagement in mathematics. As young people took on varying roles in their practice, I wanted to know how they see themselves as doers of mathematics (Martin, 2000, 2007) and what other aspects of their identities are shaped from the mathematics literacy work they participate in. The work of the YPP underscores the importance of students' critical involvement in mathematics education.

Specifically, I examine the persistence of two African American college students, whose goal was to help youth develop mathematics literacy through the Young People's Project in the city of Chicago. The two mathematics literacy workers highlighted in this chapter are referred to as college mathematics literacy workers (CMLWs). The intersection of identity, liberation, and membership within a community of practice were partially addressed through the following research questions:

1. How does the engagement of mathematics literacy workers in the Young People's Project Chicago describe a community of practice and inform the modes of belonging in identity formation of two African American CMLWs?

2. What identities and roles do African American CMLWs see themselves having in their local communities as a result of their partici-

> pation in the community of practice, specifically the Young People's Project Chicago?

The first question focuses on the socializing experiences of participation and membership in Young People's Project Chicago and how that gives rise to a community of practice. The second question considers the intersections of identity, participation, and the broader society, particularly as they emerge from the two college mathematics literacy workers' experiences as members in the organization and their evolving senses of what it means to be African American students.

The data and analysis presented in this chapter is a case study taken from my dissertation research (Brewley-Corbin, 2009). I conducted an interpretive qualitative case study of four African American mathematics literacy workers in YPP Chicago, but in this chapter, I only report on two CMLWs. Qualitative methods of inquiry are grounded in the tradition of asking questions to find meaning about reality through others' personal experiences (Winter, 2000). My goal in doing a qualitative case study was to tell an in-depth story of how membership in YPP Chicago and engagement in mathematics literacy work influenced CMLWs identity and broader community goals.

A major goal of this chapter is to provide an understanding of how being members in a community of practice, the Young People's Project Chicago, influenced how these two college mathematics literacy workers worked towards young people achieving mathematics literacy. Wenger's (1998) concepts of communities of practice and modes of belonging are employed to explain ways in which CMLWs participated in Young People's Project Chicago and subsequently how this participation influenced their identity and their work for mathematics literacy with youth. A more elaborate description of communities of practice will be provided later. First, I will begin with a discussion of the mathematics literacy issue, particularly for Black students.

THE MATHEMATICS LITERACY ISSUE

The need for complex and sophisticated mathematical knowledge and problem solving has grown over the past 100 years and is strongly attributed to technological advancement. In order for all citizens, regardless of race or educational background, to compete fully in society, they must develop an adequate level of mathematics literacy (Jablonka, 2003). They must also develop the ability to make abstractions from a given situation and determine the necessary components needed to solve problems (Dubinsky, 2000). Unfortunately, the gap between the mathematical demands on citizens and the mathematical capabilities of individuals has widened (Quantitative Literacy Design Team, 2001). While in the past one could receive employment and a living wage without complex mathematical skill, today that is less often the case. Many Black students are unable to take full advantage of careers in the sciences and technologies due to insufficient opportunities to learn mathematics.

Furthermore, mathematics literacy enables individuals to participate more fully in society by helping them make more informed economic and political decisions. Acquiring mathematics literacy also provides individuals with the opportunity to empower themselves and to broaden their perspective and awareness of issues around them (Ernest, 2002). Some scholars have argued that mathematics literacy is embedded in the right to citizenship (Kamii, 1990; Martin, 2007; Moses, 1994; Silva, Moses, Rivers, & Johnson, 1990), which refers to the rights and responsibilities of individuals participating in society for the betterment of themselves and their community.

Today, mathematics literacy has become a requirement for citizenship for many poor and minority students. This is paralleled with having reading literacy and a modest interpretation of the Constitution as a requirement for citizenship for poor Mississippi sharecroppers who fought for the right to vote in the 1960s (Moses, 1994; Moses & Cobb, 2001). Then, there was a necessity for those who were denied access to the political system to demand their right to vote, making a case that they could articulate a position on equal participation in society. Today, a similar articulation is necessary for Black students in mathematics. These students must make the case for themselves that acquiring mathematics literacy is not only a requirement for citizenship, but also a requirement for fully exercising their civil rights (Moses, 1994; Moses & Cobb, 2001). Further, these students must make the case that it should be the right of every citizen to be able to access quality mathematics education in order to develop mathematical literacy. When we prepare students for citizenship, we are preparing them to take positions on issues, utilize their voices effectively, and deal with situations critically as they arise (Rudduck, 2007). There is also a need to establish a community where students can become engaged in and excited about doing mathematics, and where they can take an active role in the teaching and learning of mathematics with their peers. The work of YPP in Chicago is a response to this call.

THE YOUNG PEOPLE'S PROJECT & YPP CHICAGO

The Young People's Project (2008) is a youth-empowerment and after-school mathematics initiative created *by* young people and *for* young people, with the goal of expanding how mathematics is experienced by youth in urban communities. There are three main objectives of YPP: 1) development of community, 2) youth leadership, and 3) academic success demonstrated through algebra completion. The YPP is an outgrowth of, and in partnership with, the Algebra Project (Moses & Cobb, 2001). Founded in Jackson,

Mississippi, in 1996, YPP's core belief is that young people can be proactive in changing their life conditions and that mathematics literacy can be an entry point in that process. Robert Moses, Algebra Project founder, believes that young people need to be inspired to fight for their own liberation. Like the Algebra Project, YPP is guided by the principle that *all children*, regardless of their racial, cultural, or socioeconomic background, can learn algebra as well as other areas of mathematics.

The YPP develops and prepares high school and college students, known as high school mathematics literacy workers[2] and CMLWs to demonstrate their mathematics skills in both after-school programs and through an Algebra Project network composed of activists, educators, parents, and teachers. Similar to the way that sharecroppers fought for their own voting rights in the 1960s, YPP gives students of color the opportunity to take ownership of their learning of mathematics. Furthermore, YPP enables students to make the argument on their own terms, and to show, through their own outreach work, that mathematics is an important skill for everyone to achieve. In the next section, I provide a brief description of YPP Chicago.

YPP Chicago was started in the summer of 2002 through local citywide partnerships. To date, the organization has prepared over 200 high school and approximately 50 college students to conduct math literacy workshops and community events. Mathematics literacy workers carry out math literacy workshops in over 20 sites in the north, south, and westside communities of Chicago (Moses, 2006). YPP Chicago seeks to develop mathematics literacy in algebra for eighth and ninth graders and develop number sense with elementary students in grades three to six. Although there are several state-funded mathematics initiatives in Chicago, many are book and worksheet focused and geared solely towards students taking and passing the Illinois State Achievement Test (ISAT). Aside from YPP, many Chicago youth are not afforded opportunities to learn mathematics outside of a school context.

COMMUNITIES OF PRACTICE, MODES OF BELONGING, AND FORMATION OF IDENTITY

Educators have questioned traditional methods of teaching and learning mathematics and the ways in which knowledge is transferred to students through those approaches. They have even suggested that types of learning situations define the types of knowledge that students acquire (Boaler, 1999; Boaler & Greeno, 2000). For mathematics, specifically, it is the "practices of learning mathematics [that] define the knowledge that is produced" (Boaler & Greeno, 2000, p. 172). In this sense, the situations and activities that arise in practice are integral to cognition, learning, (Brown, Collins, & Duguid, 1989) and even agency and identity. Social theories of learning espouse that learning occurs by doing and through social activity (Lave & Wenger, 1991; Wenger, 1998).

A community of practice (CoP) is a collective group, unified by common interests where members interact regularly in order to create and improve what they learn and share. CoP theorists espouse that learning requires extensive participation in a community whose members are engaged in a set of relationships over time. Wenger (1998) explained that CoPs exist all around us; they are an important part of our everyday lives. Through prolonged engagement, members develop common knowledge, practices, and approaches, along with a common sense of identity. Clarifying the definition of CoPs, Wenger, McDermott, and Snyder, 2002 explained:

> As [members] spend time together, they typically share information, insight, and advice. They help each other solve problems. They discuss their situations, their aspirations, and their needs. They ponder common issues, explore ideas, and act as sounding boards. They may create tools, standards, generic designs, manuals, and other documents—or they may simply develop a tacit understanding that they share. However they accumulate knowledge, they become informally bound by the value that they find in learning together....Over time, they develop a unique perspective on their topic as well as a body of common knowledge, practices, and approaches. They also develop personal relationships and established ways of interacting. They may even develop a common sense of identity. (pp. 4–5)

An integral component of any CoP and an important part of learning is construction of identities. Participants in a community build their image of how they see themselves through the positions they hold, which greatly influences their construction of identity. Wenger offers three distinct modes of belonging related to identity formation and learning: engagement, imagination, and alignment. *Engagement* describes interactions as an ongoing collaboration with members in communities that can change. Through engagement in a community, a shared reality is created that aids in co-constructing identity. Engagement embodies natural boundaries because of limitations of time to participate in the community, activities one can become involved in, and relationships that members of the community develop with one another. *Imagination* is the ability of members to create new images of themselves beyond time and space. Imagination is a creative mental process that can be thought of as the ability to dream about what is possible for oneself and what one is capable of becoming. Unlike engagement, which is concretized in creating a shared reality, imagination supersedes direct interaction and is another form of constructing a shared reality. Our imaginations allow us to locate ourselves in the social world, past, present, and future. *Alignment* is the mode of belonging in which all efforts, such as energies, actions, and practices, come together to produce coordinated activities. Alignment allows participants in a community to play their unique roles in finding common ground. Participants are able to arrange their practices within the guidelines and expectations of the community to maintain alignment. Finally, alignment necessitates very precise forms of participation and structure. These modes of belonging should be thought of as working cohesively with the others in the construction of identity.

YPP CHICAGO AS A COMMUNITY OF PRACTICE

At the time of this study, YPP Chicago was a community of approximately 130 mathematics literacy workers. Members had a set of shared values and practices that were cultivated over time. Members came to the community with backgrounds ranging from little to extensive expertise in mathematics. A common mission statement unites mathematics literacy workers. The mission statement of YPP affirms their purpose, inspires members to participate, gives members meaning and a context for their work, and guides their learning and the knowledge they produce. Members also share a commitment to improve young people's understanding of numbers through games and through other mathematics literacy activities. Mathematics literacy workers also form social bonds through prolonged interactions in this community space.

FRAMING THE INQUIRY AND ANALYSIS

This chapter draws upon data collected over a six-week period at a YPP workshop training in Chicago. Data consist of 92 hours of observations, two semi-structured interviews[3] of each participant ranging from one to two hours, student work from mathematical tasks, participant reflections, and archival data. In order to answer the research questions presented earlier in this chapter, I provide each CMLW's view of mathematics literacy, their interpretation of the YPP's mission statement, and how they sought to embody the expectations of the mission in their mathematics literacy work. The mission statement was developed by its members in 2006 and provides a framework for CMLWs' practices as they engaged in mathematics literacy work. Thematic analysis was used to shed light on how each CMLW viewed mathematics literacy and embraced the mission of YPP, and how their view of the mission statement shaped their work and identity as mathematics literacy workers within the CoP.

School Setting: Abelin High School

Data were collected at Abelin Preparatory High School,[4] where mathematics literacy workers participated in daily workshop trainings in preparation to teach mathematics games to elementary school children. Founded in 1998, Abelin Preparatory High School is a west side neighborhood charter school within the Chicago Public School system. Approximately 670 students in grades nine through 12 attended Abelin. Ninety-seven percent of the students who attended were African American, and 3% were Latino. Ninety-two percent of the student population participated in a free and reduced lunch program, and 97% of the students who attended the school resided in neighborhoods on the west side of Chicago.

Participants: Naomi and DeMarcus

Naomi was a CMLW. She was a 21-year-old African American senior who attended a local university. She had been born and raised in the South, but relocated to Chicago for school. Naomi had three majors: African American studies, gender studies, and performance studies. She had worked with non-profit organizations since she was 14 and became familiar with YPP through her involvement in other youth-oriented work. Naomi enjoyed working with young people of all ages. At the time of the study, she had been involved with YPP for about eight months, which included summer and fall semester work in the organization. Naomi returned to Abelin for a second semester to give workshop trainings. She had the official title of Instructor; however, her position could best be described as a lead CMLW in YPP. In this position, she worked closely with other CMLWs, teaching and developing curriculum for the workshop training.

DeMarcus was a 21-year-old African American college mathematics literacy worker. He was a fifth-year senior at a local university. Native to the South, he relocated to Chicago for school. DeMarcus was majoring in African American studies with a concentration in history. During the previous two years, he realized that teaching was the career he wanted to pursue. His work with YPP enabled him to improve his teaching skills and his interactions with young people. At the time of the study, DeMarcus had been in YPP for one semester and returned to Abelin High School for a second semester to give workshop trainings.

FINDINGS

The results presented in this section will focus primarily on the second research question, *What identities and roles do African American CMLWs see themselves having in their local communities as a result of their participation in the community of practice, specifically the Young People's Project Chicago?* The first question will be addressed in the section entitled: Engagement, Alignment, and Imagination in Identity.

Naomi's View of Mathematics Literacy

Naomi possessed a strong willingness to do mathematics. This was demonstrated by her overall view of mathematics, which she described as pretty positive. Naomi understood that many people, including herself, often struggle with mathematics, but she chose to look at mathematics in a positive light because she enjoyed it. Naomi believed that mathematics provided space for exploration of new ideas. For Naomi, she also saw the broadness of mathematics and believed there was still a lot to discover that was unknown about the discipline.

One important aspect of YPP outreach work is the mathematics literacy workers' understanding and development of mathematics literacy as a goal. Naomi's view of mathematics literacy included not just access to mathematics but also helping those who had been undereducated or mis-educated do mathematics in meaningful ways. Naomi believed that the work from the activities developed in YPP helped students who have difficulty with mathematics develop mathematics literacy. She explained:

> I: So how would you describe mathematics literacy?
>
> N: I mean I think math literacy is having the ability…to do math. I think when we say, "Well, these people don't have access to math," it's not necessarily like I don't have the ability to open a textbook and read it myself. Math literacy means not just the access but being able to do it.…Maybe it's two parts, like having access to it and then having the literacy to do it. So that means, you know, maybe this school is not providing that math literacy. Maybe the people who are below certain rates on their testing just get lost in the system, which we've seen over and over again. They graduate and can't do the math. So we need extra things in place. We need the Algebra Project in the schools. We need [the] YPP Project after school. We need to go out to the elementary schools. It's just sort of picking up where [they left off]. Okay, we can give access, but now we need to give literacy. We need to make sure that the students are able to do it.

In this excerpt, Naomi defines mathematics literacy as access to and the ability to execute mathematics when necessary. She also emphasizes the need to supplement what students learn in school and believed that YPP was a good complement. More importantly, she believed that through YPP's work, students get the opportunity to deepen what they know about numbers and have experiences in mathematics beyond the traditional curriculum. Reflecting on the mission statement, Naomi came up with the following explanation of its meaning:

> I: I am going to read the YPP Mission statement to you. I want you to take a minute to think about this mission statement. Tell me what it means to you.
>
> N: Okay, well, I've thought about it before, and I contextualize the entire statement, you know, radically changing the quality of education, being leaders, pushing people to their full human potential, not just your students, yourself. Everybody involved with the Project and I contextualize the entire mission statement with Bob Moses' bigger picture of being a civil rights activist and mathematics being a civil right. And me realizing that, okay, having

access to education is a *right* [italics my emphasis]. If I don't have access to education, like my government or my public or whatever should not hold that access from me. I should be able to access education so that I can exercise my full human potential. [In] a lot of minority communities and low-income communities, that's not what's happening. And so that's what I see the mission of the Young People's Project. We spread ourselves, we're like a spider.... We started off with the trainers, then the instructors, and then the CMLWs, and then we spread ourselves all around all areas of Chicago. And I think I just really think that it [was] sort of supposed to be a movement. That's how I interpret it. Like the word *radical* [italics my emphasis] to me means we are not taking this anymore. We are not going to sit here and let our students just continue to be below math literacy rates. Like we're not going to do it. That's what radical means to me. I think the [words] *radical* and *full human potential* really, you know, stick out to me in the mission.

Naomi saw herself and the other mathematics literacy workers as individuals who could be proactive in helping others understand how mathematics was perceived in their community and then do something about changing it. Naomi recognized the broader issues in education as articulated in the YPP mission, which centered on mathematics literacy and students' lack thereof. She believed that by being a part of YPP, she could make an impact in changing it.

DeMarcus' View of Mathematics Literacy

DeMarcus' desire to *do* mathematics shifted once he started high school. He explained that while he was in high school, his love for mathematics deteriorated over time. In both middle and high school, DeMarcus took several advanced mathematics courses. After attending high school, he felt he was surrounded by people that were "essentially math geniuses" and people that were going to be doing that professionally. He concluded that mathematics was something not meant for him. DeMarcus believed that in order to do mathematics effectively, you had to be in the right frame of mind, which was often a challenge for him. Once he entered college, DeMarcus decided to give up on mathematics altogether because he did not believe he could do the work. To uncover his view of mathematics, DeMarcus provided the following comments that exemplified his view of mathematics:

I: So, how would you describe your educational background as it relates to mathematics?

D: I guess I would see other people and see what they were doing with math, and due to the fact that I didn't see myself capable of doing that, I was like, "all right, this clearly isn't for me." And so I kind of gave up on math in college. But at least I used to be good at it in high school.

Although DeMarcus held this view of himself doing mathematics, he still recognized the importance of mathematics. He believed that a good grasp of mathematics and the confidence to engage in mathematics would sharpen critical thinking skills. He also believed that the skills that can be developed from doing mathematics were necessary for problem solving. The work that DeMarcus did in the YPP Chicago also helped in shaping how he approached mathematics. As he described his view of the mission statement, his goal for the students that he worked with in YPP became apparent:

> I: I want you to take a minute to think about that mission statement. Tell me what it means to you.
>
> D: I essentially feel as though the mission with the Young People's Project is essentially my mission in working with youth....My goal in education is to empower....I want to help build self-sufficient students. And I mean self-sufficient in that I want them to be of the character where they can go into a situation with as much confidence in their heart and success or failure to have the strength to take it on. And I feel as though with the Young People's Project they are doing something along the same lines. What I see as having the confidence to take on any challenge, success or failure, is what I kind of feel....The YPP mission defines as its full human potential. 'Cause I feel as though if you acknowledge your full human potential, then you won't fear anything that inhibits you from reaching that full human potential. And with YPP, it just so happens that they're doing it through work with math literacy when it comes to the lives of the highschoolers and elementary school kids.

The interpretation that DeMarcus has of the YPP mission statement was focused on building both confident and competent students. By this he was focused on helping the students that he worked with develop the courage to take on challenges, including being able to do mathematics whenever that opportunity arose.

IDENTITY IN THE CONTEXT OF MATHEMATICS LITERACY WORK

There were two identities that were common themes between Naomi and DeMarcus: *agent of change* and a *doer of mathematics*. Here, I define an agent of change as someone who purposely works toward creating some kind of social, cultural, or behavioral change in society or in others through his or her work or actions. I also define doer of mathematics as someone willing to engage in thinking deeply about mathematical ideas. The following discussion explicates how these identities emerged through discussion with Naomi and DeMarcus.

Naomi as an Agent of Change

Naomi described her role in the Young People's Project as improving the welfare of others, working for social justice, and being an agent of change. This role was linked to her future pursuits in education. Naomi said that she had a long-term goal of becoming a humanitarian and wished to continue working with non-profit organizations. She also demonstrated a level of social consciousness and awareness of the problems in schools and how she could play an active role in addressing them.

> I: What are some of your sort of broad short term and long term plans?
>
> N: Whether it be I start teaching again through other...nonprofit organizations or working for human rights campaigns, or working for social change, social policy, but as long as it is something fun...I just have to make changes. It like hurts me so much...all of this stuff that people can be doing. How can you not be doing something? So I already know, my life has proven over and over again....I know I want to make a change, I know I need to be there. I know I can't ignore it. I can't sit here knowing that students aren't graduating. Knowing that...thousands of kids are being held back....I can't do it. So I know that I am gonna be involved in some way.

Naomi also embodied being an agent of change in that she understood that her work as a CMLW was broader than workshop training. She believed that her role was to encourage other students to take ownership of the issues they saw around them (outside of the training) and in their classrooms so that they could create an impact in their communities. She explained:

> N: I feel like my purpose, I guess on a micro-level, is to make sure that Abelin can survive....I can leave, and Abelin will still be a functioning site with students who are impacted in a positive way, that they felt like they gained something and that they also gave something back to the community. So I guess on a broader scale, it's being able to affect those students in a way that they're inspired to do something in their community, or they feel like they have done something to their community.

Naomi also explained that although her work in YPP was fun and enjoyable for the mathematics literacy workers and elementary children, she was working for the broader purpose of empowerment through mathematics literacy. Naomi described a realization that she had, which was echoed in her work in YPP. She ultimately believed that she had some impact on the high school mathematics literacy workers at Abelin. She believed that the experiences they had in the workshop training with YPP from the previous semester inspired them to return the following semester. Naomi's broader mission for herself was to improve the lives of others.

DeMarcus as an Agent of Change

DeMarcus saw himself as an agent of change when he described his past school experiences and his motivations for his work in YPP. He used the inadequacies he found in his schooling as fuel for his mission to make changes:

> I: It is quite interesting to me that you're doing this work, centered around mathematics and it is nowhere at all related to your field of interest. I want you to talk a little bit about that for me.
>
> D: I took a couple of economics courses here [at New University] that involved things like social justice and finding economic solutions to social inequalities. Combine that with a sociology course that I was taking at the time, and I found myself being drawn more and more and developing more and more of an interest in fighting for social justice and for striving for social equality, which is what eventually...led me to being an African American studies major. It was African American studies courses that also reinforced my belief in these social inequalities that I felt needed to be remedied...I'd take all of these history courses, and witness how education was essentially used as a tool against Black people for so long, what they didn't know was literally hurting them....I was experiencing difficulties at New that I felt were the result of my study skills and my educational background not being strong and toward the caliber of many of my peers. So I felt almost as though my education was somewhat of an injustice. To know that all...these peers of mine were apparently achieving higher than I was due to the fact that they were built on a better foundation. I won't even say better, that they were built on a different foundation than me....So the combination for striving for social equality and improving the lives and the achievement potential for kids like me, those were the two of the biggest things that led me to be sold on trying to be an educator.

DeMarcus's goal of becoming an educator was one way that he felt he could create social change. He went on to explain that after he graduated, he was leaning towards being a teacher with Chicago Public Schools. DeMarcus also described the role he believed he had in YPP as a model of change. He viewed himself as the change he wished to see in the high school students for which he worked. In light of the flaws he recognized in his own education, DeMarcus was determined to come up with new and creative ways of teaching that would be more stimulating to high school students.

The CLMWs saw themselves as agents of change. For Naomi, it was improving the welfare of others by promoting consciousness in the community. For DeMarcus it was giving children a better educational foundation. In looking at the theme agent of change between participants, there was a range in how they viewed what they wanted to change. The participants' identities as agents of change could be viewed from a localized perspective (changing how a specific thing is experienced, e.g. mathematics) and a broader perspective (changing the agency of others, e.g. helping others to take ownership and promoting awareness of issues in the community).

IDENTITIES AS DOERS OF MATHEMATICS

Naomi as a Doer of Mathematics

Naomi possessed an inherent comfort level in doing mathematics. She talked about possessing a very strong disposition for engagement in mathematics from her schooling experience before she began her work in YPP. She believed that she could do mathematics easily and that she was good at the subject. Working in the YPP workshop training provided her further interest to engage in mathematics. In high school, Naomi did well in mathematics as demonstrated by a national mathematics competition she competed in and won. Through opportunities that were provided to her by winning a competition, she understood that possessing strong mathematics skills was an invaluable tool that afforded her opportunity and access.

> I: How would you describe your educational background as it relates to mathematics?
>
> N: It's like so [important] for minorities to be skilled in that area because it allows them to excel....It opens a lot of doors. Like if I hadn't won that competition. I got a two thousand dollar scholarship. I got a laptop....Just so many doors, so many opportunities, just for, you know, my math skills. And so, it's really important.

Naomi also made the point that she did not hesitate to do mathematics even if it required more thinking time for her when compared to others. She stated, "I really think...I enjoy math, and I like doing math, and I love discovering it, and I love figuring stuff out....And I think coming from that standpoint...the fact that it takes me longer to do math, and I'm still not intimidated by it." In this vignette, Naomi echoes the necessity of embracing struggle as part of the mathematics learning process.

DeMarcus as a Doer of Mathematics

DeMarcus possessed a comfort level in doing mathematics that developed over time. In many respects, he was an increasingly confident doer of mathematics as a result of his work in the YPP workshop training. DeMarcus attributed some of the challenges he had with mathematics to how he saw himself, his own perceived ability, and his enthusiasm. He also indicated that, due to prior schooling experiences, he had some apprehension about engaging in mathematics because he perceived his peers to be better at the subject than he was. His desire to do mathematics evolved in a positive way the more he engaged in mathematics through the workshop training. He realized that, as a result of his work in YPP, he had a renewed interest to do mathematics problems that he found challenging. DeMarcus provided the following insight about doing mathematics:

> I: What previous experiences have you had in mathematics that influences or shape the mathematical goals you have?
>
> D: I consider myself a person [that is] not too good at math. And I consider myself a person unwilling to do math for a large portion of my life. I feel as though in that respect, I can relate a lot to the kids that I work with.... Like [the workshop training has] improved my willingness to do math in everyday situations. I can definitely say before this program, I'd see [certain] number[s] and refuse to touch it. And I felt as though, [if the workshop training] was able to have this effect on me, then it would be possible for it, the kids that I work with, to have this effect [on them]. So, I guess one of the bigger influences on me is knowing this is possible.

DeMarcus recognized that although his disposition to do mathematics had evolved, some of the young people he worked with still struggled with developing a desire to do mathematics. While both of the participants were considered to be doers of mathematics, Naomi could be considered a confident doer of mathematics who was intrinsically motivated. Naomi believed that mathematics was a challenging discipline to study but chose to engage in it anyway. Naomi also expressed a general sense of enjoyment when she did mathematics. DeMarcus' increased confidence in doing mathematics was developed over time and grew from his initial uneasiness, which came from earlier experiences in his schooling. DeMarcus made a decision that he would be more confident in his approach to mathematics, and he attributed this change with his work in YPP workshop trainings.

ENGAGEMENT, ALIGNMENT, AND IMAGINATION IN IDENTITY

The results and analysis presented in this section will focus on the first research question, *How does the engagement of mathematics literacy workers in the Young People's Project Chicago describe a community of practice and inform the modes of belonging in identity formation of two African American CMLWs?*

Engagement

Nasir (2002) points out, "Learning, then, involves coming to new ways of engaging in practice. As participants' engagement shifts, so does the nature and practice of self" (p. 228). The identities of CMLWs were influenced by the levels of engagement they put forth in the YPP workshop training practices. Engagement is an ongoing process of negotiation of meaning, formation of trajectories, and construction of shared histories of practice. Engagement is an important process in how we come to see ourselves, and the way in which one engages and defines belonging. The level at which participation occurs determines similarities, differences, and shared repertoires in how things are done in practice. In YPP workshop trainings, CMLWs worked together with a common purpose of doing mathematics differently and building a community among themselves so that they could rely on one another when working with other youths at Abelin High School and at elementary schools.

The CMLWs saw themselves as individuals who could make a difference in how mathematics was taught to others both in workshop trainings and at elementary schools. Although they shared the collective identity of "mathematics literacy worker," they interpreted this position differently and took on unique roles doing and delivering mathematics. The CMLWs' level of engagement varied substantially because their priorities in the workshop training were different. The CMLWs had the responsibility of preparing the high school students during training as part of their practice. The CMLWs various identities emerged as a result of their differing priorities. The ways in which CMLWs negotiated meaning of their work and overall purpose in YPP were distinct. They spoke about what they learned about themselves because of the mathematics they practiced. The CMLWs also reflected deeply and were able to articulate both their ability and inability to engage in mathematics, referring consistently to prior schooling experiences. They then used this knowledge to improve their practice when working with high school mathematics literacy workers. Both Naomi and DeMarcus knew that the high school students they worked with would turn to them for clarification and guidance when they were unsure about the mathematics content. As a result, both Naomi and DeMarcus made it a priority to develop a deep level of understanding of the mathematics used in the workshop training so that they could be seen as experts and exemplars.

Imagination

Imagination is a critical component in identity construction because it transcends practice. Imagination is an essential element of our experience in the world and how we come to make sense of our place in it. It is the ability to create images of the world and of the self we hope to become by imagining something new and different for ourselves. Wenger (1998) explains that "imagination requires the ability to dislocate participation and reification in order to reinvent ourselves, our enterprises, our practices, and our communities" (p. 185). In this sense, identity rests not only on practice but is an expansion of time and space and can take on new dimensions beyond engagement. Imagination can create affinities to social causes, organizations, and groups of people. CMLWs saw themselves as creating new experiences for others in mathematics and in education. They saw themselves as agents of change and imagined themselves actualizing that identity in numerous ways.

The imagination of the CMLWs allowed them to connect their mathematics literacy work to the broader mission of the YPP. The CMLWs linked their practices in workshop trainings to a broader community of mathematics literacy workers in Chicago and also to the national efforts of the YPP and the Algebra Project. Both CMLWs saw themselves as being part of a social movement in mathematics education that empowered youth in communities near and far. This view can be seen in the interpretations that Naomi and DeMarcus held of the YPP mission statement and their alignment with the work of Bob Moses, both on a local and national level. It can also be seen in their explanations of the impetus for becoming a part of the YPP. Naomi wanted to better understand the students she worked with, and DeMarcus wanted to acquire experience in teaching. The CMLWs also connected the importance of doing their work to specific social problems they witnessed in school mathematics, like students failing standardized mathematics examinations and students not being taught mathematics effectively. The CMLWs used imagination to create life trajectories (Nasir, 2002) for themselves of what they hoped to become once they graduated from college. Specifically, Naomi explained that through engagement in her work with youth, both in the YPP and through other organizations, she knew that she wanted to be "some sort of humanitarian" or social activist. DeMarcus explained that as a way of combating inequities in education he saw himself becoming an educator and was leaning towards becoming a teacher in the Chicago Public Schools. Both Naomi and DeMarcus believed that their experience in the YPP was a stepping-stone to the work they wanted to accomplish later in their lives. They also believed that to create change in schools and society, they had to be a part of *some* change effort now.

Alignment

Alignment is the ability to take visions of self and transform them into relevant practice. There were varied levels of alignment among the CMLWs. For the CMLWs, alignment occurred when they worked toward their goals of becoming a humanitarian and an educator. Subsequently, the CMLWs structured their identities, practices, and behaviors to align with these goals. For instance, because Naomi wanted to become a humanitarian and an activist, she structured her practice in the training to raise the social consciousness of other mathematics literacy workers through social justice activities. She also structured her practice to include the broader purpose of the YPP, reinforcing the need for mathematics literacy in the local community and highlighting the importance of children thinking critically about numbers.

The aspiration that DeMarcus had to become an educator helped to structure his practice in the workshop training as being sensitive to the differing personalities and learning styles of the high school students he worked with so that he could be effective in his facilitation of lessons with them. He also structured his practice in guiding literacy workers when difficulty arose in the mathematics content. Finally, as a doer of mathematics, DeMarcus structured his practice in building confident students by working one-on-one with MLWs when the need arose.

MATHEMATICS LITERACY FOR LIBERATION AND LIBERATION IN MATHEMATICS LITERACY

The CMLWs in this chapter demonstrated persistence in working towards mathematics empowerment. None of the participants had a mathematics concentration in school, but they chose to engage in an after-school initiative centered on mathematics because they understood the urgency for this kind of work in their community. The CMLWs spent several hours in school each day, including their stay for YPP training, which further demonstrated their commitment and persistence.

Naomi and DeMarcus are examples of Black students who combat dominant discourses through their participation in YPP. Naomi possessed a rather strong mathematics identity and believed that possessing mathematics literacy would improve an individual's life chances. She reaped the benefits by doing well in mathematics and gave examples of this from her prior schooling when she spoke of the prize awarded to her for winning a mathematics competition. Naomi also spoke with conviction about the importance of not standing on the sidelines while young people in Chicago continued to do poorly in education, in general, and in mathematics, in particular. Naomi reaffirmed that the broader, minority and working-class communities in Chicago do not always have access to quality education, and it was up to all stakeholders to ensure that education is a right for all. As a result, her efforts in YPP attempted to mitigate these effects through her work with youth. Naomi also helped to create a counter-narrative of agency by aiding communities she worked in to circumvent violence by creating knowledge and power. Naomi embraced the YPP mission when she suggested that one of the goals of the YPP was to "push people to their full human potential."

As DeMarcus reflected on his prior schooling experiences, he realized that he had endured many inequities and injustices. He attributed these inequities to poor academic preparation and poor study skills in comparison to his college-going classmates. Although there was nothing he could do to go back and change his own schooling experiences, DeMarcus did express a commitment to improving the lives and the achievement potential of the children who participated in YPP. This commitment was also shown when

DeMarcus, despite the negative mathematics identity he developed earlier in life, continued to do mathematics. However, as Martin (2007) noted, the ability to do mathematics is only one facet of someone's life. A mathematics identity does not develop on its own, separate from other important identities people make for themselves. As a consequence of our experiences, all of our identities taken together manifest in new and different ways. The accounts of schooling that DeMarcus shared, primarily about mathematics, provided one example of a young person who gave up on a specific aspect of his academic life only to reclaim it when he found new purpose. His own struggle to understand the mathematics used in the YPP workshop training demonstrated his perseverance to perform well.

CONCLUSION

Based on the analysis presented in this chapter, there are several implications for future practice with African American students in youth empowerment initiatives that are connected to mathematics. The analysis suggests that student-led initiatives like YPP Chicago show great promise for developing mathematics literacy, developing positive identities in young people within CoPs, and evoking the brilliance of Black youth in urban settings like Chicago. Moreover, highlighting Naomi and DeMarcus reifies the necessity to search for students who are very capable and motivated in doing mathematics, but who otherwise may have other academic interests.

The Young People's Project provides a different context where the mathematics education reform movement can be actualized. It is not top-down like many reform efforts, but bottom-up; created for young people by young people. Young people not only engage in mathematics in the YPP, but they also build a social awareness of problems that exist within their communities and engage in work to help transform them. Because of the reflective nature of the work mathematics literacy workers participate in, the YPP provides a space for young people to build social skills that enable them to gain a stronger sense of who they are and who they hope to become through prolonged engagement with each other. By participating in a CoP like the YPP, students learn what it means to explore mathematics and become mathematically literate in their community and in a broader context.

Although the focus of the chapter was a case study of two CMLWs, Naomi and DeMarcus, there are inherent limitations in making generalizations. However, highlighting their stories in YPP Chicago does provide contribution to the literature and an opportunity for growth in our knowledge of student-led initiatives in mathematics. Although this chapter offered snapshots of two CMLWs, there are countless of others in YPP and beyond who are helping to change the lives of Black children in their community by using their own brilliance to evoke the brightness, skills, and voice of other Black youth in mathematics. These are young Black men and women, young Black boys and girls, who are defying discourses of Black youth difficulties in mathematics and redefining what it means to do mathematics and be mathematically literate. They are just doing good work and more of their stories should be heard.

ACKNOWLEDGMENTS

I would like to acknowledge and thank the members of YPP Chicago for giving me the opportunity to work with them to collect data for this research study.

NOTES

1. The term *Black children* is used to describe the collectiveness when referring to Black youth of a variety of ages within the diaspora.
2. High school mathematics literacy workers are also referred to as MLWs.
3. In the vignettes from the interviews that follow, the same questions were asked of each participant. The "I" signifies the question asked by the interviewer and it is followed by an "N" or "D" for each participant's response. In order to capture each participant's flow of thought, I was not confined to a rigid interview protocol.
4. This and all the names in this chapter are pseudonyms.

REFERENCES

Boaler, J. (1999). Participation, knowledge and beliefs: A community perspective on mathematics learning. *Educational Studies in Mathematics, 40,* 259–281.

Boaler, J. & Greeno, J. G. (2000). Identity, agency, and knowing in mathematics worlds. In J. Boaler (Ed.), *Multiple perspectives on mathematics teaching and learning* (pp. 171–200). Westport, CT: Ablex.

Brewley-Corbin, D. N. (2009). *Case study analysis of mathematics literacy workers' identity and understanding of numbers within a community of practice.* Unpublished dissertation, University of Georgia, Athens, GA.

Brown, J. S., Collins, A., & Duguid, P. (1989). Situated cognition and the culture of learning. *Educational Researcher, 18*(1), 32–42.

College Board. (1999). *Reaching the top: A report of the National Task Force on Minority High Achievement.* New York, NY: Author.

Dubinsky, E. (2000). Mathematical literacy and abstraction in the 21st century. *School Science and Mathematics, 100*(6), 289–297.

Ernest, P. (2002). Empowerment in mathematics education. *Philosophy of Mathematics Education Journal, 15.* Retrieved from http://people.exeter.ac.uk/PErnest/pome15/ernest_empowerment.pdf

Giroux, H., Lankshear, C., McLaren, P., & Peters, M. (1996). *Counternarratives: Cultural studies and critical pedagogies in postmodern spaces.* New York, NY: Routledge.

Gutiérrez, R. (2008). A "gap-gazing" fetish in mathematics education? Problematizing the research on the achievement gap. *Journal for Research in Mathematics Education, 7,* 357–364.

Gutiérrez, R. & Dixon-Román, E. J. (2011). Beyond gap gazing: How can thinking about education comprehensively help us (re)envision mathematics education? In B. Atweh, M. Graven, W. Secada, & P. Valero (Eds.), *Mapping equity and quality in Mathematics Education* (pp. 21–34). New York, NY: Springer.

Jablonka, E. (2003). Mathematical literacy. In A. J. Bishop, M. A. Clements, C. Kietel, J. Kilpatrick, & F. K. S Leung (Eds.), *Second international handbook of mathematics education* (pp. 75–102). Dordrecht, The Netherlands: Kluwer Academic Publishers.

Jencks, C. & Phillips, M. (Eds.). (2001). *The Black-White test score gap.* Washington, DC: Brookings Institution.

Kamii, M. (1990). Opening the algebra gate: Removing obstacles to success in college preparatory mathematics. *Journal of Negro Education, 59,* 392–406.

Lave, J. & Wenger, E. (1991). *Situated learning: Legitimate peripheral participation.* Cambridge, UK: Cambridge University Press.

Martin, D. B. (2000). *Mathematics success and failure among African American youth: The roles of sociohistorical context, community forces, school influences, and individual agency.* Mahwah, NJ: Erlbaum.

Martin, D. B. (2007). Mathematics learning and participation in the African American context: The co-construction of identity in two intersecting realms of experience. In N. S. Nasir & P. Cobb (Eds.), *Improving access to mathematics: Diversity and equity in the classroom* (pp. 146–158). New York, NY: Teachers College Press.

Moses, O. (2006). *The Young People's Project: Strategy for organizational expansion.* Chicago, IL: Young People's Project.

Moses, R. P. (1994). Remarks on the struggle for citizenship and math/science literacy. *Journal of Mathematical Behavior, 13,* 107–111.

Moses, R. P. & Cobb, C. E., Jr. (2001). *Radical equations: Civil rights from Mississippi to the Algebra Project.* Boston, MA: Beacon Press.

Nasir, N. (2002). Identity, goals, and learning: Mathematics in cultural practice. *Mathematical Thinking and Learning, 4*(2&3), 213–247.

Perry, T. (2003). Up from the parched earth: Toward a theory of African American achievement. In T. Perry, C. Steele, & A. Hilliard (Eds.). *Young, gifted, and Black: Promoting high achievement among African American students* (pp. 1–84). Boston, MA: Beacon Press.

Quantitative Literacy Design Team. (2001). The case for quantitative literacy. In L. A. Steen (Ed.), *Mathematics and democracy: The case for quantitative literacy* (pp. 1–22). Princeton, NJ: National Council on Education and the Disciplines.

Rudduck, J. (2007). Student voice, student engagement, and school reform. In D. Thiessen & A. Cook-Sather (Eds.), *International handbook of student experience in elementary and secondary school* (pp. 587–610), Netherlands: Springer.

Silva, C. M., Moses, R. P., Rivers, J., & Johnson, P. (1990). Algebra Project: Making middle school mathematics count. *Journal of Negro Education, 59,* 375–391.

Walker, E. N. (2006). Urban high school students' academic communities and their effects on mathematics success. American Educational Research Journal, 43(1), 43–73.

Wenger, E. (1998). Communities of practice. Cambridge, UK: Cambridge University Press.

Wenger, E., McDermott, R., & Snyder, W. M. (2002). Cultivating communities of practice: A guide to managing knowledge. Boston, MA: Harvard Business School.

Winter, G. (2000). A comparative discussion of the notion of 'validity' in qualitative and quantitative research. The Qualitative Report, 4(3 & 4). Retrieved August 28, 2005, from http://www.nova.edu/ssss/QR/QR4-3/winter.html

Young People's Project. (2008). The young people's project: Math literacy and social change. Chicago, IL: Author.

CHAPTER 13

UNPACKING BRILLIANCE

A New Discourse for Black[1] Students and Successful Mathematics Achievement

Nicole M. Russell

INTRODUCTION

What does it mean to be brilliant in mathematics, and what does race have to do with it? In answering that question philosophically, one should look back into the origins of Greek philosophy where the scholarship and work of great philosophers such as Plato and Aristotle dominate the mainstream academic knowledge that is taught and constructed in most secondary schools, colleges, and universities. By far, Plato's *The Republic* and Aristotle's *Politics* are prototypes of the kind of text that is considered part of the canon. It is understood that Plato stands with Socrates and Aristotle as one of the shapers of the Western intellectual tradition; few other writers have exploited so effectively the epistemology of philosophy, logic, science, and mathematics. Plato's *Republic* is essentially a treatise on education, and his goal of education was to identify and nurture that part of an individual's psyche that distinguishes him or her by *nature*. In Plato's *Republic*, Blacks (along with Greek women) were considered merchants, the lowest class,

The Brilliance of Black Children in Mathematics:
Beyond the Numbers and Toward New Discourse, pages 295–319.
Copyright © 2013 by Information Age Publishing
All rights of reproduction in any form reserved.

and Plato argues that the knowledge constructed by these individuals is unreliable because by nature, their appetites trump reason. Philosopher kings were trained in formal mathematics; thus only an elite few had the capacity to transcend the cerebral rigor and would be considered what many understand today as brilliant.

Today, brilliance in mathematics continues to be defined through the lens of meritocracy and hegemonic principals, values, and ideals. It is clear from the structure of most mathematics classrooms in the U.S. that arranging students by ability (or nature to use Plato's words) is still valued.[2] What are the underlying assumptions and expectations of teachers, professors, the mathematics community, and other stakeholders about *who* a mathematician is, what mathematics is all about, and who can be successful? These assumptions unintentionally promote the myth that mathematics is a discipline of complete objectivity and that White males represent the "prototype" of true mathematicians (see Stinson, Jett, & Williams, this volume). African Americans, both male and female, are rarely viewed as successful in mathematics. This view gets constructed in part because the majority of scholarship about African American learners and mathematics achievement rarely focuses on success, but rather centers on failure, and the explanations for their underachievement are prolific—including genetic differences between African Americans and European Americans (Jensen, 1969), economic disparities (Ornstein & Levine, 1989), unstable families, lack of parental involvement, and negative peer pressure within the African American community (Irvine & York, 1993; Ogbu & Matute-Bianchi, 1986), tracking (Oakes, Joseph, & Muir, 2004) and differences in achievement tests (National Center for Educational Statistics, n.d.).

Since the 1970s, this work has become known as achievement gap research, and it is understood by "virtually everyone that this does not refer to a gap between Africans and Asians or a gap between Africans and Latinos or a gap between Africans and anyone else other than Europeans" (Hilliard, 2003, p. 137). This is the knowledge that gets produced about African American students and mathematics. Martin (2009) contends that "it is through the knowledge production process, characterized by the continuing reliance on a race-comparative paradigm, that African American children come to be constructed as mathematically illiterate and that White children and their behavior come to be normalized" (p. 29). This lens places groups in opposition with each other, which implies that one's gain is the other's loss and sends an unintended message that Black students are not worthy of study in their own right—that a comparison group is necessary (Gutierrez, 2008). Research on the achievement gap does not capture the history nor the context of learning that has produced failing outcomes (Gutierrez, 2008; Ladson-Billings, 2006), thus conducting more contextualized studies is important work.

This volume is about unpacking brilliance—an effort of mathematics education scholars to move beyond the numbers and toward a new discourse—that Black students are brilliant in mathematics. The purpose of this chapter is to align my research and scholarship with an established community of scholars who reject research and policy perspectives that ignore Black student success (Leonard, 2008; Martin, 2009; Moody, 2001; Stinson, 2008). The findings reported here extend from my mixed-methods dissertation that used historical analysis, 69 student surveys, and 12 in-depth interviews as it explored the question "What are the factors that influence or contribute to the successful math achievement of Black students raised in the United States?" This chapter uses the in-depth interview to analyze and describe important factors influencing the mathematics experiences, education, and performance of Black students. I view these factors as part of any student's *micro* (present) background that they bring to the K–12 mathematics classroom. However, the findings of my study suggested that while these factors can influence mathematics achievement for all students, these same factors become *complex* for Black students because of the distinct schooling experiences associated with broader historical and social circumstances and contexts. Consequently, Black students also bring what I term their *macro* milieu—the mathematics experiences, learning, and achievement associated with the adults and peers in their networks as well as all of the generations that precede them (Margo, 1990; Martin, 2000). This macro milieu includes the ramifications and circumstances of 300 years of deliberate educational, political, and social actions taken to systemically deny Blacks the ability to fully participate in the study of and access to mathematics. I sought to re-envision mathematics success for African Americans given the long standing educational inequities[3] they have faced as a collective group. I re-think analysis of Black student achievement from achievement gap language to more complex descriptions of success—constructs such as resiliency, agency, and persistence (see McGee; Jett, Stinson & Williams, this volume).

This chapter is divided into three major sections. Section one presents a theoretical framework into which my study was grounded. To make sense of the literature connecting socio-cultural aspects of mathematics education and African American student experiences, I abstracted three themes for framing the theoretical perspectives in the literature: identity, social capital, and teacher-student relationships. The second section provides the methodology and results of the study. The objectives of the study were to 1) document and analyze the ways that Black students experience mathematics—in and out of school, 2) describe Black mathematically successful students, and 3) gather data that may help empower Black families and mathematics educators to move to praxis (Freire, 2002). The third section draws conclusions from and makes inferences based on the results of the research study. The goal of this section is to help create a dialogue among mathematics educators and policymakers about the impact Black students' in-and out of school experiences can have on their mathematics education.

THEORETICAL GROUNDING

There is no single body of literature that robustly integrates race in analyzing mathematics experiences;[4] thus in framing ideas for a *critical perspective* on what factors influence success for Black students socialized in the U.S, this study is grounded in Critical Race Theory (CRT) and what I call the tradition of multicultural education. CRT originates from the legal field (Delgado & Stefancic, 2001) but has found its tenets appropriate in education. Ladson-Billings (1998) argues that this relationship is critical because schools are nothing more than facsimiles of the broader society. Consequently, Ladson-Billings and Tate (1995) have advocated for the utilization of the CRT framework in education because it "asks such questions as: what roles do schools themselves, school processes and school structures play in helping to maintain racial, ethnic and gender subordination?" (Solorzano & Yosso, 2000, p. 40).

CRT is a *theory* that centers race, authorizing researchers to examine, investigate, and interpret racial inequalities in schools (Ladson-Billings & Tate, 1995; Ladson-Billings, 1998; Zamudio, Russell, Rios, & Bridgeman, 2010). Among several other tenets, CRT privileges the voices of people of color to critique society through counter-storytelling in order to illuminate and explore experiences of racial oppression. CRT can also be conceptualized as a *perspective* (Gillborn, 2006) because it is a set of beliefs about the importance of race and how it operates in society, especially in the U.S.

Multicultural education means different things to different people. The field includes educational researchers from a wide variety of backgrounds; thus different viewpoints should be expected. Some scholars conceptualize multicultural education as a philosophy of cultural pluralism within the educational system (Baptiste, 1979), while others view it at as acquiring knowledge about various groups that oppose oppression and exploitation by studying the artifacts and ideas that emanate from their efforts (Sizemore, 1981). Still other advocates think about multicultural education as a type of education that is concerned with various groups in American society that are victims of discrimination because of their unique cultural characteristics—examining key concepts such as prejudice, identity, conflicts, and modifying school practices and policies to reflect an appreciation for diversity in the U.S. (Banks, 2004). This study combines these various conceptualizations to think about the notion of a *tradition of multicultural education*. While there are differences within the schemes, they are more semantic than substantive, and an underlying assumption of the different viewpoints is that multicultural education centers the *educational process* that is to be built upon the ideas of democracy, equity, justice, equality, and human dignity (Banks, 2004, 2007; Gay, 1994; Grant, Elsbree, & Fondrie, 2004; Sleeter, 2005).

Multicultural education can take place in schools and informs all subject areas and other aspects of the curriculum and learning process as contexts and activities. It confronts social issues involving race, ethnicity, socioeconomic class, and gender and their intersections, while emphasizing reflective intervention, empowerment, and social action in changing the structures of educational institutions. Important aspects of Banks' (2004) typology of multicultural education include an equity pedagogy, knowledge construction, and empowering school cultures and structures. Also within the tradition of multicultural education is Gay's (2000) notion of culturally responsive instruction. Culturally responsive instruction uses instructional strategies that validate, empower, and transform the learning process for students of culturally and racially diverse backgrounds. I posit that CRT and multicultural education complement each other in that together they provide a theoretical and practical framework for analyzing and addressing issues of equity in schools.

RELEVANT LITERATURE

Factors influencing the mathematics achievement of any student can include areas such as identity development, social capital, and teacher-student relationships; however, when we intersect race with these factors, I argue that what we think we understand about mathematics achievement becomes increasingly complex. Critical Race Theory allows for this claim of complexity because the narratives of the students in this study represent their ways of learning mathematics, having access to mathematics, and negotiating the schooling process as Black students.

In this section, I discuss these three areas and how each might relate to Black students' experiences in math classrooms and their math achievement. While identity development, social capital, and teacher-student relationships are not necessarily interrelated, they are connected to the overarching frameworks of critical race theory and multicultural education. Identity development, social capital, and teacher-student relationships can be viewed as aspects of the learning and socialization process in which Black students participate. Multicultural education is concerned about the learning processes of people of color and promotes these processes to be equitable. These ideas can be interrogated, using CRT, to illuminate the strategies the Black students in this study put into place to self-define their success in math classrooms, which resulted in more empowerment. These themes were chosen from the literature because they represent themes found in the data analysis.

Identity Development

The notion of identity has become increasingly prominent in the mathematics education literature. Identity is a concept that is important to consider in mathematics education research because it connects elements that are integral to our understanding of mathematics contexts and learning. In a broad sense, identity can be thought of as how individuals author dynamic narratives, both by how they identify and name themselves and how they are recognized and looked upon by others. Identity is a construct that examines the extent to which individuals develop a sense of affiliation with and have come to see value in mathematics as it is realized in their classrooms (Cobb & Hodge, 2007). Identity can be positional and refers to the ways in which people understand and enact their positions in the worlds in which they live (Boaler & Greeno, 2000; Holland, Lachicotte, Skinner, & Cain, 1998). Positional identities have to do with the everyday experiences related to power, deference, and entitlement. Conceptualizing identities as positional speaks to one explanation for how Black students raised in the U.S. formulate or develop mathematics identities (Lim, 2008; Martin, 2000, 2007).

Some scholars ground their perspective on identity in a Deweyian framework of motivation, which posits that understanding who students are and who they are becoming in math classrooms is connected to what motivates them, which is connected to culture and identity (Cobb & Hodge, 2007). Dewey (1913) stated that motivation "expresses the extent to which the end foreseen is bound up with an activity with which the self is identified" (as cited in Cobb & Hodge, 2007, 170). When conceptualizing identity in this way, the evolution of students' interests can be viewed as a deeply cultural process. The identity-motivation connection implies that the cultivation of students' interests in engaging in mathematical activity should be an explicit goal of mathematics instruction. Identities are also developed in cultural activities. Cultural activities, such as playing dominoes, can be examined for understanding ways in which individuals engage in mathematics problem solving and cognitive reasoning, and can inform mathematics teaching and learning (Nasir, 2005). In her study of strategy shifts in the game of dominoes over time from elementary, high school, to adult participants, Nasir (2005) found that the set of practices and organization of activities in the game of dominoes provided a context that supported learning in a variety of ways. Players were positioned as competent and more experienced players offered consistent feedback, both of which facilitated increased access to positive identities as a learner and a participant. This study reveals how understanding who Black students are culturally and how they participate in cultural activities can be connected to positive mathematics identities. Mathematics and personal identities can be merged to garner ideas and strategies for creating learning experiences that can contribute to increased success for African American students.

Social Capital

During the last 30 years, scholars have defined, described, and developed the idea of social capital, and it has become one of the most "popular exports from sociological theory to everyday language" (Portes, 1998, p.2). Social capital has been conceptualized as social networks and resources (Stanton-Salazar, 1997) and social support and reciprocity (Siddle-Walker, 1996; Williams, 2005). Social capital is a resource that individuals draw upon to enhance productivity (Coleman, 1988), facilitate educational progress (DiMaggio & Mohr, 1985) and realize economic benefits (Lin, 2001). Coleman (1988) claimed that a primary function of social capital is to enable an individual to gain access to human, cultural, and other forms of capital as well as to institutional resources and support.

Coleman (1988) and Bourdieu (1986) offer two different constructs of social capital. Coleman (1988) emphasizes the role of social capital in communicating the norms, authority, and social controls that an individual must understand and adopt in order to succeed. In contrast, Bourdieu (1986) emphasizes the ways in which some individuals are advantaged because of their membership in particular groups. Bourdieu suggests that the amount of social capital an individual may gain access to through social networks depends on the size of the networks. Bourdieu (1986) views social capital as a mechanism that the dominant class uses to maintain its dominant position.

Despite their differences, both Coleman (1988) and Bourdieu (1986) recognize that social capital consists of resources embedded in social relations and social structures that can be mobilized when an individual wants to increase the likelihood of success in a specific action. Beadie (2008) points out that much of the literature in the history of education that invokes the idea of social capital is rooted in Black history. Franklin (2002) examines the role of Blacks in providing financial and other material resources for the support of public and private schools established in Black communities in the United States from the antebellum period to the 1960s. He describes the ways in which Black communities drew upon their own resources to create institutions and opportunities in contexts in which they were excluded. In this way, they used the social capital present in Black communities to establish a "wide-spread system of private self-supported schools" (Williams, 2005, p. 200).

Historical and contemporary structural barriers (i.e., differential access to institutional resources) have imposed limitations to social and cultural capital (see Tolley, 2003), particularly in mathematics classrooms. School success in mathematics classrooms is a function of linguistic, social, and cultural background (Zevenbergen, 2000), which can be viewed as social capital. The practices of mathematics can be exclusory for some students, particularly those whose everyday language is not consistent with the formal mathematics found within schools and classrooms. This is not to suggest a deficit in the language of the students, but to strongly advocate that there are differences between the language and experiences of the students and common school practices. The specialized vocabulary of mathematics, the semantic structure, and the lexical density of mathematics represent a form of language for the discipline, a language students must learn in order to be able to participate productively and effectively in mathematics (Zevenbergen, 2000).

Math classroom talk has its own internal rules that are not made explicit to students but form the basis for communication in the classroom, referred to as the "culture of power" (Delpit, 1988, p. 282). The culture of power exists in the educational environment, and those in power are frequently least aware of its existence, while those with less power are keenly aware of its existence (Delpit, 1988). Students who have not been exposed to the language of mathematics may be disadvantaged in the classroom as knowledge is assumed but not taught. When teachers teach all students the explicit and implicit rules of power, they take a first step toward a more just society and access to social capital (Darder, 1991; Delpit, 1988).

Teacher-Student Relationships

Interpersonal relationships are associations between two or more individuals, which can be based on commonalities. Interpersonal relationships are formed in several different contexts including social, cultural, political, or other influences; and the context can vary from family, kinship, to teachers and friends. Interdependence is also a part of interpersonal relationships. Interpersonal relationships usually involve some level of interdependence. People in a relationship tend to influence each other, share their thoughts and feelings, and engage in activities together. Because of this interdependence, most things that change or impact one member of the relationship will have some level of impact on the other member. Specifically, teacher interactions with their individual students can either strengthen or weaken the established relationship (Baker, 1999; Hughes, Cavell, & Willson, 2001). The notion of *interdependence* is relevant to this discussion because several studies have shown that African American students tend to have a field-dependent cognitive style, which can influence school success (Aschenbrenner, 1972; Clark & Halford, 1983; Cohen, 1969; Ramirez & Price-Williams, 1974; Shade, 1982). While we must be careful not to essentialize, cognitive style, as coined by Witkin (1967), has shown that cultural values reflected in socialization practices affect development of cognitive styles in children. A field-dependent cognitive style, as described by these scholars, emphasizes family and religious authority, group identity, relies more on external visual cues, and is more socially oriented, while field-independent styles characterize encouragement to question convention, individualism, reliance on bodily cues within themselves, and are generally less oriented toward social engagement with others.

When teachers establish and build relationships with their students, ideas about cognitive style (how students think and process) are important because they can provide insight into how to build effective relationships with students. Social-cultural approaches to building relationships emphasize that individual units (the teacher-student dyad) cannot be separated from their school and classroom contexts (Cobb, 1986; Rogoff, 1996). Thus, socio-cultural approaches to student–teacher relationships might examine the role of standards and norms, including the types of activities students and teachers participate in, as well as what it means to "be engaged" in learning or to have "appropriate" interactions (Davis, 2003). Social constructivist perspectives on student–teacher relationships posit that knowledge in classrooms is jointly constructed within the context of relationships (Cobb & Yackel, 1996). From this perspective, teachers and students engage in the process of negotiating meaning about both cognitive activities (e.g., learning mathematics) as well as social cognitive activities (e.g., use of humor). This includes negotiating language and power in the classroom (Cazden, 2001).

METHODS

I interviewed 12 Black high school juniors and seniors across four high schools, which I call West, Woodson, Gates, and DuBois in the Dunbar School District located in the Pacific Northwest. Each participant selected a pseudonym to be used during the study. The purpose of interviewing is not to generalize to a population, but to access the *context* of individuals' behavior and thereby provide a way for researchers to understand the meaning of that behavior (Seidman, 1998). A basic assumption of interviewing methods is that the meanings people make of their experience affects the ways they carry out that experience, which helps researchers understand individuals' actions. However, the 12 African American students had characteristics that would speak to other Black students' experiences in mathematics classrooms, such as being the sole African American student in an advanced mathematics course, such as IB calculus.[5]

Dunbar School District has 23 elementary schools, seven middle schools, and four comprehensive high schools. The total number of students enrolled is approximately 22,000, of which 2,882 or 13.1% are African American, and a total of 43.8% of all students in the district receive free or reduced-price lunch. Dunbar's database included 391 Black juniors and seniors during the academic year 2009–2010. Juniors and seniors were chosen because they have been a part of the mathematics education pipeline and are able to describe their mathematics experiences and achievement over time. I collaborated with principal contacts at each of the four high schools and established visitations to discuss the research project. Out of the 391 students, 192 attended the research presentation. Sixty-nine students signed consent forms and agreed to participate in Phase I (questionnaires) of the study. From the 69 students, 12 were selected to do interviews for Phase II of the study. These students were selected through a combination of self-interest in the project, course-taking patterns, state assessment scores, and college aspirations. A major tension in this process was the difficulty of locating a critical mass of Black mathematics students based on what is publicly legitimized as "successful"—meeting or exceeding standard on standardized mathematics tests. Consequently, the definition of success had to be problematized, and the interviews became even more essential for constructing a more complete picture of success, which included multiple possibilities like persistence, resilience, agency, course taking, grades, and fulfilling college requirements.

The data were collected during the 2009–2010 school year and interviews were conducted on mutually agreed upon sites including home and school sites. I chose to interview the students in places they felt most comfortable in describing their experiences in mathematics classrooms. I interviewed each participant once, and the interviews ranged from 30 to 60 minutes. Questions focused on teacher relationships, family and peer networks, agency, and motivation. I proceeded through multiple steps including preparing, exploring, and analyzing the data, and then describing and interpreting the results. Once the transcripts were completed, each question and the corresponding responses were read several times. I then took notes and eventually developed a qualitative codebook—the codebook documented the initial codes, their labels, and interrelating themes (Miles & Huberman, 1994). I identified twelve codes, and six central themes emerged from the data including 1) social networks and their facilitation of success, 2) mediating identities, 3) positive teacher-student relationships, 4) motivation, 5) assertive personalities, and 6) human agency. I report on three of those themes in this chapter. Miles and Huberman (1994) point out that because data overload is a chronic problem in qualitative research, selectivity is essential for data collection and analysis. The conceptual framework and the research questions were the best defense against this overload. Consequently, while the findings were represented in both the discussion of themes, I made subjective choices. Building evidence for the themes found in the interview data was the main way this study approached validity of the qualitative data.

FINDINGS

The findings reported in this chapter derive from a more extensive study, which examined the factors that influence successful mathematics achievement for Black students raised in the United States. This section describes Black students who have exercised agency, including resilience to achieve mathematics, academic, and personal success (see Noble, this volume). I use narratives and vignettes to focus my interpretive analysis on how these students view themselves as individuals and learners of mathematics, as well as the beliefs they hold about the importance of mathematics, their motivation to persist in mathematics, and the actions they take to achieve and maintain their success. Data that emerged from students' interviews in the form of quotations were carefully selected to show how the students articulated their experiences around the respective themes. I describe the findings around themes: teacher-student relationships, social capital, and identity.

Connections between Teacher-Student Relationships and Learning Mathematics

Authentic teacher-student relationships, high expectations, and role modeling can influence mathematics success among Black students. Participants conceptualized a "good teacher" as an individual with whom they had a good relationship, who had the ability to teach mathematics in multiple ways, and who exercised patience, had high expectations, and promoted openness for students to ask questions. An authentic relationship can include mutual trust, feeling comfortable with one another to share personal stories, making connections, and being willing to go the extra mile. Brianna (BT)pointed out that she has had both good and bad math teachers, but was grateful for both types of experiences:

> BT: I think I've had the spectrum. I've had the really, really good teachers that pushed really hard and put effort and are there whenever you need help. Then I've had teachers who are just like, "well you don't get it, that's your problem." So I really can't generalize them all. I think that I liked the fact that I've had, you know, a little bit of everything; therefore I know how to identify a good teacher from a bad teacher really quickly and know how to get in and out of those classes. Because for me if the teacher isn't willing to take the time and actually explain it [math] and put it in terms that I can understand there's no way I'm going to get it.

Brianna also discussed how her relationship with her math teacher eased any discomfort when she did not quite understand the particular math topic the class was studying:

> BT: I had a really good relationship with them. The teacher I had for about four years, she was really...our class was more like a...it was a math class, but it was more like a um, I don't want to call it a friendship cause it sounds really cliché....She was really open with us. She was very...because we'd known her

for four years, so it made it really easy for her and most of us to like share. Um, there was really no embarrassment when you didn't get something.

Having an authentic teacher-student relationship can influence students' desire to open up, share, and decrease any embarrassment when they do not understand specific math topic areas.

Cindy (CJ) said that her ninth-grade algebra I teacher was one of her role models. She identified this math teacher's instructional moves as something for all teachers to aspire:

CJ: Only one I see as a role model…my ninth-grade year. I still talk to her.

NR: So what did she do that made you say, yes—that's a role model for me.

CJ: Cause like she first started the class and she saw…she observed on how we learned. Every class did different things. Like it wasn't the same test, it wasn't the same skill building projects. So like our class was the last period of the day, she knew that it was a time when we did not want to pay attention so she like figured out ways that we can do math. We did like a lot of activities. You know how most teachers are supposed to go by the book and if you don't get it you are supposed to come after school?

NR: Yes.

CJ: We spent a whole month on just one lesson because just nobody got it and she wanted us to grasp it. She was like "I don't care about the schedule of the year, if you guys don't get this then we're gonna work on it." And we worked outside a lot. Like she let it, like when it was sunny she like did activities outside and we did slopes outside.

When I asked Tiana (TJ) if she viewed any of her math teachers as role models, she described her personal definition of a role model and then said that both individuals in the relationship have a role and responsibility. She also pointed out that asking questions is a less intimidating task when she has a relationship with her math teachers:

NR: And do you see your math teachers as role models?

TJ: Mmm…yes. I really do. They're someone that…they're easy to go to and I can ask any question I want and they are very patient with me. I just had to learn that I need to take initiative and go to my teachers and get the help I need.

NR: Ok. And would you say that if you had to name one thing, that's the one thing that really helps you to start being more successful in your math classes? Asking questions and taking more initiative?

TJ: Yes. I think that's the main thing.

NR: Do you think that you would still be failing or struggling had you not done that?

TJ: I think so, yeah 'cause especially this year the material is ten times harder than it was last year, but I'm getting it more because I have a relationship with my teacher and I can go and ask questions.

These student comments suggest that positive teacher-student relationships can help students feel comfortable asking questions, which facilitates support in math learning. When teachers create a space for students to question the text or give explanations of mathematics content, they shift authority and help students co-construct knowledge, an important idea of multicultural education (Banks, 2004). While this instructional technique could be viewed as beneficial for any student, because Black students have been positioned as intellectually inferior in mathematics, particularly in the U.S., it is subsequently more important that these types of experiences are created for Black learners. Teachers who develop fluid relationships with Black students can increase students' comfortability of raising questions and engaging in mathematics learning as competent contributors.

Social Networks Make it Possible to Achieve Certain Ends

I learned through the interviews I had with participants that although the students felt their families could not help them directly with math homework, their families were very aware of how taking as much math as possible was important for college admissions and potential future economic success. As a result of this awareness, the families of the students provided guidance and high expectations. Family included mothers and fathers, but also extended and fictive kin, such as aunts, uncles, cousins, and mentors.

Most families were not intentional about teaching their children the ways in which they used math on a daily basis. The students pointed out that in addition to their families, they received a wealth of information through their social network of friends. All of them, except one, were in advanced math courses beyond algebra and learned from their peers about various prestigious opportunities and resources. Briana discussed how her mother was the impetus for choosing to go to college:

BT: Honestly I don't have much of a choice. My mother made it very clear to me that she had a plan and nothing I or anyone else does is going to get in the way of that. She made it very clear that I had to go to college.

NR: Is your mother a college graduate?

BT: Yeah.

NR: So do you think if she were not on top of you like that, and that there was a choice, you would still choose college?

BT: I think I would, but it would have taken me a lot longer to get to "Ok, I need to go to college." 'Cause it's always something....I need to go to college...I have to go, there is no other option for me. I think once I got to a certain point where I realize there is nothing better in jobs or the military, I was going to get it together and go to college, 'cause that's what she did.

Tiana viewed her mother and aunt as resources to support her math learning and achievement as well as the inspiration to stay on track. She connected the fact that while her mom didn't have the knowledge of calculus, she still provided capital by getting her a math tutor:

> TJ: She [my mother] hangs over my shoulder like a hawk, but it's a good thing though. I'm happy that I have that support...especially with math class, when I don't understand something, even though she has no idea how to do math cause she only took algebra and that was it for her, but she is very supportive. She got me a tutor when I didn't understand something and plus my aunt in South Carolina, she is a principal so she took some math, so I get help from her, too.

Jerome (JJ) highlighted his grandfather and older brother as integral to his success in learning mathematics:

> JJ: My Grandpa, like, every time he sees me he tells me I need to make sure I'm still reading books and just doing math. Talking about the three R's...you know like back in the day and all that stuff...My older brother, when I was in second grade, I think it was during the summer and he'd teach me division and multiplication and stuff. So I pretty much knew stuff before most other kids my age did. He told me he was there to help me.

Exposing students to mathematics early can be a powerful strategy to support their future learning in more challenging mathematics courses.

In addition to having her parents and uncle for math support, Chantel (CR) utilized her peers in the Advancement via Individual Determination (AVID) program to help support her math learning. Chantel also indicated that while math "irritates" her, she thinks that asking questions when help is needed is her own responsibility:

> CR: My mom's in school so we used to do math together. My stepdad's an assistant principal, so if I need help he can help me. More like in the beginning of high school and middle school and elementary, that's when they really helped me, but now if I don't ask for help it's like I don't really get it. But I know I need to ask.

> NR: Were there other family and friends that talked to you?

> CR: My uncle, he's in school too, so he helps me. If I need help from...'cause I'm in AVID, and AVID helps me. I bring the question in, and we have tutors to help. So they could help me, but I chose not to bring math questions most

of the time because I don't even want to think about math. That's why I don't like math. Sitting in the math class just irritates me a lot.

Tiana said that because she comes from a family of educators, key messages about the importance of mathematics are pervasive:

TJ: My mom's an English teacher though. But where she really did help was to get me a tutor, and she also buys me a lot of math books, which helps give steps on how to study and how to do the problems, so she did help me with that even though she doesn't know any math herself. She does take the initiative to try and get me the help that I need.

An interesting point Tiana made was that her family pushed her to pursue mathematics even if she wasn't going to pursue a math-related career. Since Tiana was surrounded by educated family members, they understood that knowing mathematics is critical, not just for college admissions, but also for life in general:

TJ: Yeah. Pretty much like my family's been telling me do not drop your math classes even if you're not going to pursue a specialty, which I am, but if you're not going to pursue a major with math then you still do need math.

These students had many social resources to learn and gain support from and to provide them with information regarding the importance of mathematics. Securing resources for the purposes of supporting mathematics learning was important to the families of the students in this study. This strategy reveals that Black families do value mathematics education and believe that their children can be brilliant in mathematics if given the opportunity to learn; and they provide a counter-narrative to the conventional wisdom about Black parents not valuing education. These quotes exemplify the importance of social capital for exposing students to mathematics and encouraging mathematics achievement.

Mediating Mathematics Identities

One reason mathematics identities are important is that they can influence Black students' participation levels in math communities. The mathematics classroom is a community, and there are implicit and explicit messages about what it means to do, learn, and achieve in mathematics. At the high school level, grades are an important factor because they influence students' grade point averages, which is one of the main indicators colleges consider for admission.

It can be difficult to locate a reason to develop and nurture a mathematics identity when one grapples with either the content or the process of mastering certain math courses, skills, or concepts. Several of the participants noted that during elementary and middle school, math was their favorite subject. They related to mathematics because they understood the concepts and had more one-on-one time with the teacher. However, when they got to high school, mathematics became somewhat taxing. Queenniya (QW) discussed her frustration with math:

> QW: I don't like it. I struggle tremendously. It frustrates me a lot because there's so many different formulas and everything you have to remember and West doesn't have the best math teachers and it seems like some of them just don't care enough, but I don't know. I've just always, well I haven't always struggled with math. Seventh and eighth grade I got A's in math and when I hit high school it was just...wow.

Chantel, too, enjoyed elementary and middle school mathematics, but discussed how in high school, it became a challenge for her to get higher than a C in her more advanced math courses:

> CR: All my math...middle school I did really good in math. Elementary school I loved math it was my favorite subject, but then high school was like, wow all this is really confusing. I don't know. It's hard to get over a C for me. I had a D last year, but that was like one semester. And the others I just had Cs, and it was just like wow, what am I doing? But I know the things I need to do like study more, but it just takes so much work.

These student responses suggest that because high school mathematics can be more challenging and less enjoyable, it can affect students' beliefs about their mathematical ability and pose challenges to developing positive mathematics identities.

Although Briana and Michelle (MS) were well aware of the importance of mathematics to their potential future careers, both pointed out that they would not take mathematics courses if they were not required for their graduation:

> BT: ...because I don't like math (laughs). I don't like it and I know that in some areas either the business or broadcast journalism I am going to have to take math, but if I can bypass that by anyway, I'm going to; just because it's not my cup of tea. I've got to have that math type of mindset, so, yea.
>
> NR: And why do you think that?
>
> BT: Couldn't understand it. And if you don't understand something you get frustrated with it, and I'm just not going to take the time to work at something that no one else is helping me to get. Cause it's the blind leading the blind, running around in circles.
>
> MS: Well...I don't think I would take it, but there is a side of me that says I should even though I probably wouldn't want to just because I would feel like I'm missing some type of skill other than just reading, science, and English.

Being aware that mathematical knowledge is necessary for future careers can contribute to the development of a mathematics identity by prompting students to take mathematics courses that they otherwise would not take.

Mathematics can be enjoyable even when it is challenging. Phillip indicated that he really enjoys doing math even though it is challenging sometimes:

> Well, I'm up to calculus right now...fingers crossed; I'm taking statistics next year....I'm pretty confident with math. I've made it this far. It's kind of you know, it gets harder as it goes, but other than that I actually enjoy doing math.

Billy (BB) stated that math just comes easy for him and that it is a subject he is very passionate about. He also pointed out that it was time and effort that helped him to develop his love for mathematics:

BB: Um, yeah 'cause I like math.

NR: Ok, what do you like about it?

BB: I don't know it just comes easier to me than like the other subjects.

NR: Ok. Um, what are the subjects that are you passionate about?

BB: Math and science, like chemistry. Cause that's kind of like another math class.

NR: And would you say the reason is because like you said earlier you feel like it comes easy to you?

BB: Yeah.

NR: Did you feel that way when you were in elementary and middle school? Did you still like those subjects, math and science, or was that something that developed later in high school?

BB: In elementary school math was kind of easy but it was kind of difficult at the same time, but we had flashcards and it helped me and I just put a lot of time and effort and it just came easy to me.

NR: So, this question asks what subjects do you think you're skilled at, just comes naturally—math and science?

BB: Yeah.

NR: Ok. And how do you think you have developed those skills?

BB: By just putting a lot of time into studying the type of math I was learning.

Some students developed what I called a *temporary* mathematics identity because their choice to take advanced mathematics courses was connected to an important end—potential future careers. Briana and Michelle were clear in stating that they probably would not take additional mathematics courses if they believed not taking advanced mathematics courses was not tied to future opportunities.

DISCUSSION

This study is important because it provided insights into the experiences of Black students who self-identified as being successful in mathematics. Black students are important to study because successful stories are scarce in the mathematics education literature (Berry, 2008; Martin, 2000), and new findings add new voices and theories on Black students' learning. Few studies have privileged student voice in exploring factors that influence successful mathematics achievement, and this study addressed that gap. The findings in this study are central to placing Black students at the center of their own realities as students in math classrooms and support the importance of counter-stories as a method of empowerment in U.S. public schools. I use the student narratives as critical readings to talk back (hooks, 1994), challenge, and interrogate the mainstream image of the mathematics domain founded upon and buttressed by the paradigmatic status of Platonic and Aristotelian scholarship. The stories expressed in this study demonstrate the complexities of being Black in mathematics classrooms in the U.S. This is consistent with CRT, which focuses on the importance of race and how race operates in society.

The findings revealed that positive teacher relationships were important to mathematics learning and success. Cindy described how one of her math teachers provided opportunities for non-traditional learning such as examining slopes of lines outside of the classroom, which she found effective in her mathematics learning. The students in this study defined a "good teacher" as one whom they more than likely had a good relationship with because good teachers were role models, exercised patience, held high expectations, and promoted openness. Students viewed this openness as a safety net for questioning in the public space when they experienced cognitive dissonance, as well as opportunities to share personal aspects of their lives. The students situated their understanding of teacher relationships within ways that help us understand features of culturally responsive instruction. Through positive student relationships, teachers can demonstrate ambitious and appropriate expectations and exhibit more one-on-one support for students in their efforts toward mathematics brilliance. A few ways that this can be done include providing resources and personal assistance, modeling positive self-efficacy beliefs, and celebrating individual and collective accomplishments (Gay, 2000). Other studies have also shown that teacher care can influence mathematics success because having positive relationships with teachers can increase teacher efforts in promoting opportunities for more students to excel in mathematics (Berry, 2008). A key aim of multicultural education is to create empowering educational opportunities for students from racially and culturally diverse backgrounds that increase learning. Establishing and promoting positive relationships with Black students in mathematics classrooms for the purpose of helping students access more mathematics learning is an instructional practice that empowers students.

Social capital was an important factor that Black students in this study viewed as being influential on mathematics success. Jerome, Tiana, and Briana all commented on how their family members and fictive kin supported their mathematics learning in indirect ways. The support was indirect because many of their parents and other family members had not taken advanced mathematics to provide more direct support with homework, for example. A parent not having taken advanced math courses as students, thereby being unable to contribute to their children's math learning in more direct ways, is an illustration of what I mean by the macro background influencing the mathematics achievement of Black students. In his ethnographic study, Martin (2006) documented the experiences of African American parents and found that they struggled for mathematics literacy and situated that struggle within race-based frameworks that they believed characterized their lives and their children's lives as both African Americans and learners of mathematics. However, given this fact, many parents of the students in this study still understood the importance of mathematics education and provided resources and support such as tutors. Consistent with the literature, the students indicated that family and community members who made the expectations of academic and mathematics success explicit enabled them to be more successful in mathematics (Stinson, 2008).

The Black students in this study wrestled with mediating mathematics identities in their quest for mathematics success. Several participants mentioned that during elementary and middle school, math was their favorite subject and they identified with mathematics, but then later on in high school, mathematics was viewed as more challenging. Students identified experiences such as having to memorize formulas, non-caring teachers, and limited opportunities for one-on-one teacher time as some reasons for why high school math was more challenging. These types of experiences can influence Black students' beliefs about their mathematics ability and raise challenges to positive identity development as mathematics learners. Consistent with other studies, developing positive mathematics identities can be shaped by school level and community factors (Martin, 2000; Nasir, 2005). These school level factors influencing identity can be situated in broader contexts around race (Martin, 2006, 2009).

Because the students were aware that taking three to four years of mathematics was a requirement for most universities, some of the students developed temporary mathematics identities. They were temporary because their decisions to take advanced mathematics courses was tied to personal utility—a certain end—mainly graduation requirements and college admission, and not necessarily linked to an intrinsic "love of mathematics." In his study of African American adults' mathematics learning and participation,

Martin (2007) also found that the attraction to pursuing advanced mathematics courses lay more in its "relationship to forging a stronger African American identity than for occupational or socioeconomic advancement" (p. 158). These findings reveal that African American students' choices to take advanced mathematics can be understood within a more complex context in connecting to a larger personal endeavor.

CONCLUSION

Achievement gap language and race-comparative studies are prolific in the mathematics education literature, and this analysis (along with every chapter in this volume) attempts to show that these frameworks are limiting and insufficient in describing mathematics success for Black students. It is evident that race is endemic in U.S. society as CRT suggests. Standardized test scores position Black students within a deficit model and perpetuate racial inequalities. Therefore, there is a need for prolific scholarship that provides counter-narratives. This volume provides counter-narratives and provides examples of the brilliance of Black students in mathematics. The students' comments in this study, as well as others included in this volume, provide a more contextualized notion of success, including the barriers they had to overcome to achieve that success.

For many mainstream mathematics education scholars, the mathematics community, policy writers, and the larger society in general, objectivity in analyzing mathematics success for all learners means using standardized measures and codified guidelines. These types of accountability measures legitimate a prescribed form of social capital that is in turn replicated in schools. Analyzing mathematics success in this way is problematic because it positions Black students in a deficit model, embodies meritocracy and hegemonic ideals, and does not necessarily capture the complexities of being both Black and a mathematics learner. For these reasons, rethinking success and notions of brilliance is important work that needs to be advanced in the field, and this volume contributes to that scholarship.

The results of this study and others in the volume can be used to inform policy regarding the mathematics education of Black students. Specifically, these results further our understanding of multicultural education because they help us think more critically about how to educate Black students in mathematics in ways that allow them to fully participate in American life, not by being forced to attend to "hollow, inane, decontexualized" (Delpit, 1995, p. 45) experiences, but rather within the context of meaningful endeavors that embrace their intellectual and cultural legacies as mathematics knowers, learners, and doers.

Policy stakeholders might consider the implications of how changing extant measures of mathematics success can positively influence Black students' mathematics education and achievement. An important idea is to build upon the teacher effectiveness framework to include observation protocols for evaluation around these key ideas. In order to redress past injustices and account for intentional or unintentional limiting of access to study mathematics, the racialized experiences of Black students must be challenged and ameliorated in mathematics classrooms to account for multiple forms of social capital (Martin, 2000; Stinson, 2008). The consideration of these multiple forms of social capital in policy formation as well as in schools is required to be able to account for success across characteristics such as race, class, or gender and create equity in our classrooms.

NOTES

1. Black refers to individuals who are multigenerationally born and raised in the United States; a group of people whose families have for generations identified themselves as African American, and for whom that identification is a crucial part of their sense of themselves, their families, and their communities. Black people have a distinct identity that has been shaped in large measure by a common history of slavery and by the political struggle of the Civil Rights Movement. Multigenerational African Americans have been enculturated in how the United States socially constructs race and ethnicity (Clark, 2010).
2. See the scholarship of Jeannie Oakes, a progenitor of research on mathematics tracking in the United States.
3. See James D. Anderson, *The Education of Blacks in the South, 1860–1935* and Heather Andrea Williams, *Self-Taught: African American Education in Slavery and Freedom.*
4. Danny Martin (2009) is considered a progenitor in this new emerging body of literature examining mathematics achievement and performance as racialized experiences.
5. The International Baccalaureate (IB) program is very prestigious and rigorous.

REFERENCES

Aschenbrenner, J. (1972). Extended families among Black Americans. *Journal of Comparative Family Studies, 3,* 257–268.

Baker, J. A. (1999). Teacher-student interaction in urban at-risk classrooms: Differential behavior, relationship quality, and student satisfaction with school. *The Elementary School Journal, 100*(1), 57–70.

Banks, J. A. (2004). Multicultural education: Historical development, dimensions, and practice. In J. A. Banks & C. A. M. Banks (Eds.), *Handbook of research on multicultural education* (2nd ed., pp. 3–29). San Francisco, CA: Jossey-Bass.

Banks, J. A. (2007). Multicultural education: Characteristics and goals. In J. A. Banks & C. A. M. Banks (Eds.), *Multicultural education: Issues and perspectives* (6th ed., pp. 3–30). Hoboken, NJ: John Wiley & Sons, Inc.

Baptise, H. P. (1979). *Multicultural education: A synopsis.* Washington, DC: University Press of America.

Beadie, N. (2008). Education and the creation of capital: Or what I have learned from following the money. *History of Education Quarterly, 48*(1), 1–29.

Berry, R. Q. (2008). Access to upper level mathematics: The stories of successful African American middle school boys. *Journal for Research in Mathematics Education, 39*(5), 464–488.

Boaler, J. & Greeno, J. G. (2000). Identity, agency, and knowing in mathematical worlds. In J. Boaler (Ed.), *Multiple perspectives on mathematics teaching and learning* (pp. 171–200). Westport, CT: Ablex.

Bourdieu, P. (1986). The forms of capital. In J. G. Richardson (Ed.), *Handbook of theory and research for the sociology of education* (pp. 241–258). Westport, CT: Greenwood Press.

Cazden, C. (2001). *Classroom discourse: The language of teaching and learning.* Portsmouth, NH: Heinemann.

Clark, H. D. 2010. *We are the same but different: Navigating African American and deaf cultural identities.* Unpublished doctoral dissertation, University of Washington, Seattle, WA.

Clark, L. A. & Halford, G. S. (1983). Does cognitive style account for differences in scholastic achievement? *Journal of Cross-Cultural Psychology, 14*(3), 279–296.

Cobb, P. (1986). Contexts, goals, beliefs, and learning mathematics. *Journal for the Learning of Mathematics, 6,* 2–9.

Cobb, P. & Hodge, L. L. (2007). Culture, identity, and equity in the mathematic classroom. In N. S. Nasir & P. Cobb (Eds.), *Improving access to mathematics: Diversity and equity in the classroom* (pp. 159–171). New York, NY: Teachers College Press.

Cobb, P. & Yackel, E. (1996). Constructivist, emergent, and socio-cultural perspectives in the context of developmental research. *Educational Psychologist, 31,* 175–190.

Cohen, R. A. (1969). Conceptual styles, culture conflict and nonverbal tests of intelligence. *American Anthropologist, 71*(5), 828–856.

Coleman, J. S. (1988). Social capital in the creation of human capital. *American Journal of Sociology, 94,* 95–120.

Darder, A. (1991). *Culture and power in the classroom: A critical foundation for bicultural education.* Westport, CT: Greenwood Publishing Group.

Davis, H. A. (2003). Conceptualizing the role and influence of student-teacher relationships on children's social and cognitive development. *Educational Psychologist, 38*(4), 207–234.

Delgado, R. & Stefancic, J. (2001). *Critical race theory: An introduction.* New York, NY: University Press.

Delpit, L. D. (1988). The silenced dialogue: Power and pedagogy in educating other people's children. *Harvard Educational Review, 58,* 280–297.

Delpit, L. D. (1995). *Other people's children: Cultural conflict in the classroom.* New York, NY: The New Press.

Dewey, J. (1913). *Interest and effort in education.* Cambridge, MA: The Riverside Press.

DiMaggio, P. & Mohr, J. (1985). Cultural capital, educational attainment, and marital selection. *American Journal of Sociology, 90*(6), 1231–1261.

Franklin, V. P. (2002). Introduction: Cultural capital and African American education. *The Journal of African American History, 87,* 175–181.

Freire, P. (2002). *Pedagogy of the oppressed* (M. Bergman Ramos, Trans.). New York, NY: Continuum.

Gay, G. (1994). *A synthesis of scholarship in multicultural education.* Oak Brook, IL: North Central Regional Educational Laboratory. (ERIC Document Reproduction Service No. ED378287)

Gay, G. (2000). Culturally responsive teaching: Theory, research, and practice. New York, NY: Teachers College Press.

Gillborn, D. (2006). Critical race theory and education: Racism and anti-racism in educational theory and praxis. *Discourse: Studies in the Cultural Politics of Education, 27*(1), 11–32.

Grant, C. A., Elsbree, A. R., & Fondrie, S. (2004). A decade of research on the changing terrain of multicultural education research. In J. A. Banks & C. A. M. Banks (Eds.), *Multicultural education: Issues and perspectives* (6ᵗʰ ed., pp. 184– 207). Hoboken, NJ: John Wiley & Sons, Inc.

Gutierrez, R. (2008). A "gap-gazing" fetish in mathematics education? Problematizing research on the achievement gap. *Journal for Research in Mathematics Education, 39*(4), 357–364.

Hilliard, A. G. (2003). No mystery: Closing the achievement gap between Africans and excellence. In T. Perry, C. Steele, & A. Hilliard (Eds.), *Young gifted and Black: Promoting high achievement among African-American students* (pp. 131– 165). Boston, MA: Beacon Press.

Holland, D., Lachicotte, W., Skinner, D., & Cain, C. (1998). *Identity and agency in cultural worlds.* Cambridge, MA: Harvard University Press.

hooks, B. (1994). *Teaching to transgress: Education as the practice of freedom.* New York: Routledge.

Hughes, J. N., Cavell, T. A., & Willson, V. (2001). Further support for the developmental significance of the quality of teacher-student relationship. *Journal of School Psychology, 39*(4), 289–301. Doi ISSN 0022-4405, 10.1016/S0022-4405(01)00074-7

Irvine, J. J. & York, D. E. (1993, April). Differences in teacher attributions of school failure for African American, Hispanic, and Vietnamese students. Paper presented at the meeting of the American Educational Research Association.

Jensen, A. R. (1969). How much can we boost IQ and scholastic achievement? *Harvard Educational Review, 39,* 1–123.

Ladson-Billings, G. (1998). Just what is critical race theory and what's it doing in a nice field like education? *International Journal of Qualitative Studies in Education, 11*(1), 7–24.

Ladson-Billings, G. (2006). From the achievement gap to the education debt: Understanding achievement in U.S. schools. *Educational Researcher, 35*(7), 3–12.

Ladson-Billings, G. & Tate IV, W. F. (1995). Toward a critical race theory of education. *Teachers College Record, 97*(1), 47–68.

Leonard, J. (2008). *Culturally specific pedagogy in the mathematics classroom: strategies for teachers and students.* New York, NY: Routledge.

Lim, J. H. (2008). The road not taken: two African-American girls' experiences with school mathematics. *Race, Ethnicity & Education, 11*(3), 303–317.

Lin, N. (2001). Building a network theory of social capital. In N. Lin, K. Cook & R. S. Burt (Eds.), *Social capital: Theory and research* (pp. 3–29). New York, NY: Aldine DeGruyter.

Margo, R. A. (1990). *Race and schooling in the South, 1880–1950: An economic history.* Chicago, IL: The University of Chicago Press.

Martin, D. B. (2000). *Mathematics success and failure among African American youth: The roles of sociohistorical context, community forces, school influence, and individual agency.* Mahwah, NJ: Lawrence Erlbaum.

Martin, D. B. (2006). Mathematics learning and participation as racialized forms of experience: African American parents speak on the struggle for mathematics literacy. *Mathematical thinking and learning, 8*(3), 197–229.

Martin, D. B. (2007). Mathematics learning and participation in the African American context: The co-construction of identity in two intersection realms of experience. In N. S. Nasir & P. Cobb (Eds.), *Improving access to mathematics: Diversity and equity in the classroom* (pp. 146–158). New York, NY: Teachers College Press.

Martin, D. (2009). Liberating the production of knowledge about African American children and mathematics. In D.B. Martin (Ed.), *Mathematics teaching, learning, and liberation in the lives of Black children* (pp. 3–36). New York, NY: Routledge.

Miles, M. B. & Huberman, A. M. (1994). *Qualitative data analysis* (2nd ed.). Thousand Oaks, CA: Sage.

Moody, V. (2001). The social constructs of the mathematical experiences of African-American students. In B. Atweh, H. Forgasz, & B. Nebres (Eds.), *Sociocultural research on mathematics education* (pp. 255–278). Mawah, NJ: Lawrence Erlbaum Associates.

Nasir, N. S. (2005). Individual cognitive structuring and the sociocultural context: Strategy shifts in the game of Dominoes. *The Journal of the Learning Sciences, 14*(1), 5–34.

National Center for Educational Statistics. (n.d). The nation's report card: Mathematics 2011. Retrieved from http://nces.ed.gov/nationsreportcard/pubs/main2011/2012458.asp

Oakes, J., Joseph, R., & Muir, K. (2004). Access and achievement in mathematics and science: Inequalties that endure and change. In J. A. Banks & C. A. M. Banks (Eds.), *Handbook of Research on Multicultural Education* (2nd ed., pp. 69–90). San Francisco, CA: Jossey-Bass.

Ogbu, J. & Matute-Bianchi, M. E. (1986). Understanding socio-cultural factors: Knowledge, identity, and school adjustment. In Bilingual Education Office (Ed.), *Beyond language: Social and cultural factors in schooling language minority students* (pp. 73–143). Los Angeles, CA: Evaluation, Dissemination and Assessment Center.

Ornstein, A. C. & Levine, D. U. (1989). Social class, race, and school achievement: Problems and prospects. *Journal of Teacher Education, 40*(5), 17–23.

Plato. (1987). *The Republic.* (D. Lee, Trans.). New York, NY: Penguin

Portes, A. (1998). Social capital: Its origins and applications in modern sociology. *Annual Review of Sociology, 24*, 1–24.

Ramirez, M. & Price-Williams, D. R. (1974). Cognitive styles of children of three ethnic groups in the United States. *Journal of Cross-Cultural Psychology, 5*, 212–219.

Rogoff, B. (1996). Developmental transitions in children's participation in sociocultural activities. In A. J. Sameroff & M. M. Haith (Eds.), *The five to seven year shift: The age of reason and responsibility* (pp. 273–294). Chicago, IL: University of Chicago Press.

Seidman, I. (1998). *Interviewing as qualitative research: A guide for researchers in education and social sciences* (2nd ed.). New York, NY: Teachers College Press.

Shade, B. J. (1982). Afro-American cognitive style: A variable in school success? *Review of Educational Research, 52*, 219–244. DOI: 10.3102/00346543052002219

Siddle-Walker, V. (1996). *Their highest potential: An African American school community in the segregated South.* Chapel Hill, NC: The University of North Carolina Press.

Sizemore, B. A. (1981). The politics of multiethnic education. *Urban Educator, 5*(2), 4–11.

Sleeter, C. E. (2005). *Un-standardizing curriculum: Multicultural teaching in the standards-based classroom.* New York, NY: Teachers College Press.

Solorzano, D. & Yosso, T. (2000). Toward a critical race theory of Chicana and Chicano education. In C. Tejeda, C. Martinez, & Z. Leonardo (Eds.), *Demarcating the border of Chicana(o)/ Latina(o) education* (pp. 35–65). Cresskill, NJ: Hampton Press.

Stanton-Salazar, R. D. (1997). A social capital framework for understanding the socialization of racial minority children and youth. *Harvard Educational Review, 67*(1), 1–40.

Stinson, D. W. (2008). Negotiating sociocultural discourses: The counter-storytelling of academically (and mathematically) successful African American male students. *American Educational Research Journal, 45*(4), 975–1010.

Tolley, K. (2003). *The science education of American girls: A historical perspective.* New York, NY: RoutledgeFalmer.

Williams, H. A. (2005). *Self-taught: African American education in slavery and freedom.* Chapel Hill, NC: The University of North Carolina Press.

Witkin, H. A. (1967). A cognitive style approach to cross-cultural research. *International Journal of Psychology, 2*(4), 233–250.

Zamudio, M., Russell, C., Rios, F., & Bridgeman, J. L. (2010). *Critical race theory matters: Education and ideology.* New York, NY: Routledge.

Zevenbergen, R. (2000). "Cracking the code" of mathematics classrooms: School success as a function of linguistic, social, and cultural background. In J. Boaler (Ed.), *Multiple perspectives on mathematics teaching and learning* (pp. 201–223). Stamford, CT: Ablex.

SECTION V

PREPARING TEACHERS TO EMBRACE THE BRILLIANCE OF BLACK CHILDREN

CHAPTER 14

TAPPING INTO THE INTELLECTUAL CAPITAL OF BLACK CHILDREN IN MATHEMATICS

Examining the Practices of Pre-service Elementary Teachers

Shonda Lemons-Smith

Can I commit myself to work hard over time if I know that, no matter what I or other members of my reference group accomplish, these accomplishments are not likely to change how I and other members of my group are viewed by the larger society, or to alter our caste like position in the society? I still will not be able to get a cab. I still will be followed in department stores. I still will be stopped when I drive through certain neighborhoods. I still will be viewed as a criminal, a deviant, and an illiterate. (Perry, 2003, p. 5)

INTRODUCTION

Perry's quote is a stark articulation of the complexities surrounding Black people in the United States. It is a multifaceted reality that lies at the heart

The Brilliance of Black Children in Mathematics:
Beyond the Numbers and Toward New Discourse, pages 323–339.
Copyright © 2013 by Information Age Publishing
323

of educational underachievement broadly, and within the discipline of mathematics education specifically. Though it might be tempting to want to separate schools and schooling from the realism of the macro society, schools do not exist in a vacuum, and their policies and practices are influenced by the broader context. If you asked most mathematics teachers if they value Black students, their response would likely be a resounding, "Of course! I value all the kids in my classroom." I would argue that this declaration often comes from a place of sincerity and good intentions. However, the lens that frames teachers' conceptions of students, particularly Black students has been inexplicably shaped by a myriad of social, historical, and political forces. These forces promote what Perry (2003) calls "taken-for-granted" notions of intellectual inferiority. She asserts that these notions are going largely unnoticed because we are in a post-civil rights age where "an illusion of openness and opportunity" exists. It would seem that the 2008 election of the first African American president has likely added fuel to Perry's assertion. In the post-*Brown vs. Board of Education* era many people have embraced a "we are beyond that" stance that I contend is critically flawed. In the four decades after *Brown vs. Board of Education,* much of the educational discourse portrayed children of color as deficient and at-risk. While this discourse has lessened somewhat in recent years, notions of deficiency very much endure and are marketed to K–12 schools under the veil of facilitating understanding about certain student demographic groups (e.g., Ruby Payne's Framework of Understanding Poverty). Such perspectives contribute to the construction of how Black learners are viewed in mathematics classrooms.

This chapter explores the notion of tapping into the intellectual capital of Black children in mathematics through examining the practices of pre-service elementary teachers. In this chapter I will discuss the following: a) rethinking the mathematics achievement gap, b) promoting the brilliance of Black children, c) the development of pre-service teachers, and d) highlights of a study of the math practices of pre-service teachers.

RETHINKING THE MATHEMATICS ACHIEVEMENT GAP

In mathematics education, the heavily posited "closing the gap" has emerged as the new descriptor dividing the mathematics academic achievement of White, Black, and Hispanic learners. Closing the mathematics achievement gap is a heavily discussed topic in both K–12 and higher education communities. The intensity is evidenced simply by conducting a Google search on the topic and observing a staggering number of hits. A huge number of articles on the topic have been written (e.g., Lubienski, 2008, 2002; Lubienski & Crockett, 2007; Tate, 1997). Gutierrez (2008) raises concerns about the prominence of the achievement-gap dialogue, stating, "I see it as a moral imperative to move beyond this 'gap-gazing' fetish" (p. 357). Gutierrez

(2008) further contends, "These dangers [of gap-gazing] include offering little more than a static picture of inequities, supporting deficit thinking and negative narratives about students of color and working-class students, perpetuating the myth that the problem (and therefore solution) is a technical one, and promoting a narrow definition of learning and equity" (p. 357). Similarly, Martin (2009) maintains that framing schooling outcomes in terms of racial achievement gaps inherently purports the intellectual inferiority of Black children. Understanding the full magnitude of gap-oriented discourse requires stepping back and considering the implications for Black students. The proliferation of deficit language can negatively impact how Black students view themselves as learners of mathematics (Williams & Lemons-Smith, 2009). Martin (2000) asserts that students' mathematical identities are influenced by social, cultural, historical, and political forces. These aforementioned mathematics education scholars have pushed the dialogue surrounding the achievement gap and learning experiences of Black students in mathematics. How the achievement gap is conceptualized in the mathematics education literature is critical because it informs the preparation of pre-service mathematics teachers, and hence their instructional practices.

PROMOTING THE BRILLIANCE OF BLACK CHILDREN

Although many scholars have advocated for equity and tapping into the intellectual potential of Black children, the pervasive belief that only certain demographic groups are capable of learning mathematics still seems to exist. Hence, the critical question remains, "Why do Black children, by and large, continue to receive inequitable mathematics education despite educational reforms and policies?" That is, what prevents schools from providing cognitively demanding, high-quality mathematics instruction to Black children? Why do schools continue to miss the mark? Hilliard (1995) maintains that a commitment to "releasing the genius" of Black children is in direct contradiction to the existing structural and institutional ideology of U.S. schools. He argues that schools must make a substantive shift in the underlying ideological and pedagogical paradigm as it relates to the academic excellence of Black students. Teaching and learning must be conceptualized in a way that affirms and values Black students' personal and intellectual capital. At the core of such a pedagogical stance is the rejection of the notion that Black children come to school as blank slates or that their out-of-school experiences lack value.

The "Mathematics Beyond the School Walls" project is one pedagogical example of affirming the lived experiences of Black children (Lemons-Smith, 2009). This family outreach project took place in a high-poverty elementary school and applied the Funds of Knowledge framework (Moll & Gonzalez, 2003).[1] The project consisted of two primary components: a) collecting family visual and written artifacts, and b) analyzing and using artifacts in the mathematics teaching and learning process. Students and their families were asked to document the student's out-of-school mathematics experiences. They were asked to document the experiences by either a) taking photographs of images that (they feel) reflect how the student experiences or interacts with mathematics in his/her home and/or community environment, or b) cutting out images from magazines, newspapers, and other print media that (they feel) reflect how the student experiences or interacts with mathematics in his/her home and/or community environment. Families were also asked to jot down the reason(s) why they chose each artifact and why it is relevant to the mathematical learning of the student. After collecting the student-generated artifacts, the teachers explored the mathematical content embedded in the artifacts, mapped them to state mathematics standards, and developed corresponding tasks. Nearly all of the artifacts could be utilized as a context for teaching multiple mathematics concepts. The created mathematical tasks were aligned with the district mathematics scope and sequence and subsequently implemented. The project afforded students opportunities to infuse their cultural identities in learning activities, something that is not commonplace in mathematics classrooms. By utilizing artifacts from students' homes, families, and communities to communicate mathematical ideas, these teachers explicitly rejected the pervasive notion that students of color and students living in poverty do not bring valuable out-of-school experiences and mathematics capital to the teaching and learning process.

In the book *Culturally Specific Pedagogy in the Mathematics Classroom*, Leonard (2008) also highlights instructional practices that tap into students' cultural experiences. Leonard features several teachers, including Ms. Cho and Ms. Baker. She describes how their classrooms reflect a sense of shared learning community, high expectations, and a valuing of students' contributions to the mathematics teaching and learning process. For example, Leonard recounts how Ms. Cho and Ms. Baker used the text *Sweet Clara and the Freedom Quilt* as a springboard for introducing the idea of quilting. The freedom quilt functioned as the cultural connection and spurred the learning of area, perimeter, symmetry, congruence, measurement, and geometric shapes.

THE DEVELOPMENT OF PRE-SERVICE TEACHERS

This chapter strongly argues for tapping into the intellectual capital of Black children in mathematics. Therefore, it is critical to consider how teacher preparation programs prepare pre-service teachers to work with Black children. In order to promote the brilliance of Black children, teachers must have critical requisite knowledge, skills, and dispositions. In the landmark report "Saving the African American Child," Hilliard and Sizemore (1984) attributed the quality-of-service gap to teachers' lack of preparedness to teach culturally diverse student populations, stating that: a) teachers are unfamiliar with essential culturally relevant academic subject matter information, b) they have not been exposed to the work of many educators who have worked successfully with Black learners, c) they are unfamiliar with important professional literature specific to the education of Black students, and d) they tend to believe that strange new teaching methods have to be invented in order to teach our children so that they can meet general academic standards. Two decades later similar sentiments are still being echoed. Howard and Aleman (2008) outline three elements comprising teacher capacity for teaching diverse populations: a) subject matter and pedagogical content knowledge, b) knowledge of effective practice about teaching in diverse settings, and c) development of a critical consciousness. The conversation surrounding preparing teachers to work with students of color often centers on interpersonal relations and less on curriculum and content. By and large, teacher preparation programs rely on a standalone multicultural course as a mechanism for addressing issues of diversity.

Banks, Cochran-Smith, Moll, Richert, Zeichner, LePage, Darling-Hammond, Duffy, and McDonald (2005) argue that this is a cursory approach and advocate for integrated experiences throughout the professional preparation program. To develop mathematics teachers who are capable of enacting high quality, culturally relevant instruction, the traditional mathematics methods, mathematics content courses, and relevant pedagogy courses must be revisited and reshaped to embrace an equity-oriented approach to teaching and learning. Notions of diversity, equity, and social justice must be central to and evident in all aspects of the course readings, discussions, assignments, lesson plans, and field experiences. Typically pre-service teachers draw on their own mathematics experiences and have distinct views of what constitutes "good" mathematics teaching, what counts as mathematical knowledge, and who is good at mathematics. Therefore, pre-service teachers must engage in provocative exercises that challenge those ideas. They must be encouraged to adopt instructional practices that reflect an asset perspective and explicitly focus on valuing and tapping into the mathematics capital that *all* students bring to the classroom.

What, then, is the role of teacher preparation in preparing teachers who have the knowledge, skills, and dispositions to promote the brilliance of Black children and affirm their intellectual capital? I contend that teacher preparation programs must cultivate pre-service teachers who fundamentally believe in the brilliance of Black children. Webster's dictionary defines believe as: "To exercise belief in; to regard or accept as true; to place confidence in; to think; to suppose." Sadly, the brilliance of Black children is not accepted as an absolute truth. Black children's intelligence is not acknowledged unless there is a preponderance of academic evidence. Even then, children are considered outliers rather than representing the core community. A critical issue at the heart of the conversation regarding the brilliance of Black children is the notion of what counts as valuable mathematical knowledge and who possesses it. Martin (2000, 2006, 2009) has eloquently outlined this argument. Martin (2009) asserts, "In many mathematics classrooms, teachers and students participate in a range of practices in which they develop, contest, and internalize beliefs about what counts as math literacy and who is mathematically literate, contributing to the construction of these classrooms as highly racialized spaces" (p.10). Hence, pre-service teachers must be provided counter experiences and narratives that challenge long held paradigms regarding Black children and the brilliance they embody. They must be provided scaffolding and support in their efforts to operationalize mathematics practices that foster the brilliance of Black children.

STUDYING THE MATHEMATICS PRACTICES OF PRE-SERVICE TEACHERS

In order to enact real change, it is not enough for pre-service teachers to merely espouse positive views about Black students, but those views must be reflected in their mathematics instructional practices. The next portion of the chapter highlights a case study conducted by the author. The study focuses on pre-service teachers' perceptions of Black students as well as students living in poverty situations.

Framing the Study

Ladson-Billings' (1995) theory of culturally relevant pedagogy (CRP) served as the framing perspective of the study. Ladson-Billings asserts that culturally relevant teachers exhibit the following broad qualities with respect to the underlying propositions:

1. Conceptions of Self and Others—Suggests that culturally relevant teachers hold high expectations for all students and believe all students are capable of achieving academic excellence;
2. Social Relations—Assumes that culturally relevant teachers establish and maintain positive teacher-student relationships and classroom learning community as well as are passionate about teaching and view it as a service to the community; and
3. Conceptions of Knowledge—Suggests that culturally relevant teachers view knowledge as fluid and facilitate students' ability to construct their own understanding.

Given the research question guiding the study and emphasis on instructional practices, the dimension Conceptions of Knowledge was the primary focal point. That is, the study looked explicitly at whether urban teachers used students' backgrounds, families, communities, lived, or out-of-school experiences to anchor the mathematics. I refer to this pedagogical approach as contextual anchoring. Teachers engaged in contextual anchoring draw upon children's informal knowledge and experiences, make connections to formal mathematics, and use those connections to help children understand mathematical concepts.

METHOD

Context

The qualitative study was conducted in a large urban city in the southeast United States. The participants were pre-service mathematics teachers placed in high-poverty, urban elementary schools. The context for the study was a mathematics methods course and accompanying student teaching experience. The course is embedded in an alternative certification program for individuals who have an undergraduate degree in an area other than education. Unlike many alternative certification programs this program is not a "boot camp." Rather, individuals matriculate full-time for one academic year (summer, fall, and spring semesters). During that time they take coursework required for P-5 certification and complete 900 hours of student teaching. Another feature that makes the certification program unique is its explicit focus on preparing teachers for urban, high-poverty elementary schools. Diversity, equity, and culturally relevant pedagogy function as the cornerstone and are integrated throughout all aspects of the program including the methods course. Hence, these principles are evident in the mathematics methods course readings, discussions, assignments, and lesson plans. It's fair to say that the course is a significant departure from the historically traditional field of mathematics education.

Data Sources/Participants

A student perceptions questionnaire, math lesson plans, and math teaching observations served as the data sources for the study. All of the data were collected during the last semester of the certification program (spring semester). Seventeen prospective teachers enrolled in the mathematics methods course comprised the larger sample. Each of them completed the student perceptions questionnaire. However, a subset of seven individuals was selected as focal participants. These individuals were selected because of my dual role as their mathematics methods instructor and student teaching supervisor. For each of the focal participants a student perceptions questionnaire, three math lesson plans, and three math teaching observations were collected as data. The questionnaire consisted of two open-ended questions and twenty Likert-scale questions that documented participants' perspectives on mathematics teaching and learning as it relates to students of color and students living in poverty. The open-ended questions were:

- Education statistics indicate that White students in the U.S. score significantly higher on mathematics standardized tests and are more likely to enroll in advanced level mathematics courses. Why do you believe this disparity exists?
- Education statistics also indicate a disparity in the mathematics achievement of low income and middle/high income students in U.S. schools. Why do you believe this disparity exists?

On the Likert component, students were asked to indicate whether they strongly agree, agree, disagree, or strongly disagree. Questions reflected three domains: achievement, out-of-school experiences, and teacher ideology. That section posed statements such as:

- Black students are capable of attaining the same level of mathematics achievement as White students.
- Students living in poverty bring valuable out-of-school experiences and mathematics informal knowledge to the teaching and learning process.
- A good teacher values the out-of-school experiences of all students and utilizes them as a springboard for teaching and learning mathematical ideas.

The math lesson plans included the typical components of objectives, materials, procedures, and assessment. The three math teaching observations lasted thirty minutes each and were documented using a departmental observation rubric.

Data Analysis

The two open-ended questions were coded using the following descriptors:

- Positive—Response was free of deficit language; children's families, communities, or cultures were not devalued or blamed; low student expectations not expressed.
- Moderate—Response had subtle hints of deficit language; subtle devaluing or blaming of children's families, communities, or cultures; subtle low student expectations expressed.
- Negative—Response had deficit language; children's families, communities, or cultures were explicitly devalued; low student expectations explicitly expressed.

For the Likert-scale items the following numeric scores were assigned: 4 = strongly agree, 3 = agree, 2 = disagree, and 1 = strongly disagree. A higher numeric score indicates positive views towards students of color and students living in poverty as it relates to mathematics teaching and learning. The possible range of scores for the questionnaire is 20 to 80. Participants scoring 64 to 80 were designated as reflecting positive views (responded strongly agree on 80% of the questions), 48 to 63 as moderately positive (responded agree on at least 80% of the questions), and 47 or below as minimally positive.

The participants' lesson plans and teaching observations were coded using the following descriptors:

- Substantive—Uses students' backgrounds, families, communities, lived, or out-of-school experiences to *anchor* the mathematics. The anchor draws upon children's informal knowledge and experiences, makes connections, and helps them understand the mathematical concept.
- Cursory—Uses students' backgrounds, families, communities, lived, or out-of-school experiences to *engage* student interest in the lesson. For instance, it uses a scenario that is familiar to children (e.g., using the store concept of Chuck E. Cheese's because there's a Chuck E. Cheese's in the children's community), or superficial (e.g., inserting children's names into story problems), or generic (e.g., using ice cream in a story problem because all kids like ice cream).
- No Attempt—Makes no attempt to uses students' backgrounds, families, communities, lived, or out-of-school experiences to facilitate learning.

FINDINGS AND DISCUSSION

Aggregate Participant Results—Student Perceptions Questionnaire (Likert)

Descriptive data are shown to report the findings of the student perceptions questionnaire on the Likert-scale items. The results of the entire sample of seventeen pre-service teachers are shown in Table 14.1.

This summary provides a broad snapshot as to how the focal participants' perceptions compare to their counterparts. The mean score of the larger sample is 76. Hence, three focal participants scored above the mean (Chelsea, Karen, Bonita) and four below the mean (Kelli, Mary, Nancy, Alicia).

TABLE 14.1. Larger Sample Summary of Data on Likert-Type Items

Entire Class	Likert Score
Student 1	80
Student 2	80
Student 3	80
Student 4	80
Student 5	79
Student 6	79
Student 7	79
Student 8	78
Student 9	76
Student 10	75
Student 11	74
Student 12	74
Student 13	73
Student 14	72
Student 15	71
Student 16	70
Student 17	65

Focal Participant Results—Student Perceptions Questionnaire (Likert/Open-Ended)

Descriptive data are shown to report the findings of the student perceptions questionnaire on the Likert-scale items. The results of the seven focal participants are shown in Table 14.2. As evidenced, all seven focal participants were designated as positive. That is, their responses to the Likert-scale questions reflect affirming views towards students of color and students living in poverty.

Next, data are shown to report the findings on the open-ended items on the student perceptions questionnaire. Table 14.3, reveals patterns for the focal participants.

Focal Participants' Espoused Perceptions

The student perceptions questionnaire reflected two areas of the study framework: Conceptions of Self and Others and Social Relations. Although all focal participants' Likert scores reflected positive views, contradictions were evident with three participants—Karen, Bonita, and Alicia. Although

TABLE 14.2.   Focal Participants Summary of Data on Likert-Type Items

Focal Participant	Likert Score	Designation
Chelsea	79	POS
Karen	79	POS
Bonita	78	POS
Kelli	75	POS
Mary	74	POS
Nancy	74	POS
Alicia	71	POS

Note: POS: Positive views; MOD: Moderately positive views; MIN: Minimally positive views

Karen had the second highest Likert score of 79, her narrative responses were coded as negative (narrative #1) and positive (narrative #2). A negative rating indicates that Karen's narrative #1 had deficit language, children's families, communities, or cultures were explicitly devalued, and/or low student expectations explicitly expressed. Although Bonita posted the third highest Likert score of 78, only one of her narrative responses were coded as positive (narrative #1). Her narrative #2 response was coded as moderate. A moderate rating indicates that Bonita's narrative #2 had subtle hints of deficit language, subtle devaluing or blaming of children's families, communities, or cultures, and/or subtle low student expectations expressed. Unlike Karen and Bonita, whose Likert scores top the focal participants, Alicia had the third lowest Likert score of 71. Her narrative #1 was denoted as positive, while her narrative #2 was denoted as moderate. A moderate rating indicates that Alicia's narrative #2 had subtle hints of deficit language, subtle devaluing or blaming of children's families, communities, or cultures, and/or subtle low student expectations expressed. While there is a disconnect between Karen, Bonita, and Alicia's open-ended and

TABLE 14.3.   Focal Participants Summary of Data on Open-Ended Items

Focal Participant	Narrative 1	Narrative 2
Chelsea	POS	POS
Karen	NEG	POS
Bonita	POS	MOD
Kelli	POS	POS
Mary	POS	POS
Nancy	POS	POS
Alicia	POS	MOD

Note: POS: Positive narrative; MOD: Moderate narrative; NEG: Negative narrative

Likert-scaled responses, it is worth noting that it isn't a total disconnect. That is, each of them had at least one narrative response that was designated as positive.

Another interesting observation is that two of the three less positive responses came from narrative #2. Narrative #2 asked participants why they believe a disparity exist between the mathematics achievement of low income and middle/high income students in U.S. schools. Narrative #1 posed a parallel question regarding Black and White students. Therefore, the responses were less positive regarding socioeconomic class than race.

Focal Participant Results—Lesson Plans and Teaching Observations

Lastly, data are shown to report the findings on the lesson plans and teaching observations. Table 14.4 reveals patterns for the focal participants.

Focal Participants' Practice

The participants' lesson plans and teaching observations reflected the third area of the study framework, Conceptions of Knowledge. While the focal participants espoused positive views on the student perceptions questionnaire, only two individuals had lesson plans and teaching episodes that reflected a substantive effort, Karen and Nancy. Two out of three of Karen's lesson plans and teaching episode were designated as substantive, while one out of three of Nancy's were. It is encouraging to note, however, that all of the other five participants attempted to engage in contextual anchoring and received at least one designation of cursory. Specifically, all three of Mary's lesson plans and observations were categorized as cursory. Chelsea, Bonita, and Alicia each had two out of three, while Kelli had one out of three.

TABLE 14.4. Focal Participants Summary Lesson Plan and Teaching Observation Data

Focal Participant	Lesson 1	Lesson 2	Lesson 3	Teaching 1	Teaching 2	Teaching 3
Chelsea	N	C	C	N	C	C
Karen	C	S	S	C	S	S
Bonita	N	C	C	N	C	C
Kelli	N	C	N	N	C	N
Mary	C	C	C	C	C	C
Nancy	C	S	C	C	S	C
Alicia	N	C	C	N	C	C

Note: S: Substantive anchor; C: Cursory anchor; N: No attempt to anchor

In both the course and field experience, participants had guidance and support in drawing upon children's informal knowledge and experiences, making connections, and using those connections to help students understand mathematical concepts. Some might suggest that the inability to engage in contextual anchoring speaks to the overall inexperience of the participants. However, I would caution against diminishing the findings because the participants were able to demonstrate other indicators of effective mathematics teaching. On the contrary, they were generally successful in addressing various mathematical topics, differentiating learning, and using math centers, manipulatives, and the like. So the question is why were the focal participants unsuccessful at contextual anchoring?

Gay (2010) posits a similar point asking, "Can prospective teachers recognize specific beliefs embedded in particular teaching decisions and behaviors?" (p. 144). She further asserts,

> Many prospective teachers do not think deeply about their attitudes and beliefs toward ethnic, cultural, and racial diversity; some even deliberately resist doing so. When my students are asked about them, they make declarations about being "beyond race, ethnicity, and culture," colorblind, or advocates of racelessness. Yet they are hard pressed to articulate with any depth of thought what these ideologies mean, why they support them, and how they translate into teaching behaviors. (p. 145)

The ideas expressed by Gay are salient in considering why the participants may have been unsuccessful at using students' backgrounds, families, communities, lived, or out-of-school experiences to anchor the mathematics. I would argue that teachers' ability or lack thereof to engage in contextual anchoring speaks to the very notion of what it means to value students. The notion of valuing is one that heavily permeates literature related to teaching diverse student populations. However, what valuing "looks like" with regard to content-specific instructional practices isn't as widely addressed.

Earlier I stated that if you ask most mathematics teachers if they value Black students, their response would likely be a resounding, "Of course! I value all the kids in my classroom." While such declarations are likely sincere, those good intentions often do not translate into action. The participants in this study verbalized generally affirming perceptions of Black students and students living in poverty. However, only two of them experienced any level of success at mining students' backgrounds, families, communities, lived, or out-of-school experiences to anchor their mathematics lessons. It should also be noted that there was some resistance to engaging in contextual anchoring. I observed this both in class and in the field. I contend that this resistance strongly reflects "what a person says he believes" versus "what he *really* believes." This idea is akin to the old adage "actions speak louder than words."

The findings of the study speak to the need for teacher preparation programs to further advance the critical consciousness of pre-service teachers. Howard and Aleman (2008) contend that critical consciousness "a central part of teacher capacity is examining one's own ideas and how these influence the work" (p. 166). Moll and Arnot-Hopffer (2005) echo this position and suggest teachers must have "ideological clarity" about the work they do and acknowledge that teaching is a political enterprise. Programs must engage pre-service teachers in curricular activities and experiences that encourage them to confront the implicit, often unspoken, notion that only the experiences of some students are valuable and reflect mathematical knowledge.

The study also highlights the need for prospective teachers to be pushed further in their development with regard to Conceptions of Knowledge, as defined by Ladson-Billings. I contend that teacher education programs seem to focus more on encouraging candidate growth in the areas of Knowledge of Self and Others and Social Relations. Programs must vigorously prepare teachers to teach mathematics, science, literacy, and social studies in a way that encourages students to use their personal intellectual capital to engage in learning and construct their own understanding. By personal intellectual capital, I mean students' lived experiences that they bring to the mathematics learning community (i.e., their backgrounds, families, communities, frames of reference, and out-of-school experiences). Accomplishing this type of teacher development requires that diversity, equity, and culturally relevant pedagogy are integral components of methods courses and reflected in course readings, discussions, and assignments. Further, programs must be purposeful in their design of field experiences, for field experiences and mentor teachers play pivotal roles in pre-service teacher development.

While teacher preparation programs must ensure that pre-service teachers possess caring and affirming views of Black students, those affective attributes must be coupled with culturally relevant instructional practices. Good intentions are simply not enough. Teachers must be able to plan and implement mathematics lessons that use students' backgrounds, families, lived, or out-of-school experiences to anchor the mathematics and facilitate powerful mathematical learning.

FINAL THOUGHT

This chapter explored the notion of tapping into the intellectual capital of Black children in mathematics through examining the practices of pre-service elementary teachers. The highlighted study is an example of how teachers can engage in pedagogical practices that promote the brilliance of Black students and affirm the experiences they bring to the mathematics teaching and learning process. Schools, teachers, and teacher preparation programs must aggressively challenge the underlying ideological and pedagogical paradigms related to the mathematics excellence of Black students. In order to illuminate the brilliance of Black children in mathematics, schools, teachers, parents, teacher educators, policy makers, and other stake holders must accept the brilliance as an absolute truth rather than an anomaly. The existing rhetoric and slogans that often surrounds Black children should not be confused as promoting brilliance. Rather, it often denotes average or below average notions of capability. In this chapter I call for embracing brilliance and engaging in purposeful action that facilitates the accomplishment of that in which we say we believe. Notions of what counts as valuable mathematical knowledge and who possesses it must be approached from the perspective that Black children *are* brilliant...in mathematics and beyond. As Dr. Asa Hilliard affirms, "First, we must believe that the genius is there. Do we really believe that it is?" (Hilliard, 1991, p. 34). I close with one of Hilliard's poignant quotes,

> The risk for our children in school is not a risk associated with their intelligence. Our failures have nothing to do with IQ, nothing to do with poverty, nothing to do with race, nothing to do with language, nothing to do with style, nothing to do with the need to discover new pedagogy, nothing to do with the development of unique and differentiated special pedagogies, nothing to do with the children's families. All of these are red herrings. The study of them may ultimately lead to some greater insight into the instructional process; but at present they serve to distract attention from the fundamental problem facing us today. We have one and only one problem: Do we truly will to see each and every child in this nation develop to the peak of his or her capacities? (1991, p. 36)

NOTE

1. Moll & Gonzalez' (2003) framework, Funds of Knowledge, reflects a sociocultural perspective and is built on anthropological methods. The basic tenet of Funds of Knowledge is that students' home and social contexts provide powerful informal knowledge. That informal knowledge, in turn, can be used to facilitate school based teaching and learning. Funds of Knowledge explicitly challenges the notion that low-income households do not possess valuable knowledge and skills, and provides a structure for considering how children's knowledge might be accessed, identified, and documented.

ACKNOWLEDGEMENT

The contents of this publication were developed under the PACT+ grant funded by the Transition to Teaching and Teacher Quality Programs and the Office of Innovation and Improvement (OII), U.S. Department of Education. However, those contents do not necessarily represent the policy of the Department of Education, and you should not assume endorsement by the Federal Government.

REFERENCES

Banks, J., Cochran-Smith, M., Moll, L., Richert, A., Zeichner, K., LePage, P.,McDonald, M. (2005). Teaching diverse learners. In L. Darling-Hammond & J. Bransford (Eds.), *Preparing teachers for a changing world: What teachers should learn and be able to do* (pp. 232–274). San Francisco, IL: Jossey-Bass.

Gay, G. (2010). Acting on beliefs in teacher education for cultural diversity. *Journal of Teacher Education, 61*(1-2), 143–152.

Gutierrez, R. (2008). On "gap gazing" fetish in mathematics education? Problematizing research on the achievement gap. *Journal for Research in Mathematics Education, 39*(4), 357–364.

Hilliard, A. (1995). Mathematics excellence for cultural "minority" students: What is the problem? In Iris M. Carl (Ed.), *Prospects for School Mathematics* (pp. 99–113). Reston, VA: The National Council of Teachers of Mathematics.

Hilliard, A. (1991). Do we have the will to educate all children? *Educational Leadership, 49(1)*, 31–36.

Hilliard, A. & Sizemore, B. (1984). *Saving the African American child: A report of the Task Force on Black Academic and Cultural Excellence.* Washington, DC: National Alliance of Black School Educators.

Howard, T. C. & Aleman, G. R. (2008). Teacher capacity for diverse learners: What do teachers need to know? In M. C. Smith, S. Feiman-Nemser, D. J. McIntyre & K.E. Demers (Eds.), *Handbook of research on teacher education* (pp. 157–174). New York, NY: Routledge.

Ladson-Billings, G. (1995). Towards a theory of culturally relevant pedagogy. *American Educational Research Journal, 32*(3), 465–491.

Lemons-Smith, S. (2009). Mathematics beyond the school walls project: Exploring the dynamic role of students' lived experiences. In D. Y. White, & J. S. Spitzer (Eds.), *Mathematics for every student: Responding to diversity, grades Pre-K–5* (pp. 129–136). Reston, VA: National Council of Teachers of Mathematics.

Leonard, J. (2008). *Culturally specific pedagogy in the mathematics classroom.* New York, NY: Routledge.

Lubienski, S. (2008). On "gap gazing" in mathematics education: The need for gaps analyses. *Journal for Research in Mathematics Education, 39*(4), 350–356.

Lubienski, S. (2002). A closer look at Black-White mathematics gaps: Intersections of race and SES in NAEP achievement and instructional practices data. *Journal of Negro Education, 71*, 269–287.

Lubienski, S. & Crockett, M. (2007). NAEP mathematics achievement and race/ethnicity. In P. Kloosterman & F. Lester (Eds.), *Results from the ninth mathematics assessment of NAEP* (pp. 227–260). Reston, VA: National Council of Teachers of Mathematics.

Martin, D. (2009). Researching race in mathematics education. *Teachers College Record, 111*(2), 295–338.

Martin, D. (2006). Mathematics learning and participation in African American context: The co-construction of identity in two intersecting realms of experience. In N. Nasir & P. Cobb (Eds.), *Diversity, equity, and access to mathematical ideas* (pp. 146–158). New York, NY: Teachers College Press.

Martin, D. (2000). *Mathematics success and failure among African American youth.* New York, NY: Routledge.

Moll, L. & Arnot-Hopffer, E. (2005). Sociocultural competence in teacher education. *Journal of Teacher Education, 56*(3), 242–247.

Moll, L. & Gonzalez, N. (2003). Engaging life: A funds-of-knowledge approach to multicultural education. In J. Banks & C. Banks (Eds.), *Handbook of research on multicultural education* (2nd ed., pp. 699–715). San Francisco, CA: Jossey-Bass.

Perry, T. (2003). Up from the parched earth: Toward a theory of African-American achievement. In T. Perry, C. Steele, & A.G. Hilliard III (Eds.), *Young, gifted, and Black: Promoting high achievement among African-American students,* (pp. 1–10). Boston: Beacon Press.

Tate, W. F. (1997). Race-ethnicity, SES, gender, and language proficiency trends in mathematics achievement: An update. *Journal for Research in Mathematics Education, 28,* 652–679.

Williams, B. A. & Lemons-Smith, S. (2009). Perspectives on equity and access in mathematics and science for a 21st century democracy: Re-visioning our gaze. *Democracy & Education, 18*(3), 23–28.

CHAPTER 15

DEVELOPING TEACHERS OF BLACK CHILDREN

(Re)Orienting Thinking in an Elementary Mathematics Methods Course

Tonya Gau Bartell, Mary Q. Foote, Corey Drake,
Amy Roth McDuffie, Erin E. Turner, and Julia M. Aguirre

INTRODUCTION

The research reported in this chapter is part of a larger research project, Teachers Empowered to Advance CHange in Mathematics (TEACH MATH), aimed at transforming preK–8 mathematics teacher preparation so that new generations of mathematics teachers will be equipped with powerful tools and strategies to increase mathematics learning and achievement in our nation's increasingly diverse public schools (Turner et al., 2012). This is not to say that the efforts of this project, or even efforts within an entire elementary mathematics methods course, are sufficient for the preparation of highly qualified teachers. Learning to teach is a complex process. Expertise in teaching does not develop within a short teacher

The Brilliance of Black Children in Mathematics:
Beyond the Numbers and Toward New Discourse, pages 341–367.
Copyright © 2013 by Information Age Publishing

education program (Berliner, 1994; Hammerness, Darling-Hammond, & Bransford, 2005), but rather develops over time as teachers learn from various sources, including their own teaching experiences (Hiebert, Morris, Berk, & Jansen, 2007). Nonetheless, we focus on teacher preparation as we believe it is important to intervene early in supporting mathematics teachers' development.

One of the project goals is to develop a series of instructional modules for elementary mathematics methods courses that support prospective teachers (PSTs) in developing competencies related to accessing and connecting to children's mathematical thinking and children's community and cultural knowledge in instruction. This chapter focuses on PST engagement with and reflection on three activities from the modules: an individual interview with shadowing experience and a mathematics problem solving interview from *The Mathematics Learning Case Study Module*, and a community mathematics exploration experience from *The Community Mathematics Exploration Module*. (For more information on one or more of the three instructional modules, see Aguirre et al. (2012); Bartell et al., 2010; Turner et al., 2012.)

The TEACH MATH project is not focused exclusively on supporting PSTs in becoming more effective teachers of Black students. The goals are broader and include a focus on supporting PSTs in becoming more effective teachers of children from marginalized groups more generally. We believe, however, that it is important to examine the ways in which our project is able to support the development of PSTs specifically with regard to mathematics instruction for Black students. Historically, Black students have not been well served by public schools (Nieto, 2004). As recently as the 1960s, theories of cultural deprivation and cultural deficiency (Valencia, 1997) followed upon the Moynihan (1965) report's unfortunate labeling of the Black family as a "tangle of pathologies." This report supported the infusion of notions of deviance and deficiency into folk myths about the Black community and Black families. We live still today with this racist legacy. The assumptions held by many in the dominant society about the Black community affect the beliefs that many teachers hold about the learning potential of Black children and the resources in the homes of Black students and constitute nothing less than cultural bias against Black children in United States schools (King & Wilson, 1990; Ladson-Billings, 1994).

In this chapter, then, we speak to ways in which our course activities supported PSTs' development specifically with regard to teaching Black students. Our participants include 17 non-Black PSTs who each worked with one Black case study student from grades K–5 over the course of a semester-long mathematics methods course. The activities in which they engaged were intended to (re)orient PSTs from talking about what children (in the case of this report, Black children) *cannot* do, to consider instead what Black children *can* do by having the PSTs attend to the competencies, knowledge, and skills Black children bring to the mathematics classroom (Martin, 2007). We address the question of how these activities support the initial preparation of effective mathematics teachers of Black children so that they can be positioned to recognize and support the brilliance of Black children.

BACKGROUND PERSPECTIVES

Many teachers, including but not limited to the predominantly White, middle-class females at the elementary school level (Howard, 1999; Nieto, 2004), all too often have negative stereotypes and deficit perspectives of children of color as learners, of their culture, and of their communities (Artiles & McClafferty, 1998; Groulx, 2001; Larson & Ovando, 2001; Milner, 2005). These perceptions have implications for curriculum, instruction, and children's learning (Schofield, 1989) and place children of color at a disadvantage (Larson & Ovando, 2001). One goal of involving PSTs in the mathematics methods course activities of interviewing and shadowing a child and exploring parts of the community is to (re)orient PSTs toward a resource perspective as opposed to this deficit perspective. We do not mean to suggest that all PSTs enter our course with a deficit orientation toward Black children. Rather, PSTs have varied prior experiences with diverse populations and communities in general and with Black children and Black communities specifically. Regardless of prior background and experience, however, PSTs are still developing the knowledge, skills, and dispositions required for effective teaching and can benefit from (re)orienting in particular ways toward what children *can* do for the purpose of supporting students' mathematics learning.

THEORETICAL PERSPECTIVES

To conceptualize "effective mathematics teaching," we look to research that has long documented the practices of effective teachers of Black children. This research suggests that effective teachers of Black children value students' perspectives, identities, status, competencies, and prior knowledge (Martin, 2000, 2006); provide space for their voices (Malloy, 2009; Milner, 2003); engage consistently in activities and experiences that allow them to know themselves and their students (Milner, 2003); explicitly reject deficit perspectives of children (Foster, 1990); and uphold their "responsibility to be certain that every child 'reached his or her highest potential'" (Siddle-Walker, 1996, p. 158). More specifically, effective teachers of Black children employ pedagogies centered on children's cultural capital and community contexts (Gay, 2010; Irvine & Irvine, 1983; Ladson-Billings, 1994; Leonard, 2008; Martin, 2007, 2009; Matthews, 2003). This latter research argues that effective instruction means teachers understand and draw upon children's cultural, linguistic, and community-based knowledge and experiences (Ladson-Billings, 1994; Silver & Stein, 1996; Turner, Celedón-Pattichis & Marshall, 2008).

Although work of this type may have applicability across content areas, it is particularly important to do this work in the field of mathematics education, as mathematics as a discipline is often thought to be culture free and accessible to all in the same way (Ladson-Billings, 1997; Nasir, Hand, & Taylor, 2008). We do not discount the importance of a solid understanding of mathematics content and of children's mathematical thinking. We contend that additional experiences that help prospective mathematics teachers understand the mathematical knowledge and practices of students and their communities can enhance teachers' ability to provide effective mathematics instruction (see also Gay, 2010; Ladson-Billings, 1994, 2001; Leonard & Guha, 2002; Nieto, 2004).

Linking Culture to Teaching

Leonard (2008), in discussing culturally specific pedagogy in the mathematics classroom, described this as teachers becoming "students of the students" (p. 15), learning about children's cultures and backgrounds so as to create curriculum that "engage[s] students in culturally specific mathematics tasks" (p. 49). Ladson-Billings (1994) described culturally relevant pedagogy as a "pedagogy that empowers students intellectually, socially, emotionally, and politically by using cultural referents to impart knowledge, skills, and attitudes" (pp.17–18). Culturally relevant teachers acknowledge the link between culture and learning and view the cultural capital of all students as beneficial to school success (Howard, 2003). To be effective, teachers need to recognize and build upon the knowledge, skills, competencies, and resources each child in the classroom brings. Gay (2010) described culturally responsive pedagogy as a bridge between children's home and school cultures; teachers build on the cultural knowledge and prior experiences of diverse students—or students' strengths—to make learning more appropriate and effective for them. "It is based on the assumption that when academic knowledge and skills are situated within the lived experiences and frames of reference of students, they are more personally meaningful, have higher interest appeal, and are learned more easily and thoroughly" (Gay, 2002, p. 106). To be effective, teachers need to understand children's perceptions and experiences so as to design instruction based on the knowledge and experiences children bring to the classroom.

Even attempts to position teachers as "students of the students" is a complex process requiring engagement with, for example, issues of race, class, culture, and power (Bartell, 2011). Activities that involve PSTs in interacting with students and their communities, however, hold potential as an important initial step in teacher development because they provide an avenue for PSTs to gain knowledge of the contexts of Black children's lives, to see the rich resources available to them in their communities, and to see their competencies and, ultimately, their brilliance. Experiences learning about specific children or engaging with children's communities can support PSTs in appreciating the complexity of schooling with respect to negotiating school and community values and the importance of understanding children and their families more generally (Duesterberg, 1998). Further, such experiences can support teachers in beginning to develop and adapt instruction and curriculum based on knowledge of individual children and the community context (Dunkin, Welch, Merritt, Phillips, & Craven, 1998; Gay, 2010; Ladson-Billings, 2001; Leonard, 2008).

PARTICIPANTS

As noted previously, the participants in the study were 17 PSTs enrolled in mathematics methods courses as part of their teacher education programs. Nine of these PSTs took their mathematics methods course as part of an undergraduate teacher education program at a university in the mid-Atlantic region of the United States. An additional eight PSTs took their mathematics methods course as part of a graduate teacher education program at a university in the Northwest region of the United States. As is the case with most elementary teacher education programs, our participants were predominantly White, middle-class females. Thirteen of the 17 participants were female and four were male and, based on self-description, 13 were White, one was Chinese American, one was Brazilian, and two did not specify. Furthermore, drawing on university teacher education program demographic data for PSTs, and PSTs' self-reported data when available, at least 75% of these PSTs have middle-class socioeconomic backgrounds. Additional information about PSTs' self-described experiences with diverse populations and communities in general and with Black children and Black communities specifically was not information explicitly requested from PSTs. Whenever available, however, PSTs' experiences are shared when the PSTs are introduced to the reader.

INSTRUCTIONAL ACTIVITIES

Each PST, as part of *The Mathematics Learning Case Study Module*, conducted two individual interviews with a child and shadowed that child for part of a school day. Additionally, as part of *The Community Mathematics Exploration Module*, PSTs conducted one or more visits to specific locations in the child's community.

Because research demonstrates that teachers have an easier time developing relationships with students who are like them (Spindler & Spindler,

1982), PSTs were asked to choose a child (the same child for all of these activities) who was different from them in one or more socio-cultural ways (i.e., gender, race, socioeconomic status, home language) so as to (re)orient PSTs to recognize the resources and knowledge all children bring to the mathematics classroom.

Activity 1: Individual Interview and Shadowing

In the first individual interview and shadowing experience, the goal was for PSTs to learn about the interests, competencies, and rich experiences of a child both in school and in their home and community spaces. PSTs were provided with an interview protocol created by the researchers/methods instructors that included questions in three areas aimed to support the PSTs in understanding the contexts in which the student lived and learned. The first group of questions aimed to support PSTs in finding out more about the child, including their interests (e.g., "Tell me five things I don't know about you"), activities they engaged inside and outside of school (e.g., "What kinds of things do you like to do after school or at home?"), and what they identified as activities at which they excel (e.g., "What are your favorite school subjects?"). The second group of questions focused on identifying places, locations, and activities in the community that were familiar to the child (e.g., "If I was going to walk from the school to your house, what are some things/places that I would see?") and finding out what the child knew about the potential mathematical activity in those settings (e.g., "Can you think of any places in your community where people do math or use math?"). The third and final group of questions aimed to support PSTs in finding out more about the child's ideas, attitudes and/or dispositions toward mathematics (e.g., "What are some things in math that you are really good at?" "What about math do you not like, if anything?").

PSTs also engaged in at least four hours of "shadowing" their case study student during the school day in both academic and non-academic settings (e.g., hallway, lunchroom, recess). The goal of this activity was to support PSTs in attending to the ways the child demonstrated specific skills, competencies, and knowledge across settings.

Activity 2: Mathematics Problem Solving Interview

In the second individual interview, PSTs conducted a mathematics problem solving interview with their case study students to support PSTs in learning more about their students' mathematical thinking, including their strategies for solving problems and their understanding of particular content areas. Questions for the mathematics problem-solving interview were provided by the researchers/methods instructors. For children in grades K–2, addition, subtraction, multiplication, and division story problems that varied with respect to difficulty were provided (e.g., different problem types were provided such as join, start unknown or join, result unknown [Carpenter, Fennema, Franke, Levi, & Empson, 1999] and could be further differentiated by selecting one-, two-, or three-digit numbers). In addition, for children in grades three through five, story problems also focused on children's knowledge of tens (e.g., "If 10 donuts fit on a plate, how many plates would we need for 87 donuts?") and rational number concepts (e.g., "There are four children playing together. They have seven brownies to share for a snack. How much will each child get if they each get an equal share?").

Upon completion of these two interviews and the shadowing experience, PSTs were asked to write a guided reflection on what they began to learn about the child through these brief encounters. These reflections serve as the primary data source for analysis in relation to this activity. We must keep in mind that these are beginning activities, not an exhaustive examination of the child. Such activities provide only a small window into the experiences of the student. But is it an important window. Teachers often spend an entire academic year with a student and still know little about their lives and interests.

Activity 3: Community Exploration

The goal of the community exploration activity was for PSTs to identify mathematical practices and mathematical funds of knowledge in children's communities and build on these practices and funds of knowledge in a standards-based mathematics lesson. Pairs or small groups of PSTs conducted one or more visits to a specific location in the child's community. The choice of the location was informed by the first individual interview as well as other discussions in the classroom and school setting, and included locations such as parks, community health centers, professional offices, construction sites, corner stores, restaurants, bakeries, and so forth. The assignment given to PSTs required them to look for and document examples of mathematics, (e.g., observing people using mathematics, talking to individuals who work/play/shop in the setting about how they use mathematics) and formulate a series of questions and data sources about the context that could be mathematized. Upon completion of the community exploration portion of this activity, PSTs designed a mathematics lesson that aimed both to deepen children's mathematical understanding of a particular concept and to connect to the community context that they learned about in their community walk.

As with the prior activities, this is a beginning activity and not an exhaustive examination of a local community; it serves to (re)orient PSTs to seeing mathematics in the community and challenging a point of view often prevalent in schools that mathematics, itself, is school based (Schoenfeld, 1992).

Moreover, using contexts in lessons, gained from an examination of spaces that are familiar to children, supports children in having easier access to the concepts being studied (Gay, 2002).

PSTs also wrote reflections about their community exploration experience, focusing on what they began to learn about the community, about the children and their families, and about themselves. Moreover, PSTs reflected on the experience of designing a mathematics lesson based on the community exploration experience, including the benefits and challenges of teaching that draws on and connects to the mathematical funds of knowledge in children's communities. These reflections serve as the primary source of data for analysis with respect to the community engagement experience.

DATA COLLECTION AND ANALYSIS

As noted in the sections above, the data sources for this chapter included 17 PSTs' written reflections on the individual interview and shadowing experience, the mathematics problem-solving interview, and the community exploration experience. The first two authors analyzed data using open coding (Corbin & Strauss, 2008) and analytic induction (Bogdan & Biklen, 1992) to identify patterns of similarities and/or differences as well as emerging themes across the 17 participants' data. Differences in interpretation were discussed until consensus was reached (Miles & Huberman, 1994). Next, using a narrative inquiry approach (Clandinin & Connelly, 2000), we constructed narrative accounts of the participants' interactions with their case study children across instructional activities. We then added our interpretations of these stories. Drafting these narratives allowed us to further illuminate themes, narrative threads, and patterns within participants' stories with greater nuance and detail.

RESULTS AND DISCUSSION

In this section, we provide a brief overview of the three central themes that emerged from our analysis of the data across the 17 participants in order to contextualize the narratives that form the more significant portion of our results and discussion. Following this, we provide extended narratives for selected participants to serve as representative cases reflecting these themes. We blend our analyses of these narratives with the narrative results themselves in order to offer the reader a more integrated presentation of results and discussion with respect to the overarching themes and their significance with respect to the mathematics education of Black children.

Themes

The first theme that emerged from our data analysis is that most PSTs (13 of the 17) explicitly stated that the opportunity to interact with a case study student through interviews, shadowing, and community exploration was a positive experience (see Figure 15.1). This is important, as Leonard (2008) argues that it is a critical task for PSTs to see value in learning about Black children, their culture, and their communities in an effort to improve mathematics instruction for these children. Second, close interactions with a Black child supported many PSTs (11 of the 17) in reframing their interpretations of that child or their communities. For example, some PSTs reframed their interpretations of children from viewing them as struggling with mathematics or lacking in some way to viewing them as exhibiting strengths and competencies with mathematics. Other PSTs reflected that these activities supported them in beginning to confront negative or deficit assumptions about Black children's communities, allowing them to see, for example, the resources that communities could provide with respect to informing mathematics instruction.

Third, and lastly, while in many cases PSTs were able to confront assumptions they held, deficit perspectives also surfaced (for 7 of the aforementioned participants). We now turn to the narratives, which form the more significant part of our results and discussion.

Narratives: A (Re)orienting Experience

For many PSTs, close interactions with Black children supported them in reframing their interpretations of those children or their communities. We present two extended narratives in the following section as representative of this (re)orienting.

The case of Monica. Monica was a White, female PST for whom learning mathematics had never come easily. She felt that throughout her schooling she struggled to keep up with her classmates in spite of time that her father and eventually a tutor spent working with her at home. She nonetheless came to appreciate mathematics, in part because of her mother, who tried

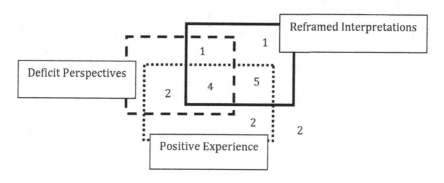

FIGURE 15.1. Themes

to pique her interest in math by doing such things as buying her Scieszka's (1995) book, *Math Curse*. This appreciation of mathematics is something she hoped to instill in her students. Monica also noted that, in grades K–12, she attended schools with predominantly White, middle-class student populations and that the activities she participated in outside of school, such as church, sports, and band, also consisted primarily of White, middle-class participants.

Monica worked with an African American female student, Aniya, who was a first grader in a school where the student body was identified as 54.7% White, 19.9% Hispanic, 10.1% Asian/Pacific Islander, and 7.7% Black with 5.2% of the student body being identified as biracial or multiracial. Approximately half (52.1%) of the students received free or reduced lunch. Monica described Aniya as outgoing or "spunky," even "loud," and as demonstrating middle to high performance in mathematics. Aniya loved to play with Legos and hoped to build houses and buildings when she grew up. Although prior to the course activities Monica thought of Aniya as somewhat "disruptive at times" and as "seeming to observe more than actively participate," she came to see that perhaps what was interpreted as inattention and inability was actually the result of boredom and lack of appropriate intellectual challenge.

During the mathematics problem-solving interview, Monica noticed that Aniya was easily able to directly model or count-on to arrive at solutions to typical subtraction problems, including those where the numbers were large for a first grader (e.g., I had 50 dogs. 13 ran away. How many dogs do I have now?). Furthermore, she documented that Aniya solved problems that Carpenter and colleagues (1999) suggested are more complicated for students, as the unknown quantity is not the result of the operation (e.g., $50 - 13 = x$), but the change between quantities (e.g., $7 - x = 4$ or Jason has seven cookies. He eats some of his cookies. He now has four cookies. How many cookies did he eat?).

During the shadowing experience, Monica observed that Aniya did not join her peers in shouting out answers in mathematics class. While initially interpreting this as inability, Monica looked closer and checked Aniya's paper, seeing that it was "perfect." Monica reflected on a similar experience observing Aniya in the context of reading where she seemed to "flip through the pages of her book," seemingly "not reading at all," only to find when she asked Aniya to read a book to her that "she quickly read through the book without any difficulty" and "had actually been reading the books and getting new ones when she finished." These observations, coupled with knowledge gained in the mathematics problem-solving interview, supported Monica in reorienting herself to see Aniya as competent, yet bored. Monica reflected, "through this observation I realized that students aren't always quiet because they don't understand, they might be bored like Aniya and may need a challenge to do something more."

In the extended case presented here, we see that close interactions with a Black child, Aniya, supported Monica in reframing her interpretations of that child away from seeing how the child was lacking (e.g., disruptive or non-participatory) to instead recognizing knowledge and competence (e.g., bored, intelligent, needing more intellectual challenge). All too often, and despite evidence to the contrary, *meta-narratives* about African American, Latino, and other historically marginalized populations about lack of parental involvement or an innate lack of mathematical intelligence (Giroux, Lankshear, McLaren, & Peters, 1996) serve to frame teachers' perceptions of children of color in their mathematics classrooms (DiME, 2007). Supporting PSTs in moving away from and challenging these framings serves to "facilitate a shift in identity," producing "narratives about *who can do mathematics*" (Battey & Chan, 2010, pp. 143–144, emphasis in original). We see this as an important initial step in teacher development toward recognizing the brilliance of Black children, as a focus on what children *can* do allows PSTs to build on a child's knowledge and resources to develop classroom practices that support the students' mathematics learning and achievement.

While in Monica's case, as with other PSTs, this reorienting experience occurred with respect to their interpretations of the individual child, for other PSTs these instructional activities supported them in also reorienting toward the communities of Black children. The case of Mark, which we turn to next, is illustrative of both of these ideas.

The case of Mark. Mark was a White, male PST who grew up in a "wealthy, primarily White suburb of Silicon Valley." Mark described his experiences with mathematics from elementary school through his junior year in high school as initially being those of confusion about the material, which "quickly devolved into frustration, followed by embarrassment and eventually apathy." He attributed his interest as a senior in high school in fantasy sports and sports betting as providing meaning to mathematics and supporting him in appreciating and understanding mathematics in new ways. He hoped to similarly support his students in seeing themselves in the mathematics curriculum.

Mark worked with an African American male student, Robert, who was a fifth grader in a school in which 91.6% of the students received free or reduced lunch. The composition of the student body was identified as follows: 42.1% Black, 26.6% White, 20.2% Asian/Pacific Islander, 2.7% American Indian/Alaskan Native, with 7.3% of the student population being identified as biracial or multiracial. Mark described Robert as a student who struggled with basic addition and subtraction computation, but who was quite adept at using multiplication and division (which for Mark meant that Robert could procedurally solve multiplication and division problems). During the mathematics problem-solving interview, Mark realized that Robert's understanding of multiplication and subtraction was "illustrated brilliantly" as he reasoned through the following multi-step problem: "Lily has 3 boxes of chocolates. Each box has 4 pieces of chocolate. If Lily gives 5 pieces of chocolate to her sister, how many pieces of chocolate will she have then?" While we would argue that these number choices might be better suited for students in lower grades, possibly indicating deficit thinking on Mark's part with respect to Robert, it is important to also note that such problem solving did not seem to have been a central part of Robert's prior mathematics instruction. Mark noted that Robert's current mathematics class included daily timed multiplication/division or addition/subtraction number facts practice, that "solving" word problems was primarily framed as, in the words of Robert, "turning words into numbers," and that, according to Mark, prior instruction related to multiplication and division was focused on "rote memorization." Thus, the ability to solve multi-step word problems, even with low numbers, could be viewed as a more difficult task within this context. Furthermore, Mark went into the problem-solving interview thinking that Robert would not successfully solve such a problem (even with numbers better suited for children in lower grades). Instead, he saw that Robert demonstrated knowledge of subtraction (understanding both that "giving away five" suggests you "minus something" and that his answer was reasonable), as well as a "highly structured and organized approach to problem solving." In this way, Mark began to reorient himself toward Robert, recognizing the mathematical competence and knowledge Robert had.

Mark also described Robert as a student who identified mathematics as his favorite subject and who had a relatively strong grasp of how math was "present in everyday life." He noted that Robert was aware of areas where he was still working to understand the material, and that he persisted to learn the material in part because he could see the practical importance mathematics played in his life. Mark was impressed by Robert's ability to articulate the role of mathematics in his life, and found that based on his own experience, he could relate to Robert's mathematical struggle and the importance that Robert's ability to identify relevance for math in his life could play in supporting his mathematics learning. Together, these experiences supported Mark in beginning to reframe his interpretations of Robert's mathematical knowledge:

> While not a strong math student in terms of computational ability [with respect to addition and subtraction] and test scores, it seemed that he grasped the concept that math was present in everyday life and that he needed to rely on his own [reasoning] skills....What I can take away from this experience is

that even though on paper Robert it not a strong student, his grasp on the importance of the subject matter...is a pretty powerful concept.

Mark did not only reorient toward Robert, but also toward the community. Drawing on his interactions with Robert and other students in the class, Mark visited a local Boys and Girls club that many students attended with their friends and families. Mark described his experience visiting the Boys and Girls club as an "interesting, informative, and eye-opening experience" both because of the mathematics that was happening in the setting and more specifically because he was able to obtain a "glimpse of [his students'] world" outside of the school environment. Mark was impressed by the fact that Saturday morning basketball games were "an event" that brought students, extended family members, friends, and neighbors who "all seemed to know each other" together. He began to notice what he could learn from these community members for his own instruction. For example, he noted that watching how children and their parents interact around asking permission to do things (e.g., sitting with friends, going to the snack bar) could inform how he might handle similar requests in his classroom. He reflected on the experience:

> I have come to expect certain behaviors from some of my students; however, that perspective has changed since the math walk. Having seen my students outside of the school environment, really being themselves, it gives me a much better insight into what really makes them tick.

In this extended example, we see that Mark had a positive attitude toward his community exploration and toward Robert, finding value in what can be learned from the community to orchestrate mathematics instruction and in Robert's ability to see how mathematics is relevant to one's lived experiences. We also see Mark differentiating between knowledge exhibited on tests or "on paper" and what he felt is important mathematical knowledge that Robert had, thus producing new narratives about what counts as mathematical knowledge and who can do mathematics. With respect to the community exploration experience, it is unclear how much of Mark's self-described change in perspective related specifically to his prior expectations about interactions with a Black child's community.

What underlying assumptions might he have held about Black communities that reflect meta-narratives in society and thus required a change in perspective? On the one hand, his surprise that families and extended families were involved in the community center, for example, may reflect an underlying stereotype born out of a media-fed meta-narrative around African Americans and lack of parental involvement (Giroux et al., 1996). On the other hand, Mark positioned himself as a learner in a setting where Black children's parents and family members could be a valuable resource with respect to supporting their children's mathematics learning. He reflected, "It helped me be aware that the community is a valuable source for just about everything, including math." Such notable (re)orienting, as many of the PSTs in this study described these experiences to be, is important because finding interactions with Black communities a positive experience in which they discover resources they did not realize existed might be considered a first step in supporting PSTs' abilities to recognize, access, and build upon such resources in mathematics instruction in order to support the mathematical learning of Black students.

Narratives: Emergence of Deficit Perspectives

While many PSTs were able to make progress in identifying and confronting assumptions they held, deficit perspectives also surfaced. As we might expect, based on the myths in the dominant society regarding Black communities (King & Wilson, 1990; Ladson-Billings, 1994), the deficit thinking fell into two categories: a) deficit views of Black children's communities or families, and b) deficit views of the children themselves. In the two extended narratives that follow, we present representative cases of PSTs who formed good relationships with their Black case study students and found many positive points to discuss about them. In spite of this positive orientation, deficit perspectives did emerge.

The case of Tara. Tara was a White PST who saw herself as not very capable with mathematics. In spite of having a dad who was very interested in mathematics, she had several difficult experiences with it in school that fed this view of herself. However, she was determined to find some enjoyment in mathematics so that she could support her future students in having a better experience than she did, and was dedicated to improving her knowledge of mathematics. Tara worked with an African American male student, Jay, who was a first grader in a school in which 46.2% of the students received free or reduced lunch. The composition of the student body was identified as follows: 44.2% White, 19.1% Hispanic, 13.6% Asian/Pacific Islander, 12.3% Black, 1.4% American Indian/Alaska Native, with 9.3% of the student body being identified as biracial or multiracial. Tara described Jay as an active, enthusiastic, and very social child who seemed to be a bit of a perfectionist. He had some struggles in school with spelling and reading, but this was not true of mathematics, a subject he said he liked and which Tara indicated seemed to come fairly easily to him. He loved to dance and could often be seen dancing in the classroom and dancing instead of walking down the hall, something that would get him reprimanded. Tara saw Jay's enthusiasm as positive and did not pathologize this behavior as indicating that Jay was hyperactive or needed to learn to control himself. Instead she said:

Jay gets along with everyone in our class and has friends....He is quick to say sorry if he has hurt someone's feelings and eager to help another student when they need it. Jay has never acted out purposely, and his main behavior issues are his need to be active. Jay is the kind of student you want because of his lovable personality and the excitement he brings to the classroom.

Tara observed that Jay did very well in mathematics class with part-part-whole problems using bare numbers (i.e. problems that were not contextualized as story problems). When Tara worked more closely with Jay through the mathematics problem-solving interview, she found that he did not do as well with several contextualized part-part-whole type problems. Although she at first thought it was due to working with contextualized problems, she discovered that it was due to Jay's unfamiliarity with the structure of some of the problems she was posing. Tara asked Jay to solve this problem: "Tara is saving money to buy her brother a present. The present costs 7 dollars. She has 4 dollars so far. How many more dollars does Tara need to buy the present?" Jay incorrectly answered 11, but Tara noted that the computation he did for 7+4 was correct although he did not seem to understand either the problem structure or the context well enough to solve it correctly. When asked how he had arrived at 11 for an answer, Jay indicated that his understanding of the problem was that Tara had 7 dollars and then saved 4 more so the present must cost eleven dollars. Despite the fact that Jay arrived at the incorrect solution, Tara recognized that he had strength with computation that could be brought to bear on working on problems like this. When Tara went back to look at the part-part-whole problems that Jay had done so well with in math class, she found they all had the structure Carpenter and colleagues (1999) call join result unknown (the typical "addition" problem with the structure of 7+4=x as opposed to the "missing part" structure of the problem Tara had posed, 4+x=7). In her recommendations for next steps with Jay, Tara recommended using several of the missing part activities that she had encountered in her methods text. She noted that these activities used a game format that she thought would be appealing to Jay as it would allow him to work with others and move around while he worked. In this way, her ideas supported what Boykin and colleagues call movement and communalism features of African American culture that are often undervalued in schools (Boykin & Bailey, 2000a, 2000b; Boykin, Coleman, Lilja, & Tyler, 2004). With the lack of attention that schools pay to these features, they undermine the development of strong cultural identities of Black children as the culture of school validates the middle-class, White experience as the norm, often defining others' cultural experiences as deficient if not abnormal (Asante, 1991; Lareau, 2003).

In discussing her experiences interacting with Jay during the activities of the *Mathematics Learning Case Study Module*, and working with him in the classroom more generally, Tara wrote that she had learned about Jay's strengths with computation and difficulties with less routine addition and subtraction problem structures. She gained insight into various strategies that children use when they solve mathematical problems. She learned that problems are more accessible when the contexts are linked to children's experiences. She concluded, "I have learned so much more from this assignment than I initially thought I would, and am excited to see how I can use these techniques in my student teaching experience."

In this extended example, thus far, we see that Tara had a very positive attitude toward Jay and finds much to value both in the liveliness and enthusiasm he demonstrated and in his mathematical skills. She recognized that he needed instructional activities to support growth in mathematical understanding and suggested appropriate activities that would not only support this development, but would also be appealing to an active first grade boy. Despite this picture of a PST who had recognized strengths in the student and made apt suggestions for further work with him that both acknowledge academic needs and attend to preferences for movement and communalism, we can see some deficit thinking creep in as a result of her work in the community exploration activity.

As part of their preparation for visiting supermarkets in the community for their community exploration activity, Tara talked to staff at the school and students in her classroom about their experiences with grocery shopping and food. According to Tara, several of the students told her that although they knew fruits and vegetables were good for them, they did not like them, they refused to eat them, and were not required to eat them at home or ate fast food. Tara commented,

> I realize that some of this could be related to a limited budget, however I am a huge proponent of living a healthy lifestyle and would be very interested in teaching students (even at this young age) about ways to eat healthy for cheap and take care of their bodies.

This view of the home as one that does not support healthy eating may have been adopted hastily and with too little examination. Tara seemed prepared to accept what the students said about their eating habits at face value and was prepared to implicate the families, and in particular poor families, in not providing a healthy diet for their children. Tara did not suggest a plan for talking to or surveying families about their eating habits, or discussing the issue with parents to find their point of view. Whether or not it is true that students would benefit from changing their eating habits, this stance follows the savior mentality that many teachers bring to their interactions with Black communities (Martin, 2007). Tara acted with some level of arrogance in seeing herself as one who could save students from their environment and teach them how to live healthily. While some may consider this a relatively minor instance of deficit thinking about children's families, it nonetheless should be a signal for mathematics teacher educators that there may be other ways in which deficit thinking about Black children and their families and communities may linger and that more work is necessary in the methods course to bring this deficit thinking into the open so that it can be challenged. Activities such as the student case study and the community exploration activity provide an opportunity for PSTs to share their views of Black children and their families and communities. If deficit thinking exists, it is good that it surfaces within the methods class where it can be addressed and where PSTs can be supported in (re)orienting their thinking in more productive ways that will support the learning and educational attainment of Black children.

The case of Laura. Laura was a White PST who worked with Renee, an African American female student who was a first grader at a school in which 61% of the students received free or reduced lunch. The composition of the student body was identified as follows: 78% Black, 16% White, 5% Hispanic, and 1% Asian/Pacific Islander. In addition to the difference in race, Laura noted that she and her case study student were from different socioeconomic backgrounds.

As part of the community exploration activity, Laura and her group went on a walk around the community of the school. They were accompanied by her cooperating teacher (who was from the community) and the teacher's son, who was also a first grader at the school. Laura and her group learned that nearly all students at the school lived close by and walked to school. Because of the students' familiarity with the neighborhood around the school, Laura and her group developed a lesson in which students would determine the shortest distance to the school from four familiar locations. She noted:

> We got the chance to see how excited the students were to see their school, the playground they go to, or their sibling's school on the map. This got them very engaged in the lesson because they had experience walking from these places in the community in their own lives. They could easily visualize this because it was something very real to them.

Although we do not know about student performance for this lesson, we do know that engagement and participation were high. Some researchers believe that participation *leads to* learning, that students who participate more generally learn more from the lesson, and that low rates of participation can predict low achievement particularly in the early grades (Cohen, 1984; Finn & Cox, 1992). Other researchers believe that participation *is* learning (Lave & Wenger, 1991). It is our contention, then, that experiences such as the community exploration position PSTs to support participation by drawing on contexts that students recognize and find familiar, lead to increased participation and learning.

In addition to this positive work done drawing on community information, Laura also worked individually with her case study student, Renee. Despite the differences in their backgrounds, Laura saw that they both valued the school experience. Laura described Renee as a helpful and kind person who also performed well in school mathematics and claimed mathematics as her favorite school subject. She performed very well on some contextualized mathematics problems she solved with Laura, including division problems involved with sharing cookies among friends. In another instance, despite calling the problem "kind of tricky," Renee was able to solve the following multi-step problem using direct modeling: "Ra'niah has 3 packages of graham crackers. Each package has 4 graham crackers in it. If Ra'niah gives 5 graham crackers to Renee, how many will Ra'niah have left?" Laura recognized that Renee was easily solving problems that many first graders might find difficult.

Renee appeared very motivated in school and was a diligent worker. Nonetheless, Laura felt she lacked confidence, as Renee was hesitant to volunteer to answer questions in whole-group settings unless, in Laura's opinion, she was certain her answer was correct. During their initial interview, Renee chronicled for Laura various family members' educational accomplishments, leading Laura to comment, "It seems like her family values education." While this is a positive comment about Renee's family, it is one that also gives pause, as there is an unstated deficit perspective under this positive statement that not *all* families value education, and may indicate a lingering deficit view of some families. In our experience with both prospective and practicing teachers, children's families are at times labeled as *not* valuing education (despite the fact that research has shown this *not* to be the case, e.g. Stevenson, Chen, & Uttal, 1990) when situations arise such as non-attendance at parent-teacher conferences or other school events, or incomplete homework being returned. Little energy is expended in examining the reasons that underlie these situations. Parents may indeed value education and yet because of working particular shifts or multiple jobs may be unable to be present at school events. This same situation may influence the ability of parents to help with homework. In our experience (and as research indicates, e.g., Perry, 2003), Black families are disproportionately (although not exclusively) implicated as *not* valuing education, and little is done to suggest or implement plans to support the child and parents who are in these circumstances. (For a notable exception see the work of Fuson and colleagues [2000], who supported parents in identifying a family member or neighbor who could work with their child on school assignments.) The activities that Laura engaged in allowed these views to surface so that they could be addressed in the methods classroom.

In the case of Laura, we see a PST who, from the start, had a positive orientation toward Renee, recognizing that Renee performed extraordinarily well on mathematics tasks. Laura may nonetheless have jumped to conclusions about the reasons underlying the reticence to participate. It may not have been that Renee only answered when she was confident of an answer. An alternative explanation would be that Renee was uncomfortable with the dynamics of the classroom and unsure as to whether her contributions would be valued. It might be argued that many PSTs (and even many practicing teachers) are under-prepared to examine the classroom interaction patterns and participation structures to see how they support (or inhibit) Black children's participation and instead locate the "problem" of non-participation within the child. It might be, building on the work of Boykin and colleagues (2004), that a participation structure that supported communal work as opposed to the "teacher asks, student answers" structure of this classroom would have better supported Renee in making consistent oral contributions.

FURTHER COMMENTARY AND
IMPLICATIONS FOR PRACTICE

Important first steps for PSTs in recognizing and acknowledging the brilliance of Black children include a) seeing value in learning about Black children and their communities (in other words, viewing these instructional experiences as positive and valuable) (Leonard, 2008) and b) confronting meta-narratives about Black children by focusing on what children *can do*, producing different narratives about what mathematics knowledge is of value and about who can do mathematics (DiME, 2007). As noted previously, our data indicate that the instructional activities reported on in this chapter were viewed by most PSTs as positive and by many PSTs as supportive for reorienting their interpretations of Black children or their communities. Yet, despite these positive orientations and reorientations, deficit perspectives surfaced.

While we cannot be certain that the ways in which PSTs discussed confronting previously held deficit notions was not in some way influenced by the fact that students were receiving a grade from their instructor around these activities, the articulation and discussion of reoriented views in the public space of the classroom is important. Some participants took first steps toward adopting a disposition that allowed them to recognize and acknowledge the brilliance of Black children. They may serve as an important support for the (re)orienting of those PSTs who continue to hold persistently deficit views of Black children, their homes, and communities. An opportunity to be exposed in the university classroom to the (re)oriented views of other PSTs who began the semester with deficit thinking similar to their own, but who were able to begin to (re)orient their perspectives, may support (re)orientation on their part as well. It is one thing to have this point of view advocated by the professor. It is another, and arguably more effective, matter to have peers and fellow PSTs contribute discussions focused on how they have (re)oriented their thinking about their particular case study students. By extension, deficit views of all Black children are called into question, an important component of the (re)orienting process.

Mathematics teacher educators do not have to leave this to chance, but can build discussions around some of the deficit orientations that surface throughout the semester so as to support PSTs in confronting them, encouraging those whose views have shifted to share their thinking. Although we do not know if PSTs in this study were aware of their deficit perspectives coming into the methods course, what is important to our message is that some began to confront these perspectives. This positioned them to be more aware that they might hold other deficit perspectives in need of examination. Teacher education programs, more generally, need to organize spaces such as those created with the activities presented here, where negative and deficit perspectives toward children and communities can be intentionally raised, aired, and most importantly confronted.

Shifts of the kind we are hoping to support take time, are fragile, and are not always immediately sustained. While some might argue that supporting the learning of Black children cannot and should not hinge on, nor wait for, the reorientation of teachers (White teachers in particular) (e.g., Martin, 2007), the reality is that those teaching and preparing to be teachers remain largely White, middle-class females (Howard, 1999) who must be supported in exposing and challenging meta-narratives underlying their work with Black children. This is not to say that there is not urgency in this matter; it is most urgent that mathematics teacher educators and PSTs focus on the mathematics learning of Black children.

Helping PSTs make sense of what they see and observe is critical. Absent such support, PSTs may interpret what they see and do in children's communities in ways that reinforce deficit ideas about children and families. Support must continue beyond the elementary mathematics methods course, and might include more intentional placement of student teachers both in schools where their deficit thinking might be triggered and with mentor teachers who can provide the kind of supervision that can continue to expose and confront deficit notions. It is important to note, however, that even within a one-semester course, we see PSTs moving away from deficit perspectives of Black children and their families that they brought into the methods classroom.

While we cannot be certain that this move from deficit perspective in their discourse is accompanied by a shift in their beliefs and their practice, it can be seen as an early step in that shift. Some begin taking these first steps by finding interaction with Black children and communities a positive experience; others reframe their focus toward Black children and their communities such that they discover resources they did not realize existed. Having this resource perspective on Black communities positions these PSTs to be able to access those valuable resources and build on them in mathematics lessons to support the mathematical learning of Black students.

Beyond seeing the community as a resource, some of the PSTs in this study were able to identify and appreciate specific ways in which Black children were competent, even brilliant. For some this meant (re)orienting their thinking to be open to see and acknowledge that competence and brilliance. In the narratives presented, we see some PSTs beginning to value Black children's perspectives toward and prior knowledge of mathematics, and engaging in some activities and experiences that supported them in coming to know themselves and their students, at times (re)orienting them away from negative or deficit perspectives of children, all important steps in becoming effective teachers of Black children (Foster, 1990; Martin, 2000, 2006; Milner, 2003).

NOTES

1. This material is based upon work supported by the National Science Foundation under Grant No. 1228034. Any opinions, findings, and conclusions or recommendations expressed in this material are those of the author(s) and do not necessarily reflect the views of the National Science Foundation.
2. The names of all PSTs and case study students are pseudonyms. In cases where PSTs had used pseudonyms for their case study students in their course writing, we respected those decisions and use the same pseudonyms. In the cases where PSTs referred to case study students by their first names, and for the PSTs themselves, we selected pseudonyms that mirrored their names.

REFERENCES

Aguirre, J., Turner, E., Bartell, T. G., Drake, C., Foote, M. Q., & Roth McDuffie, A. (2012). Analyzing effective mathematics lessons for English learners: A multiple mathematical lens approach. In S. Celedón-Pattichis & N. Ramirez (Eds.), *Beyond good teaching: Advancing mathematics education for ELLs* (pp. 207–219). Reston, VA: NCTM.

Artiles, A. J. &McClafferty, K. (1998). Learning to teach culturally diverse learners: Charting change in preservice teachers' thinking about effective teaching. *Elementary School Journal, 98*(3), 189–220.

Asante, M. K. (1991). The Afrocentric idea in education. *Journal of Negro Education, 60*, 170–180.

Bartell, T. G. (2011). Caring, race, culture, and power: A research synthesis toward supporting mathematics teachers in caring with awareness. *Journal of Urban Mathematics Education, 4*(1), 50–74.

Bartell, T. G., Foote, M. Q., Aguirre, J. M., Roth McDuffie, A., Drake, C., & Turner, E. E. (2010). Preparing preK–8 teachers to connect children's mathematical thinking and community based funds of knowledge. In P. Brosnan, D. B. Erchick, & L. Flevares (Eds.). *Proceedings of the 32nd annual meeting of the North American chapter of the International Group for the Psychology of Mathematics Education* (pp. 1183–1191). Columbus, OH: The Ohio State University.

Battey, D. & Chan, A. (2010). Building community and relationships that support critical conversations on race: the case of cognitively guided instruction. In M. Q. Foote's (Ed.), *Mathematics teaching & learning in K–12: equity and professional development* (pp. 137–150). New York, NY: Palgrave.

Berliner, D. C. (1994). Expertise: The wonder of exemplary performances. In J. N. Mangiere & C. C. Block (Eds.), *Creating powerful thinking in teachers and students: Diverse perspectives* (pp. 161–186). Orlando, FL: Harcourt Brace.

Bogdan, R. & Biklen, S. (1992). *Qualitative research for education: An introduction to theory and methods.* Boston, MA: Allyn and Bacon.

Boykin, A. W. & Bailey, C. (2000a). *The role of cultural factors in school relevant cognition functioning: Description of home environmental factors, cultural orientations, and learning preferences.* Baltimore, MD: CRESPAR.

Boykin, A. W. & Bailey, C. (2000b). *The role of cultural factors in school relevant cognition functioning: Synthesis of findings on cultural contexts, cultural orientations, and individual differences.* Baltimore, MD: CRESPAR.

Boykin, A. W., Coleman, S., Lilja, A., & Tyler, K. (2004). *Building on children's cultural assets in simulated classroom performance environments: Research vistas in the communal learning paradigm.* Baltimore, MD: CRESPAR.

Carpenter, T. P., Fennema, E., Franke, M., Levi, L., & Empson, S. (1999). *Children's mathematics: Cognitively guided instruction.* Portsmouth, NH: Heinemann.

Clandinin, D. J. & Connelly, F. M. (2000). *Narrative inquiry: Experience and story in qualitative research.* San Francisco, CA: Jossey-Bass.

Corbin, J. & Strauss, A. (2008). *Basics of qualitative research* (3rd ed.). Thousand Oaks, CA: Sage.

DiME. (2007). Culture, race, power, and mathematics education. In F. Lester (Ed.), *Second handbook of research on mathematics teaching and learning: A project of the National Council of Teachers of Mathematics* (pp. 405–433). Charlotte, NC: Information Age.

Duesterberg, L. (1998). Rethinking culture in the pedagogy and practices of preservice teachers, *Teaching and Teacher Education, 14,* 497–512.

Dunkin, M., Welch, A., Merritt, A., Phillips, R., & Craven, R. (1998) Teachers' explanations of classroom events: Knowledge and beliefs about teaching civics and citizenship, *Teaching and Teacher Education, 14,* 141–151.

Foster, M. (1990). The politics of race: Through the eyes of African American teachers. *Journal of Education, 172,* 123–141.

Fuson, K. C., De La Cruz, Y., Smith, S., Lo Cicero, A., Hudson, K., Ron, P., & Steeby, R. (2000). Blending the best of the twentieth century to achieve a mathematics equity pedagogy in the twenty-first century. In M. J. Burke (Ed.), *Learning mathematics for a new century* (pp. 197–212). Reston, VA: National Council of Teachers of Mathematics.

Gay, G. (2002). Preparing for culturally responsive teaching. *Journal of Teacher Education, 53*(2), 106–116.

Gay, G. (2010). *Culturally responsive teaching: Theory, practice, and research* (2nd ed.). New York, NY: Teachers College Press.

Giroux, H., Lankshear, C., McLaren, P., & Peters, M. (1996). *Counternarratives: cultural studies and critical pedagogies in postmodern space.* New York, NY: Routledge.

Groulx J. G. (2001). Changing preservice teacher perceptions of minority schools. *Urban Education, 36*(1), 60–92.

Hammerness, K., Darling-Hammond, L., & Bransford, J. (with Berliner, D., Cochran-Smith, M., McDonald, M., & Zeichner, K.). (2005). How teachers learn and develop. In L. Darling-Hammond & J. Bransford (Eds.), *Preparing teachers for a changing world: What teachers should learn and be able to do* (pp. 358–389). San Francisco, CA: Jossey-Bass.

Hiebert, J., Morris, A. K., Berk, D., & Jansen, A. (2007). Preparing teachers to learn from teaching. *Journal of Teacher Education, 58,* 47–61.

Howard, G. (1999). *We can't teach what we don't know: White teachers, multiracial schools.* New York, NY: Teachers College Press.

Howard, T. C. (2003). Culturally relevant pedagogy: Ingredients for critical teacher reflection. *Theory into Practice, 42*(3), 195–202.

Irvine, R. W. & Irvine, J. J. (1983). The impact of the desegregation process on the education of Black students: Key variables. *The Journal of Negro Education, 52,* 410–422.

King, J. & Wilson, T. L. (1990). Being the soul-freeing substance: A legacy of hope in Afro humanity. *Journal of Education, 172*(2), 9–27.

Ladson-Billings, G. (1994). *The dreamkeepers: Successful teachers of African American children.* San Francisco, CA: Jossey-Bass.

Ladson-Billings, G. (1997). It doesn't add up: African American students' mathematics achievement. *Journal for Research in Mathematics Education, 28,* 697–708.

Ladson-Billings, G. (2001). *Crossing over to Canaan: the journey of new teachers in diverse classrooms.* San Francisco, CA: Jossey-Bass.

Lareau, A. (2003). *Unequal childhoods: class, race and family life.* Berkeley, CA: University of California Press.

Larson, C. L. & Ovando, C. J. (2001). *The color of bureaucracy: The politics of equity in Multicultural school communities.* Belmont, CA: Thomson Learning, Inc.

Lave, J. & Wenger, E. (1991). *Situated learning: Legitimate peripheral participation.* Cambridge, UK: Cambridge University Press.

Leonard, J. (2008). *Culturally specific pedagogy in the mathematics classroom: strategies for teachers and students.* New York, NY: Routledge.

Leonard, J. & Guha, S. (2002). Creating cultural relevance in teaching and learning mathematics. *Teaching Children Mathematics, 9*(2), 114–118.

Malloy, C. E. (2009). Instructional strategies and dispositions of teachers who help African American students gain conceptual understanding. In D. B. Martin (Ed.), *Mathematics teaching, learning and liberation in the lives of Black children* (pp. 88–122). New York, NY: Routledge.

Martin, D. B. (2000). *Mathematics success and failure among African American youth: The roles of sociohistorical context, community forces, school influence, and individual agency.* Mahwah, NJ: Lawrence Erlbaum Assoc. Inc.

Martin, D. B. (2006). Mathematics learning and participation as racialized forms of experience: African American parents speak on the struggle for mathematics literacy. *Mathematical Thinking and Learning, 8*(3), 197–229.

Martin, D. B. (2007). Beyond missionaries or cannibals: Who should teach mathematics to African American children? *The High School Journal, 91*(1), 6–28.

Martin, D. B. (2009). *Mathematics teaching, learning and liberation in the lives of Black children.* New York, NY: Routledge.

Matthews, L. E. (2003). Babies overboard! The complexities of incorporating culturally relevant teaching into mathematics instruction. *Educational Studies in Mathematics, 53,* 61–82.

Miles, M. B. & Huberman, A. M. (1994). *Qualitative data analysis: An expanded sourcebook* (2nd ed.). Thousand Oaks, CA: Sage Publications.

Milner, H. R. (2003). Teacher reflection and race in cultural contexts: History, meaning and methods in teaching. *Theory into Practice, 42,* 173–180.

Milner, H. R. (2005). Stability and change in US prospective teachers' beliefs and decisions about diversity and learning to teach. *Teaching and Teacher Education, 21*(7), 767–786.

Moynihan, D. (1965). *The Negro family: The case for national action.* Washington DC: United States Department of Labor.

Nasir, N. S., Hand, V., & Taylor, E. V. (2008). Culture and mathematics in school: boundaries between "cultural" and "domain" knowledge in the mathematics classroom and beyond. *Review of Research in Education, 32,* 187–240. doi: 10.3102/0091732X07308962

Nieto, S. (2004). *Affirming diversity* (4th ed.). Boston, MA: Pearson.

Perry, T. (2003). Up from the parched Earth: toward a theory of African-American achievement. In T. Perry, C. Steele, & A. Hilliard (Eds.), *Young, gifted, and black: promoting high achievement among African-American students* (pp. 1–10). Boston, MA: Beacon Press.

Schoenfeld, A. H. (1992). Learning to think mathematically: Problem solving, metacognition and sense making in mathematics. In D. Grouws (Ed.) *Handbook for research on mathematics teaching and learning* (pp. 334–370). New York, NY: Macmillan.

Schofield, J. W. (1989). *Black and White in school: Trust, tension, or tolerance?* New York, NY: Teachers College Press.

Scieszka, J. (1995). *Math curse.* New York, NY: Viking Juvenile.

Siddle-Walker, V. (1996). *Their highest potential: An African American school community in the segregated South.* Chapel Hill, NC: University of North Shayna Press.

Silver, E. A. & Stein, M. K. (1996). The QUASAR project: The "revolution of the possible" in mathematics instructional reform in urban middle schools. *Urban Education, 30,* 476–521.

Spindler, G. & Spindler, L. (1982). Roger Harker and Schonhausen: From the familiar to the strange and back again. In G. Spindler (Ed.), *Doing the ethnography of schooling* (pp. 20–46). Prospect Heights, IL: Waveland Press, Inc.

Stevenson, H. W., Chen, C., & Uttal, D. H. (1990). Beliefs and achievement: A study of Black, White, and Hispanic children. *Child Development 61*(2), 508–523.

Turner, E. E., Drake, C., Roth McDuffie, A., Aguirre, J. M., Bartell, T. G., & Foote, M. Q. (2012). Promoting equity in mathematics teacher preparation: A framework for advancing teacher learning of children's multiple mathematics knowledge bases. *Journal of Mathematics Teacher Education, 15*(1), 67–82. doi: 10.1007/s10857-011-9196-6

Turner, E., Celedón-Pattichis, S., & Marshall, M. E. (2008). Cultural and linguistic resources to promote problem solving and mathematical discourse among Hispanic kindergarten students. In R. Kitchen & E. Silver (Eds.), *Promoting high participation and success in mathematics by Hispanic students: Examining opportunities and probing promising practices* [A Research Monograph of TODOS: Mathematics for ALL], 1, 19-42. Washington, DC: National Education Association Press.

Valencia, R. (1997). *The evolution of deficit thinking: Educational thought and practice.* Washington DC: Falmer Press.

CONTRIBUTING AUTHORS

Julia Aguirre (PhD, mathematics education, University of California Berkeley, 2002) is an assistant professor of education at the University of Washington Tacoma. She earned a Bachelor's degree from the University of California Berkeley in psychology and a Master's degree in education from the University of Chicago. She is a mathematics teacher educator committed to transforming instructional practice to better serve historically marginalized youth. Her work in culturally responsive mathematics teaching supports a new generation of teachers to work effectively in culturally and linguistically diverse schools.

Tonya Gau Bartell (PhD, mathematics education, University of Wisconsin Madison, 2005) is an assistant professor of mathematics education at Michigan State University. She is interested in promoting equity in mathematics education for all students through examining how issues of race, culture, and power intersect with issues of mathematics teaching and learning. Her research focuses on teachers' development of mathematics pedagogy for social justice and pedagogy integrating a focus on mathematics, children's mathematical thinking, and children's community and cultural knowledge.

The Brilliance of Black Children in Mathematics:
Beyond the Numbers and Toward New Discourse, pages 367–374.

Robert Q. Berry, III (PhD, mathematics education, University of North Carolina at Chapel Hill, 2003) is an associate professor of mathematics education in the Curry School of Education at the University of Virginia. His scholarly interests focus on equity issues in mathematics education and the work of elementary mathematics specialists. Recent accomplishments include being elected to the board of directors, National Council of Teachers of Mathematics (2011–2014).

Cheryl Lewis Beverly (MA, Africana studies, California State University at Long Beach, 2002) is a retired counselor and professor at El Camino College, in Torrance, California. Other degrees include a Bachelor's of Arts degree from Michigan State University in history and secondary education, Master's of Science degree from California State University, Los Angeles, in guidance and counseling, and a certificate of advanced studies from the University of Chicago in educational psychology. She is licensed in California as an educational psychologist.

Jamie M. Bracey (PhD, educational psychology, Temple University, 2011) is an award winning scholar conducting research on using culturally and socially relevant learning environments to accelerate minority students' learning in science, technology, engineering and mathematics (STEM). Dr. Bracey is the director of STEM education, outreach and research for Temple University, and is responsible for developing and managing external community relations, policy development, education programming and collaborative research partnerships related to STEM teaching and learning. Her research emphasis explores motivation to persist for cultural and linguistic minorities.

Denise Natasha Brewley (PhD, mathematics education, University of Georgia, 2009) is an assistant professor in the School of Science and Technology at Georgia Gwinnett College. Other degrees include a Bachelor's of Science in mathematics from Spelman College, Master's of Science in applied mathematics and Master's of Business Administration in finance from Clark Atlanta University. Her current research interests include identity, understanding how communities of practice are developed in mathematics spaces, and creating significant learning experiences for students taking undergraduate mathematics courses.

Iman Chahine (Ph.D., University of Minnesota, 2008) is an assistant professor of mathematics education at the Department of Middle-Secondary Education and Instructional Technology at Georgia State University (GSU). She is the director of two study abroad programs on Ethnomathematics and Indigenous Mathematical Knowledge Systems in Morocco and South

Africa. She received her teaching diploma and B.S. degree from the American University of Beirut. Her research interests include ethnomathematics, indigenous mathematical knowledge systems, situated cognition, and multicultural mathematics. Dr. Chahine's recent accomplishments include being the recipient of the 2012 GSU Instructional Innovation Award.

Corey Drake (PhD, mathematics education, Northwestern University, 2000) is associate professor and director of teacher education at Michigan State University. Her research interests include teacher learning from and about curriculum materials, as well as preparing teachers to integrate children's multiple mathematical knowledge bases.

Mario Eraso (PhD, curriculum & instruction, Florida International University, 2007) is an assistant professor of mathematics education at Texas A&M Commerce. He also earned Master's of Science degrees from Florida International University in mathematics education and from Lehigh University in civil engineering. Dr. Eraso's research focuses on students' cognitive transition from arithmetic to algebraic reasoning and students' spatial visualization abilities. He started his career as a civil engineer and worked as a structural engineer for six years. He teaches undergraduate and graduate courses in mathematics education at Texas A&M Commerce and collaborates with others on STEM education initiatives.

Mary Q. Foote (PhD, mathematics education, University of Wisconsin Madison, 2006) is an associate professor of mathematics education in the department of elementary and early childhood education at Queens College, CUNY. Her research interests focus on cultural and community knowledge and practices and how they might inform mathematics teaching practice.

Maisie L. Gholson (BS, electrical engineering, Duke University, 2001) is a doctoral student at the University of Illinois at Chicago. She is a fellow of the National Science Foundation Graduate Research Fellows Program in STEM education. Her proposed research relates to how classroom talk develops African American children's sense of self racially and academically and mediates students' participation patterns and mathematics achievement.

Christopher C. Jett (PhD, mathematics education, Georgia State University, 2009) is an assistant professor in the department of mathematics at the University of West Georgia in Carrollton, Georgia. He also earned Bachelor's of Science and Master's of Science degrees from Tennessee State University. He teaches mathematics content courses for prospective mathematics teachers. His research interests include employing a critical race philosophical and theoretical framework to mathematics education research and investigating the experiences of successful African American male students in mathematics.

Shelly M. Jones (PhD, mathematics education, Illinois State University, 2002) is an assistant professor at Central Connecticut State University in New Britain, Connecticut. She teaches undergraduate and graduate content, curriculum and methods courses. Her interests include culturally relevant mathematics, integrating elementary school mathematics and music, and the effects of college students' attitudes and beliefs about mathematics on their success in college. She earned a Master's of Science degree from the University of Bridgeport in Connecticut and a Bachelor's of Science degree from Spelman College in Atlanta, Georgia.

Shonda Lemons-Smith (PhD, mathematics education, Indiana University, 2002) is an assistant professor of mathematics education in the department of early childhood education at Georgia State University. She earned Bachelor's of Science and Master's of Science degrees from Fort Valley State University. Her research focuses on mathematics education in urban contexts; specifically, teacher development, issues of equity, and culturally relevant pedagogy. She has 21 years of experience in the field of mathematics education at the K–12 and college/university level. In addition, she is very active in the education community at the national, state, and local levels as well as internationally.

Jacqueline Leonard (PhD, mathematics education, University of Maryland at College Park, 1997) is the director of the Science and Mathematics Teaching Center (SMTC) and professor of mathematics education in the department of elementary and early childhood education at the University of Wyoming. She is the first person of color to serve as the SMTC director in the center's 42-year history. Other degrees include a Bachelor's of Arts from St. Louis University, a Master's of Arts in teaching from the University of Texas at Dallas, and a Master's of theology from Southern Methodist University. Her research agenda focuses on teaching mathematics for cultural relevance and social justice.

Danny Bernard Martin (PhD, mathematics education, University of California at Berkeley, 1997) is Professor of Education and Mathematics at the University of Illinois at Chicago. His research has focused primarily on understanding the salience of race and identity in Black learners' mathematical experiences, taking into account sociohistorical and sociostructural forces, community forces, school forces, and individual agency. Dr. Martin is author of the book *Mathematics Success and Failure Among African Youth*

(2000, Erlbaum) and editor of *Mathematics Teaching, Learning, and Liberation in the Lives of Black Children* (2009, Routledge).

Lou Edward Matthews (PhD, mathematics education, Illinois State University, 2002) is the director of educational standards and accountability for the Bermuda Education Ministry. He has also earned Bachelor's of Science degrees in mathematics and accounting from Atlantic Union College and a Master's of Science in secondary mathematics education from Alabama A&M University. His accomplishments include being past-president of the Benjamin Banneker Association (2007–2009). His research interests include urban education and culturally relevant pedagogy.

Oren McClain (PhD, mathematics education, University of Virginia, 2011) is an assistant dean in the Office of African-American Affairs at the University of Virginia. In this role, he advises and supports undergraduate students in mathematics-intensive courses and majors. He received his Bachelor's of Science and Master's of Engineering degrees from Old Dominion University. His research interests include equity in STEM education, mathematics identity construction, and mathematics experiences of minority students in predominately White colleges and universities.

Ebony O. McGee (PhD, mathematics education, University of Illinois at Chicago, 2009) is an assistant professor of diversity and urban schooling at Vanderbilt University's Peabody College. Other degrees include a Bachelor's of Science degree from North Carolina A & T State University in electrical engineering and Master's of Science degree from New Jersey Institute of Technology in industrial engineering. She is a National Science Foundation postdoctoral fellow at Northwestern University, with the Scientific Careers and Research Development Group. More generally, her research focuses on racial stereotypes, wellness, and identity development in high-achieving marginalized students of color in STEM disciplines.

Malaika McKee (PhD, higher education policy and administration, University of Minnesota, 2007) is a visiting professor at the University of Illinois, Champaign Urbana in the department of African American studies. Other degrees include a Master's of Education from Harvard Graduate School of Education. She is a recipient of the diversity faculty fellowship at the Morgridge College of Education, University of Denver, and a research fellowship at the Mathematica Policy Institute in Princeton, New Jersey. Her research interests focus on the intersection of civic engagement and higher education with special emphasis on the experiences of students of color.

Yolanda A. Parker (PhD, mathematics education, Illinois State University, 2004) is an associate professor in the mathematics department at Tarrant County College in Fort Worth, Texas. She earned a Bachelor's of Science from Texas A & M University and a Master's of Arts from Dartmouth College. Her research interests include culturally relevant and responsive pedagogy in mathematics classrooms, Black females' experiences in mathematics, and algebra teachers' self-efficacy.

Holly H. Pinter (MEd, middle grades education, Western Carolina University, 2009) is a visiting assistant professor at Western Carolina University in the School of Teaching and Learning. Ms. Pinter also earned a BS in middle grades education from Western Carolina University in 2005. She is finishing her doctoral work to earn a PhD in mathematics education at the University of Virginia. Her research interests include teacher quality, middle school mathematics education, student discourse practices, and pre-service teacher preparation. Before entering the doctoral program, Ms. Pinter taught seventh- and eighth-grade mathematics for five years in Buncombe County Schools in Asheville, North Carolina.

Amy Roth McDuffie (PhD, mathematics education, University of Maryland at College Park, 1998) is an associate professor at Washington State University Tri-cities. Her research and teaching focus on professional learning and development for prospective and practicing teachers, with attention to equity and supporting diverse students' learning in mathematics.

Nicole M. Russell (PhD, curriculum and instruction, University of Washington, 2011) is an assistant professor in the department of curriculum and instruction and teacher education at the University of Denver. She holds a Bachelor's degree in business economics and a Master's degree in human development. She is a certificated mathematics teacher for K–12 and a National Board for Professional Standards Candidate in adolescent mathematics. Her research interests include mathematics achievement of African Americans, culturally responsive pedagogy, equity and access in mathematics classrooms, classroom discourse, the role of African American culture in learning broadly, and mathematics development.

David W. Stinson (PhD, mathematics education, University of Georgia, 2004) is an associate professor of mathematics education in the College of Education at Georgia State University. His research interests include exploring socio-cultural, -historical, and -political aspects of mathematics and mathematics teaching and learning from a critical postmodern theoretical (and methodological) perspective. He is a co-founder and current editor-in-chief of the *Journal of Urban Mathematics Education* and co-editor of the book *Teaching Mathematics for Social Justice: Conversations with Educators* (NCTM, 2012).

Danté A. Tawfeeq (PhD, mathematics education, Florida State University, 2003) is an associate professor in the department of mathematics and co-ordinator of the Math Foundations and Reasoning Program at John Jay College, a senior college of the City University of New York (CUNY). Other degrees include a Bachelor's of Science in mathematics and Master's of Education from Florida A&M University. He is a recipient of the Mathematical Association of America Dolciani-Holloran Foundation Project NExT Fellowship. He worked as an assistant for the Eisenhower Consortium for Mathematics and Science Education and as program developer for the Florida Collaborative for the Excellence in Teacher Preparation. He is a former high school teacher of mathematics.

Erin E. Turner (PhD, mathematics education, University of Texas Austin, 2003) is an assistant professor of mathematics education in the department of teaching, learning and sociocultural studies at the University of Arizona. Her research focuses on issues of equity and social justice in teaching and learning mathematics, with particular attention to teaching mathematics in culturally and linguistically diverse contexts.

Brian Williams (PhD, educational studies, Emory University, 2003) currently serves as director of the Alonzo A. Crim Center for Urban Educational Excellence and an associate professor in the department of early childhood education in the College of Education at Georgia State Education in Atlanta, Georgia. His work is situated at the intersection of science education, urban education, and education for social justice. More specifically, he is interested in the ways in which equity issues related to race, ethnicity, culture, and class influence science teaching and learning and access to science literacy. Dr. Williams also holds a Bachelor's of Science degree from Norfolk State University and Master's of Science degree from Georgia Institute of Technology. He is also a Former Ford Foundation Fellow and Spencer Fellow.

York M. Williams (PhD, urban education, Temple University, 2007) is an assistant professor in the department of special education at West Chester University. He also earned a Master's of Education degree from Temple University, Master's of Arts in philosophy of social Science from West Chester University; and the Bachelor's of Arts in philosophy from North Carolina Central University. His work focuses on recruiting and retaining culturally diverse students in gifted education, multicultural and urban education school choice, minority student achievement and underachievement, and family involvement. Additional interests include the intersection of urban school violence and achievement among African American males through the lens of social and juvenile justice.

Paul W. Yu (PhD, mathematics education, Illinois State University, 2004) is an associate professor in the department of mathematics at Grand Valley State University in Allendale, Michigan. He also earned the Master's of Science in mathematics from Illinois State University and Bachelor's of Science in mathematics from the University of Illinois, Champaign Urbana. He teaches mathematics education courses for prospective elementary and secondary mathematics teachers. His research interests include semiotics and prototype theory as theoretical frameworks to investigate the learning of mathematics among school aged (K–12) students and college pre-service mathematics teachers.